NEWTON'S PRINCIPIA

MOTTE'S TRANSLATION REVISED

SIR ISAAC NEWTON

(See Appendix, Note 1, page 627)

Sir Isaac Newton's
MATHEMATICAL PRINCIPLES
OF NATURAL PHILOSOPHY AND HIS SYSTEM OF THE WORLD

*Translated into English by Andrew Motte in 1729.
The translations revised, and supplied with an
historical and explanatory appendix, by*

FLORIAN CAJORI

Volume One: THE MOTION OF BODIES

UNIVERSITY OF CALIFORNIA PRESS
Berkeley, Los Angeles, London

Contents

Volume One

PRINCIPIA

BOOK I. THE MOTION OF BODIES

BOOK II. THE MOTION OF BODIES
(IN RESISTING MEDIUMS)

Volume Two

BOOK III. THE SYSTEM OF THE WORLD
(IN MATHEMATICAL TREATMENT)

THE SYSTEM OF THE WORLD

APPENDIX

Editor's Note to the Present Revision

PROFESSOR FLORIAN CAJORI died August 15, 1930. In May of the following year I was invited by the University of California Press to edit this work. After much delay, due in part to unavoidable circumstances and in part to the time consumed in the extraordinary care taken in reading, checking, and rereading the proofs, this edition of Newton's *Principia* is now ready to be run off the press.

The manuscript as presented to the Press contained no Preface. Much of the material that would be included in the usual Preface is contained in the first few notes of the Appendix, pages 627 ff. Professor Cajori probably intended to prepare a Preface while the book was in the process of manufacture. There being none, the customary acknowledgment of thanks to various persons who assisted him in one way or another is lacking. Lest I unknowingly omit some to whom thanks are due, I refrain from attempting any such acknowledgment on behalf of the author.

As the title page states, this is a revision of Motte's translation of the *Principia*. From many conversations with Professor Cajori, I know that he had long cherished the idea of revising Newton's immortal work by rendering certain parts into modern phraseology (e.g., to change the reading of "reciprocally in the subduplicate ratio of" to "inversely as the square root of") and to append historical and critical notes which would provide instruction to some readers and interest to all. This is his last work; one most fitting to crown a life devoted to investigation and to writing the history of the sciences in his chosen field.

R. T. CRAWFORD

Berkeley, California,
March 31, 1934.

PHILOSOPHIÆ

NATURALIS

PRINCIPIA

MATHEMATICA.

Autore *J S. NEWTON*, *Trin. Coll. Cantab. Soc.* Matheseos
Professore *Lucasiano*, & Societatis Regalis Sodali.

IMPRIMATUR.

S. PEPYS, *Reg. Soc.* PRÆSES.

Julii 5. 1686.

LONDINI,

Jussu *Societatis Regiæ* ac Typis *Josephi Streater*. Prostat apud
plures Bibliopolas. *Anno* MDCLXXXVII.

The Ode Dedicated to Newton by Edmund Halley

THIS ODE PREFIXED TO THE PRINCIPIA OF NEWTON
IS HERE TRANSLATED BY LEON J. RICHARDSON
PROFESSOR OF LATIN
IN THE UNIVERSITY OF CALIFORNIA
FROM THE VERSION AS GIVEN IN THE FIRST EDITION

TO THE ILLUSTRIOUS MAN

ISAAC NEWTON

AND THIS HIS WORK

DONE IN FIELDS OF THE MATHEMATICS AND PHYSICS

A SIGNAL DISTINCTION OF OUR TIME AND RACE

Lo, for your gaze, the pattern of the skies!
What balance of the mass, what reckonings
Divine! Here ponder too the Laws which God,
Framing the universe, set not aside
But made the fixed foundations of his work.

The inmost places of the heavens, now gained,
Break into view, nor longer hidden is
The force that turns the farthest orb. The sun
Exalted on his throne bids all things tend
Toward him by inclination and descent,
Nor suffers that the courses of the stars
Be straight, as through the boundless void they move,

But with himself as centre speeds them on
In motionless ellipses. Now we know
The sharply veering ways of comets, once
A source of dread, nor longer do we quail
Beneath appearances of bearded stars.

At last we learn wherefore the silver moon
Once seemed to travel with unequal steps,
As if she scorned to suit her pace to numbers—
Till now made clear to no astronomer;
Why, though the Seasons go and then return,
The Hours move ever forward on their way;
Explained too are the forces of the deep,
How roaming Cynthia bestirs the tides,
Whereby the surf, deserting now the kelp
Along the shore, exposes shoals of sand
Suspected by the sailors, now in turn
Driving its billows high upon the beach.

Matters that vexed the minds of ancient seers,
And for our learned doctors often led
To loud and vain contention, now are seen
In reason's light, the clouds of ignorance
Dispelled at last by science. Those on whom
Delusion cast its gloomy pall of doubt,
Upborne now on the wings that genius lends,
May penetrate the mansions of the gods
And scale the heights of heaven. O mortal men,
Arise! And, casting off your earthly cares,

Learn ye the potency of heaven-born mind,
Its thought and life far from the herd withdrawn!

The man who through the tables of the laws
Once banished theft and murder, who suppressed
Adultery and crimes of broken faith,
And put the roving peoples into cities
Girt round with walls, was founder of the state,
While he who blessed the race with Ceres' gift,
Who pressed from grapes an anodyne to care,
Or showed how on the tissue made from reeds
Growing beside the Nile one may inscribe
Symbols of sound and so present the voice
For sight to grasp, did lighten human lot,
Offsetting thus the miseries of life
With some felicity. But now, behold,
Admitted to the banquets of the gods,
We contemplate the polities of heaven;
And spelling out the secrets of the earth,
Discern the changeless order of the world
And all the aeons of its history.

Then ye who now on heavenly nectar fare,
Come celebrate with me in song the name
Of Newton, to the Muses dear; for he
Unlocked the hidden treasuries of Truth:
So richly through his mind had Phoebus cast
The radiance of his own divinity.
Nearer the gods no mortal may approach.

Newton's Preface to the First Edition

SINCE THE ANCIENTS (as we are told by *Pappus*) esteemed the science of mechanics of greatest importance in the investigation of natural things, and the moderns, rejecting substantial forms and occult qualities, have endeavored to subject the phenomena of nature to the laws of mathematics, I have in this treatise cultivated mathematics as far as it relates to philosophy. The ancients considered mechanics in a twofold respect; as rational, which proceeds accurately by demonstration, and practical. To practical mechanics all the manual arts belong, from which mechanics took its name. But as artificers do not work with perfect accuracy, it comes to pass that mechanics is so distinguished from geometry that what is perfectly accurate is called geometrical; what is less so, is called mechanical. However, the errors are not in the art, but in the artificers. He that works with less accuracy is an imperfect mechanic; and if any could work with perfect accuracy, he would be the most perfect mechanic of all, for the description of right lines and circles, upon which geometry is founded, belongs to mechanics. Geometry does not teach us to draw these lines, but requires them to be drawn, for it requires that the learner should first be taught to describe these accurately before he enters upon geometry, then it shows how by these operations problems may be solved. To describe right lines and circles are problems, but not geometrical problems. The solution of these problems is required from mechanics, and by geometry the use of them, when so solved, is shown; and it is the glory of geometry that from those few principles, brought from without, it is able to produce so many things. Therefore geometry is founded in mechanical practice, and is nothing but that part of universal mechanics which accurately proposes and demonstrates the art of measuring. But since the manual arts are chiefly employed in the moving of bodies, it happens that geometry is commonly referred to their magnitude, and mechanics to their motion. In this sense rational mechanics will be the science of motions resulting from any forces whatsoever, and of the forces required to produce any motions, accurately proposed and demonstrated. This part of mechanics, as far as it extended to the five powers which relate to manual arts, was cultivated by the ancients, who considered gravity (it not being a manual power) no otherwise than in moving weights by those powers. But I consider philosophy rather than arts and write not concerning manual but natural powers, and consider chiefly those things which relate to gravity, levity, elastic force, the resistance of fluids, and the like forces, whether attractive or impulsive; and therefore I offer this work as the mathematical principles of philosophy, for the whole burden of philosophy seems to consist in this—from the phenomena of motions to investigate the forces of nature, and then from these forces to demonstrate the other phenomena; and to this end the general propositions in the first and second Books are directed. In the third Book I give an example of this in the explication of the System of the World; for by the propositions mathematically demonstrated in the former Books, in the third I derive from the celestial phenomena the forces of gravity with which bodies tend to the sun and the several planets. Then from these forces, by other propositions which are also mathematical, I deduce the motions of the planets, the comets, the moon, and the sea. I wish we could derive the rest of the phenomena of Nature by the same kind of reasoning from mechanical principles, for I am induced by many reasons to suspect that they may all depend upon certain forces by which the particles of bodies, by some causes hitherto unknown, are either mutually impelled towards one another, and cohere in regular figures, or are repelled and recede from one another. These forces being unknown, philosophers have hitherto attempted the search of Nature in vain; but I hope the principles here laid down will afford some light either to this or some truer method of philosophy.

In the publication of this work the most acute and universally learned Mr. *Edmund Halley* not only assisted me in correcting the errors of the press and preparing the geometrical figures, but it was through his solicitations that it came to be published; for when he had obtained of me my demonstrations of the figure of the celestial orbits, he continually pressed me to communicate the same to the *Royal Society,* who afterwards, by their kind encouragement and entreaties, engaged me to think of publishing them. But after I had begun to consider the inequalities of the lunar motions, and had entered upon some other things relating to the laws and measures of gravity and other forces; and the figures that would be described by bodies attracted according to given laws; and the motion of several bodies moving among themselves; the motion of bodies in resisting mediums; the forces, densities, and motions, of mediums; the orbits of the comets, and such like, I deferred that publication till I had made a search into those matters, and could put forth the whole together. What relates to the lunar motions (being imperfect), I have put all together in the corollaries of Prop. LXVI, to avoid being obliged to propose and distinctly demonstrate the several things there contained in a method more prolix than the subject deserved and interrupt the series of the other propositions. Some things, found out after the rest, I chose to insert in places less suitable, rather than change the number of the propositions and the citations. I heartily beg that what I have here done may be read with forbearance; and that my labors in a subject so difficult may be examined, not so much with the view to censure, as to remedy their defects.

Is. NEWTON

Cambridge, Trinity College, May 8, 1686.[1]

[1 Appendix, Note 3.]

Newton's Preface to the Second Edition

IN THIS SECOND EDITION of the *Principia* there are many emendations and some additions.[1] In the second section of the first Book, the determination of forces, by which bodies may be made to revolve in given orbits, is illustrated and enlarged. In the seventh section of the second Book the theory of the resistances of fluids was more accurately investigated, and confirmed by new experiments. In the third Book the lunar theory and the precession of the equinoxes were more fully deduced from their principles; and the theory of the comets was confirmed by more examples of the calculation of their orbits, done also with greater accuracy.

Is. NEWTON

London, March 28, 1713.

[1 Appendix, Note 4.]

Cotes's Preface to the Second Edition[1]

W E HEREBY PRESENT to the benevolent reader the long-awaited new edition of *Newton's* Philosophy, now greatly amended and increased. The principal contents of this celebrated work may be gathered from the adjoining Table. What has been added or modified is indicated in the author's Preface. There remains for us to add something relating to the method of this philosophy.

Those who have treated of natural philosophy may be reduced to about three classes. Of these some have attributed to the several species of things, specific and occult qualities, according to which the phenomena of particular bodies are supposed to proceed in some unknown manner. The sum of the doctrine of the Schools derived from *Aristotle* and the *Peripatetics* is founded on this principle. They affirm that the several effects of bodies arise from the particular natures of those bodies. But whence it is that bodies derive those natures they don't tell us; and therefore they tell us nothing. And being entirely employed in giving names to things, and not in searching into things themselves, they have invented, we may say, a philosophical way of speaking, but they have not made known to us true philosophy.

Others have endeavored to apply their labors to greater advantage by rejecting that useless medley of words. They assume that all matter is homogeneous, and that the variety of forms which is seen in bodies arises from some very plain and simple relations of the component particles. And by going on from simple things to those which are more compounded they certainly proceed right, if they attribute to those primary relations no other relations than those which Nature has given. But when they take a liberty of imagining at pleasure unknown figures and magnitudes, and uncertain situations and motions of the parts, and moreover of supposing occult fluids, freely pervading the pores of bodies, endued with an all-performing subtilty, and agitated with occult motions, they run out into dreams and chimeras, and neglect the true constitution of things, which certainly is not to be derived from fallacious conjectures, when we can scarce reach it by the most certain observations. Those who assume hypotheses as first principles of their speculations, although they afterwards proceed with the greatest accuracy from those principles, may indeed form an ingenious romance, but a romance it will still be.

There is left then the third class, which possess experimental philosophy. These indeed derive the causes of all things from the most simple principles possible; but then they assume nothing as a principle, that is not proved by phenomena. They frame no hypotheses, nor receive them into philosophy otherwise than as questions whose truth may be disputed. They proceed therefore in a twofold

[1 Appendix, Note 5.]

method, synthetical and analytical. From some select phenomena they deduce by analysis the forces of Nature and the more simple laws of forces; and from thence by synthesis show the constitution of the rest. This is that incomparably best way of philosophizing, which our renowned author most justly embraced in preference to the rest, and thought alone worthy to be cultivated and adorned by his excellent labors. Of this he has given us a most illustrious example, by the explication of the System of the World, most happily deduced from the Theory of Gravity. That the attribute of gravity was found in all bodies,[1] others suspected, or imagined before him, but he was the only and the first philosopher that could demonstrate it from appearances, and make it a solid foundation to the most noble speculations.

I know indeed that some persons, and those of great name, too much prepossessed with certain prejudices, are unwilling to assent to this new principle, and are ready to prefer uncertain notions to certain. It is not my intention to detract from the reputation of these eminent men; I shall only lay before the reader such considerations as will enable him to pass an equitable judgment in this dispute.

Therefore, that we may begin our reasoning from what is most simple and nearest to us, let us consider a little what is the nature of gravity in earthly bodies, that we may proceed the more safely when we come to consider it in the heavenly bodies that lie at the remotest distance from us. It is now agreed by all philosophers that all circumterrestrial bodies gravitate towards the earth. That no bodies having no weight are to be found, is now confirmed by manifold experience. That which is relative levity is not true levity, but apparent only, and arises from the preponderating gravity of the contiguous bodies.

Moreover, as all bodies gravitate towards the earth, so does the earth gravitate again towards all bodies. That the action of gravity is mutual and equal on both sides, is thus proved. Let the mass of the earth be divided into any two parts whatever, either equal or unequal; now if the weights of the parts towards each other were not mutually equal, the lesser weight would give way to the greater, and the two parts would move on together indefinitely in a right line towards that point to which the greater weight tends, which is altogether contrary to experience. Therefore we must say that the weights with which the parts tend to each other are equal; that is, that the action of gravity is mutual and equal in contrary directions.

The weights of bodies at equal distances from the centre of the earth are as the quantities of matter in the bodies. This is inferred from the equal acceleration of all bodies that fall from a state of rest by their weights; for the forces by which unequal bodies are equally accelerated must be proportional to the quantities of the matter to be moved. Now, that all falling bodies are equally accelerated, appears from this, that when the resistance of the air is taken away, as it is under

[1 Appendix, Note 6.]

an exhausted receiver of Mr. *Boyle,* they describe equal spaces in equal times; but this is yet more accurately proved by the experiments with pendulums.

The attractive forces of bodies at equal distances are as the quantities of matter in the bodies. For since bodies gravitate towards the earth, and the earth again towards bodies with equal moments, the weight of the earth towards each body, or the force with which the body attracts the earth, will be equal to the weight of the same body towards the earth. But this weight was shown to be as the quantity of matter in the body; and therefore the force with which each body attracts the earth, or the absolute force of the body, will be as the same quantity of matter.

Therefore the attractive force of the entire bodies arises from and is composed of the attractive forces of the parts, because, as was just shown, if the bulk of the matter be augmented or diminished, its power is proportionately augmented or diminished. We must therefore conclude that the action of the earth is composed of the united actions of its parts, and therefore that all terrestrial bodies must attract one another mutually, with absolute forces that are as the matter attracting. This is the nature of gravity upon earth; let us now see what it is in the heavens.

That every body continues in its state either of rest or of moving uniformly in a right line, unless so far as it is compelled to change that state by external force, is a law of Nature universally received by all philosophers. But it follows from this that bodies which move in curved lines, and are therefore continually bent from the right lines that are tangents to their orbits, are retained in their curvilinear paths by some force continually acting. Since, then, the planets move in curvilinear orbits, there must be some force operating, by the incessant actions of which they are continually made to deflect from the tangents.

Now it is evident from mathematical reasoning, and rigorously demonstrated, that all bodies that move in any curved line described in a plane, and which, by a radius drawn to any point, whether at rest or moved in any manner, describe areas about that point proportional to the times, are urged by forces directed towards that point. This must therefore be granted. Since, then, all astronomers agree that the primary planets describe about the sun, and the secondary about the primary, areas proportional to the times, it follows that the forces by which they are continually turned aside from the rectilinear tangents, and made to revolve in curvilinear orbits, are directed towards the bodies that are placed in the centres of the orbits. This force may therefore not improperly be called centripetal in respect of the revolving body, and in respect of the central body attractive, from whatever cause it may be imagined to arise.

Moreover, it must be granted, as being mathematically demonstrated, that, if several bodies revolve with an equable motion in concentric circles, and the squares of the periodic times are as the cubes of the distances from the common centre, the centripetal forces will be inversely as the squares of the distances. Or, if bodies revolve in orbits that are very nearly circular and the apsides of the orbits are at rest, the centripetal forces of the revolving bodies will be inversely as the squares of the distances. That both these facts hold for all the planets, all astronomers agree. Therefore the centripetal forces of all the planets are inversely as the squares of the distances from the centres of their orbits. If any should object, that the apsides of the planets, and especially of the moon, are not perfectly at rest, but are carried progressively with a slow kind of motion, one may give this answer, that, though we should grant that this very slow motion arises from a slight deviation of the centripetal force from the law of the square of the distance, yet we are able to compute mathematically the quantity of that aberration, and find it perfectly insensible. For even the ratio of the lunar centripetal force itself, which is the most irregular of them all, will vary inversely as a power a little greater than the square of the distance, but will be well-nigh sixty times nearer to the square than to the cube of the distance. But we may give a truer answer, by saying that this progression of the apsides arises not from a deviation from the law of inverse squares of the distance, but from a quite different cause, as is most admirably shown in this work. It is certain then that the centripetal forces with which the primary planets tend to the sun, and the secondary planets to their primary, are accurately as the inverse squares of the distances.

From what has been hitherto said, it is plain that the planets are retained in their orbits by some force continually acting upon them; it is plain that this force is always directed towards the centres of their orbits; it is plain that its intensity is increased in its approach and is decreased in its recession from the centre, and that it is increased in the same ratio in which the square of the distance is diminished, and decreased in the same ratio in which the square of the distance is augmented. Let us now see whether, by making a comparison between the centripetal forces of the planets and the force of gravity, we may not by chance find them to be of the same kind. Now, they will be of the same kind if we find on both sides the same laws and the same attributes. Let us then first consider the centripetal force of the moon, which is nearest to us.

The rectilinear spaces which bodies let fall from rest describe in a given time at the very beginning of the motion, when the bodies are urged by any forces whatsoever, are proportional to the forces. This appears from mathematical reasoning. Therefore the centripetal force of the moon revolving in its orbit is to the force of gravity at the surface of the earth, as the space which in a very small interval of time the moon, deprived of all its circular force and descending by its centripetal force towards the earth, would describe, is to the space which a heavy body would describe, when falling by the force of its gravity near to the earth, in the same small interval of time. The first of these spaces is equal to the versed sine of the arc described by the moon in the same time, because that versed sine measures the translation of the moon from the tangent, produced by the centripetal force, and therefore may be computed, if the periodic time of the moon and its distance from the centre of the earth are given. The last space is found by experiments with pendulums, as Mr. *Huygens* has shown. Therefore by making a calculation we shall find that the first space is to the latter, or the centripetal force of the moon revolving in its orbit will be to the force of gravity at the surface of the earth, as the square of the semidiameter of the earth to the square of the semidiameter of the orbit. But by what was shown before, the very same ratio holds between the centripetal force of the moon revolving in its orbit, and the centripetal force of the moon near the surface of the earth. Therefore the centripetal force near the surface of the earth is equal to the force of gravity. Therefore these are not two different forces, but one and the same; for if they were different, these forces united would cause bodies to descend to the earth with twice the velocity they would fall with by the force of gravity alone. Therefore it is plain that the centripetal force, by which the moon is continually either impelled or attracted out of the tangent and retained in its orbit, is the very force of terrestrial gravity reaching up to the moon. And it is very reasonable to believe that this force should extend itself to vast distances, since upon the tops of the highest mountains we find no sensible diminution of it. Therefore the moon gravitates towards the earth; but on the other hand, the earth by a mutual action equally gravitates towards the moon, which is also abundantly confirmed in this philosophy, where the tides in the sea and the precession of the equinoxes are treated of, which arise from the action both of the moon and of the sun upon the earth. Hence lastly, we discover by what law the force of gravity decreases at great distances from the earth. For since gravity is noways different from the moon's centripetal force, and this is inversely proportional to the square of the distance, it follows that it is in that very ratio that the force of gravity decreases.

Let us now go on to the other planets. Because the revolutions of the primary planets about the sun and of the secondary about Jupiter and Saturn are phenomena of the same kind with the revolution of the moon about the earth, and because it has been moreover demonstrated that the centripetal forces of the primary planets are directed towards the centre of the sun and those of the secondary towards the centres of Jupiter and Saturn, in the same manner as the centripetal force of the moon is directed towards the centre of the earth, and since, besides, all these forces are inversely as the squares of the distances from the centres, in the same manner as the centripetal force of the moon is as the square of the distance from the earth, we must of course conclude that the nature of all is the same. Therefore as the moon gravitates towards the earth and the earth again towards the moon, so also all the secondary planets will gravitate towards their primary, and the primary planets again towards their secondary, and so all the primary towards the sun, and the sun again towards the primary.

Therefore the sun gravitates towards all the planets, and all the planets towards the sun. For the secondary planets, while they accompany the primary, revolve the meanwhile with the primary about the sun. Therefore, by the same argument, the planets of both kinds gravitate towards the sun and the sun towards them. That the secondary planets gravitate towards the sun is moreover abundantly clear from the inequalities of the moon, a most accurate theory of which, laid open with a most admirable sagacity, we find explained in the third Book of this work.

That the attractive force of the sun is propagated on all sides to prodigious distances and is diffused to every part of the wide space that surrounds it, is most evidently shown by the motion of the comets, which, coming from places immensely distant from the sun, approach very near to it, and sometimes so near that in their perihelia they almost touch its body. The theory of these bodies was altogether unknown to astronomers till in our own times our excellent author most happily discovered it and demonstrated the truth of it by most certain observations. So that it is now apparent that the comets move in conic sections having their foci in the sun's centre, and by radii drawn to the sun describe areas proportional to the times. But from these phenomena it is manifest and mathematically demonstrated, that those forces by which the comets are retained in their orbits are directed towards the sun and are inversely proportional to the squares of the distances from its centre. Therefore the comets gravitate towards the sun, and therefore the attractive force of the sun not only acts on the bodies of the planets, placed at given distances and very nearly in the same plane, but reaches also the comets in the most different parts of the heavens, and at the most different distances. This therefore is the nature of gravitating bodies, to exert their force at all distances to all other gravitating bodies. But from thence it follows that all the planets and comets attract one another mutually, and gravitate towards one another, which is also confirmed by the perturbation of Jupiter and Saturn, observed by astronomers, and arising from the mutual actions of these two planets upon each other, as also from that very slow motion of the apsides, above taken notice of, which arises from a like cause.

We have now proceeded so far, that it must be acknowledged that the sun, and the earth, and all the heavenly bodies attending the sun, attract one another mutually. Therefore all the least particles of matter in every one must have their several attractive forces proportional to their quantities of matter, as was shown above of the terrestrial bodies. At different distances these forces will be also inversely as the squares of their distances; for it is mathematically demonstrated, that globes attracting according to this law are composed of particles attracting according to the same law.

The foregoing conclusions are grounded on this axiom which is received by all philosophers, namely, that effects of the same kind, whose known properties are the same, take their rise from the same causes and have the same unknown properties also. For if gravity be the cause of the descent of a stone in *Europe*, who doubts that it is also the cause of the same descent in *America?* If there is a mutual gravitation between a stone and the earth in *Europe*, who will deny the same to be mutual in *America?* If in *Europe* the attractive force of a stone and the earth is composed of the attractive forces of the parts, who will deny the like composition in *America?* If in *Europe* the attraction of the earth be propagated to all kinds of bodies and to all distances, why may we not say that it is propagated in like manner in *America?* All philosophy is founded on this rule; for if that be taken away, we can affirm nothing as a general truth. The constitution of particular things is known by observations and experiments; and when that is done, no general conclusion of the nature of things can thence be drawn, except by this rule.

Since, then, all bodies, whether upon earth or in the heavens, are heavy, so far as we can make any experiments or observations concerning them, we must certainly allow that gravity is found in all bodies universally. And in like manner as we ought not to suppose that any bodies can be otherwise than extended, movable, or impenetrable, so we ought not to conceive that any bodies can be otherwise than heavy. The extension, mobility, and impenetrability of bodies become known to us only by experiments; and in the very same manner their gravity becomes known to us. All bodies upon which we can make any observations, are extended, movable, and impenetrable; and thence we conclude all bodies, and those concerning which we have no observations, are extended and movable and impenetrable. So all bodies on which we can make observations, we find to be heavy; and thence we conclude all bodies, and those we have no observations of, to be heavy also. If anyone should say that the bodies of the fixed stars are not heavy because their gravity is not yet observed, they may say for the same reason that they are neither extended nor movable nor impenetrable, because these properties of the fixed stars are not yet observed. In short, either gravity must have a place among the primary qualities of all bodies, or extension, mobility, and impenetrability must not. And if the nature of things is not rightly explained by the gravity of bodies, it will not be rightly explained by their extension, mobility, and impenetrability.

Some I know disapprove this conclusion, and mutter something about occult qualities. They continually are cavilling with us, that gravity is an occult property, and occult causes are to be quite banished from philosophy. But to this the answer is easy: that those are indeed occult causes whose existence is occult, and imagined but not proved; but not those whose real existence is clearly demonstrated by observations. Therefore gravity can by no means be called an occult cause of the celestial motions, because it is plain from the phenomena that such a power does really exist. Those rather have recourse to occult causes, who set imaginary vortices of a matter entirely fictitious and imperceptible by our senses, to direct those motions.

But shall gravity be therefore called an occult cause, and thrown out of philosophy, because the cause of gravity is occult and not yet discovered? Those who affirm this, should be careful not to fall into an absurdity that may overturn the foundations of all philosophy. For causes usually proceed in a continued chain from those that are more compounded to those that are more simple; when we are arrived at the most simple cause we can go no farther. Therefore no mechanical account or explanation of the most simple cause is to be expected or given; for if it could be given, the cause were not the most simple. These most simple causes will you then call occult, and reject them? Then you must reject those that immediately depend upon them, and those which depend upon these last, till philosophy is quite cleared and disencumbered of all causes.

Some there are who say that gravity is preternatural, and call it a perpetual miracle. Therefore they would have it rejected, because preternatural causes have no place in physics. It is hardly worth while to spend time in answering this ridiculous objection which overturns all philosophy. For either they will deny gravity to be in bodies, which cannot be said, or else, they will therefore call it preternatural because it is not produced by the other properties of bodies, and therefore not by mechanical causes. But certainly there are primary properties of bodies; and these, because they are primary, have no dependence on the others. Let them consider whether all these are not in like manner preternatural, and in like manner to be rejected; and then what kind of philosophy we are like to have.

Some there are who dislike this celestial physics because it contradicts the opinions of *Descartes,* and seems hardly to be reconciled with them. Let these enjoy their own opinion, but let them act fairly, and not deny the same liberty to us which they demand for themselves. Since the *Newtonian* Philosophy appears true to us, let us have the liberty to embrace and retain it, and to follow causes proved by phenomena, rather than causes only imagined and not yet proved. The business of true philosophy is to derive the natures of things from causes truly existent, and to inquire after those laws on which the Great Creator actually chose to found this most beautiful Frame of the World, not those by which he might have done the same, had he so pleased. It is reasonable enough to suppose that from several causes, somewhat differing from one another, the same effect may arise; but the true cause will be that from which it truly and actually does arise; the others have no place in true philosophy. The same motion of the hour-hand in a clock may be occasioned either by a weight hung, or a spring shut up within. But if a certain clock should be really moved with a weight, we should laugh at a man that would suppose it moved by a spring, and from that principle, suddenly taken up without further examination, should go about to explain the motion of the index; for certainly the way be ought to have taken would have been actually to look into the inward parts of the machine, that he might find the true principle of the proposed motion. The like judgment ought to be made of those philosophers who will have the heavens to be filled with a most subtile matter which is continually carried round in vortices. For if they could explain the phenomena ever so accurately by their hypotheses, we could not yet say that they have discovered true philosophy and the true causes of the celestial motions, unless they could either demonstrate that those causes do actually exist, or at least that no others do exist. Therefore if it be made clear that the attraction of all bodies is a property actually existing *in rerum natura,* and if it be also shown how the motions of the celestial bodies may be solved by that property, it would be very impertinent for anyone to object that these motions ought to be accounted for by vortices; even though we should allow such an explication of those motions to be possible. But we allow no such thing; for the phenomena can by no means be accounted for by vortices, as our author has abundantly proved from the clearest reasons. So that men must be strangely fond of chimeras, who can spend their time so idly as in patching up a ridiculous figment and setting it off with new comments of their own.

If the bodies of the planets and comets are carried round the sun in vortices, the bodies so carried, and the parts of the vortices next surrounding them, must be carried with the same velocity and the same direction, and have the same density, and the same inertia, answering to the bulk of the matter. But it is certain, the planets and comets, when in the very same parts of the heavens, are carried with various velocities and various directions. Therefore it necessarily follows that those parts of the celestial fluid, which are at the same distances from the sun, must revolve at the same time with different velocities in different directions; for one kind of velocity and direction is required for the motion of the planets, and another for that of the comets. But since this cannot be accounted for, we must either say that all celestial bodies are not carried about by vortices, or else that their motions are derived, not from one and the same vortex, but from several distinct ones, which fill and pervade the spaces round about the sun.

But if several vortices are contained in the same space, and are supposed to penetrate one another, and to revolve with different motions, then because these motions must agree with those of the bodies carried about by them, which are perfectly regular, and performed in conic sections which are sometimes very eccentric, and sometimes nearly circles, one may very reasonably ask how it comes to pass that these vortices remain entire, and have suffered no manner of perturbation in so many ages from the actions of the conflicting matter. Certainly if these fictitious motions are more compounded and harder to be accounted for than the true motions of the planets and comets, it seems to no purpose to admit them into philosophy, since every cause ought to be more simple than its effect. Allowing men to indulge their own fancies, suppose any man should affirm that the planets and comets are surrounded with atmospheres like our earth, which hypothesis seems more reasonable than that of vortices; let him then affirm that these atmospheres by their own nature move about the sun and describe conic sections, which motion is much more easily conceived than that of the vortices penetrating one another; lastly, that the planets and comets are carried about the sun by these atmospheres of theirs: and then applaud his own sagacity in discovering the causes of the celestial motions. He that rejects this fable must also reject the other; for two drops of water are not more like than this hypothesis of atmospheres, and that of vortices.

Galileo has shown that when a stone projected moves in a parabola, its deflection into that curve from its rectilinear path is occasioned by the gravity of the stone towards the earth, that is, by an occult quality. But now somebody, more cunning than he, may come to explain the cause after this manner. He will suppose a certain subtile matter, not discernible by our sight, our touch, or any other of our senses, which fills the spaces which are near and contiguous to the surface of the earth, and that this matter is carried with different directions, and various, and often contrary, motions, describing parabolic curves. Then see how easily he may account for the deflection of the stone above spoken of. The stone, says he, floats in this subtile fluid, and following its motion, can't choose but describe the same figure. But the fluid moves in parabolic curves, and therefore the stone must move in a parabola, of course. Would not the acuteness of this philosopher be thought very extraordinary, who could deduce the appearances of Nature from mechanical causes, matter and motion, so clearly that the meanest man may understand it? Or indeed should not we smile to see this new *Galileo* taking so much mathematical pains to introduce occult qualities into philosophy, from whence they have been so happily excluded? But I am ashamed to dwell so long upon trifles.

The sum of the matter is this: the number of the comets is certainly very great; their motions are perfectly regular and observe the same laws with those of the planets. The orbits in which they move are conic sections, and those very eccentric. They move every way towards all parts of the heavens, and pass through the planetary regions with all possible freedom, and their motion is often contrary to the order of the signs. These phenomena are most evidently confirmed by astronomical observations, and cannot be accounted for by vortices. Nay, indeed, they are utterly irreconcilable with the vortices of the planets. There can be no room for the motions of the comets, unless the celestial spaces be entirely cleared of that fictitious matter.

For if the planets are carried about the sun in vortices, the parts of the vortices which immediately surround every planet must be of the same density with the planet, as was shown above. Therefore all the matter contiguous to the perimeter of the earth's orbit[1] must be of the same density as the earth. But this great orb and the orb of Saturn must have either an equal or a greater density. For to make the constitution of the vortex permanent, the parts of less density must lie near the centre, and those of greater density must go farther from it. For since the periodic times of the planets vary as the ³⁄₂th powers of their distances from the sun, the periods of the parts of the vortices must also preserve the same ratio. Thence it will follow that the centrifugal forces of the parts of the vortex must be inversely as the squares of their distances. Those parts therefore which are more remote from the centre endeavor to recede from it with less force; whence, if their density be deficient, they must yield to the greater force with which the parts that lie nearer the centre endeavor to ascend. Therefore the denser parts will ascend, and those of less density will descend, and there will be a mutual change of places, till all the fluid matter in the whole vortex be so adjusted and disposed, that being reduced to an equilibrium its parts become quiescent. If two fluids of different density be contained in the same vessel, it will certainly come to pass that the fluid of greater density will sink the lower; and by a like reasoning it follows that the denser parts of the vortex by their greater centrifugal force will ascend to the higher places. Therefore all that far greater part of the vortex which lies without the earth's orb, will have a density, and by consequence an inertia, answering to the bulk of the matter, which cannot be less than the density and inertia of the earth. But from hence will arise a mighty resistance to the passage of the comets, such as must be very sensible, not to say enough to put a stop to and absorb their motions entirely. But it appears from the perfectly regular motion of the comets, that they suffer no resistance that is in the least sensible, and therefore that they do not meet with matter of any kind that has any resisting force or, by consequence, any density or inertia. For the resistance of mediums arises either from the inertia of the matter of the fluid, or from its want of lubricity. That which arises from the want of lubricity is very small, and is scarcely observable in the fluids commonly known, unless they be very tenacious like oil and honey. The resistance we find in air, water, quicksilver, and the like fluids that are not tenacious, is almost all of the first kind, and cannot be diminished by a greater degree of subtilty, if the density and inertia, to which this resistance is proportional, remains, as is most evidently demonstrated by our author in his noble theory of resistances in the second Book.

[¹ Appendix, Note 7.]

Bodies in going on through a fluid communicate their motion to the ambient fluid by little and little, and by that communication lose their own motion, and by losing it are retarded. Therefore the retardation is proportional to the motion communicated, and the communicated motion, when the velocity of the moving body is given, is as the density of the fluid; and therefore the retardation or resistance will be as the same density of the fluid; nor can it be taken away, unless the fluid, coming about to the hinder parts of the body, restore the motion lost. Now this cannot be done unless the impression of the fluid on the hinder parts of the body be equal to the impression of the fore parts of the body on the fluid; that is, unless the relative velocity with which the fluid pushes the body behind is equal to the velocity with which the body pushes the fluid; that is, unless the absolute velocity of the recurring fluid be twice as great as the absolute velocity with which the fluid is driven forwards by the body, which is impossible. Therefore the resistance of fluids arising from their inertia can by no means be taken away. So that we must conclude that the celestial fluid has no inertia, because it has no resisting force; that it has no force to communicate motion with, because it has no inertia; that it has no force to produce any change in one or more bodies, because it has no force wherewith to communicate motion; that it has no manner of efficacy, because it has no faculty wherewith to produce any change of any kind. There fore certainly this hypothesis may be justly called ridiculous and unworthy a philosopher, since it is altogether without foundation and does not in the least serve to explain the nature of things.[1] Those who would have the heavens filled with a fluid matter, but suppose it void of any inertia, do indeed in words deny a vacuum, but allow it in fact. For since a fluid matter of that kind can noways be distinguished from empty space, the dispute is now about the names and not the natures of things. If any are so fond of matter that they will by no means admit of a space void of body, let us consider where they must come at last.

For either they will say that this constitution of a world everywhere full was made so by the will of God to this end, that the operations of Nature might be assisted everywhere by a subtile ether pervading and filling all things; which cannot be said, however, since we have shown from the phenomena of the comets, that this ether is of no efficacy at all; or they will say, that it became so by the same will of God for some unknown end, which ought not be said, because for the same reason a different constitution may be as well supposed; or lastly, they will not say that it was caused by the will of God, but by some necessity of its nature. Therefore they will at last sink into the mire of that infamous herd who dream that all things are governed by fate and not by providence, and that matter exists by the necessity of its nature always and everywhere, being infinite and eternal. But supposing these things, it must be also everywhere uniform; for variety of forms is entirely inconsistent with necessity. It must be also unmoved;

[¹ Appendix, Note 8.]

for if it be necessarily moved in any determinate direction, with any determinate velocity, it will by a like necessity be moved in a different direction with a different velocity; but it can never move in different directions with different velocities; therefore it must be unmoved. Without all doubt this world, so diversified with that variety of forms and motions we find in it, could arise from nothing but the perfectly free will of God directing and presiding over all.

From this fountain it is that those laws, which we call the laws of Nature, have flowed, in which there appear many traces indeed of the most wise contrivance, but not the least shadow of necessity. These therefore we must not seek from uncertain conjectures, but learn them from observations and experiments. He who is presumptuous enough to think that he can find the true principles of physics and the laws of natural things by the force alone of his own mind, and the internal light of his reason, must either suppose that the world exists by necessity, and by the same necessity follows the laws proposed; or if the order of Nature was established by the will of God, that himself, a miserable reptile, can tell what was fittest to be done. All sound and true philosophy is founded on the appearances of things; and if these phenomena inevitably draw us, against our wills, to such principles as most clearly manifest to us the most excellent counsel and supreme dominion of the All-wise and Almighty Being, they are not therefore to be laid aside because some men may perhaps dislike them. These men may call them miracles or occult qualities, but names maliciously given ought not to be a disadvantage to the things themselves, unless these men will say at last that all philosophy ought to be founded in atheism. Philosophy must not be corrupted in compliance with these men, for the order of things will not be changed.

Fair and equal judges will therefore give sentence in favor of this most excellent method of philosophy, which is founded on experiments and observations. And it can hardly be said or imagined, what light, what splendor, hath accrued to that method from this admirable work of our illustrious author, whose happy and sublime genius, resolving the most difficult problems, and reaching to discoveries of which the mind of man was thought incapable before, is deservedly admired by all those who are somewhat more than superficially versed in these matters. The gates are now set open, and by the passage he has revealed we may freely enter into the knowledge of the hidden secrets and wonders of natural things. He has so clearly laid open and set before our eyes the most beautiful frame of the System of the World, that if King *Alphonso* were now alive, he would not complain for want of the graces either of simplicity or of harmony in it. Therefore we may now more nearly behold the beauties of Nature, and entertain ourselves with the delightful contemplation; and, which is the best and most valuable fruit of philosophy, be thence incited the more profoundly to reverence and adore the great Maker and Lord of all. He must be blind who from the most wise and excellent contrivances of things cannot see the infinite Wisdom and Goodness of their Almighty Creator, and he must be mad and senseless who refuses to acknowledge them.

Newton's distinguished work will be the safest protection against the attacks of atheists, and nowhere more surely than from this quiver can one draw forth missiles against the band of godless men. This was felt long ago and first surprisingly demonstrated in learned English and Latin discourses by *Richard Bentley,* who, excelling in learning and distinguished as a patron of the highest arts, is a great ornament of his century and of our academy, the most worthy and upright Master of our *Trinity College.* To him in many ways I must express my indebtedness. And you too, benevolent reader, will not withhold the esteem due him. For many years an intimate friend of the celebrated author (since he aimed not only that the author should be esteemed by those who come after, but also that these uncommon writings should enjoy distinction among the literati of the world), he cared both for the reputation of his friend and for the advancement of the sciences. Since copies of the previous edition were very scarce and held at high prices, he persuaded by frequent entreaties and almost by chidings, the splendid man, distinguished alike for modesty and for erudition, to grant him permission for the appearance of this new edition, perfected throughout and enriched by new parts, at his expense and under his supervision. He assigned to me, as he had a right, the not unwelcome task of looking after the corrections as best I could.

ROGER COTES

Fellow of Trinity College,
Plumian Professor of Astronomy
and Experimental Philosophy.

Cambridge, May 12, 1713.

Newton's Preface to the Third Edition

IN THIS THIRD EDITION, prepared with much care by *Henry Pemberton*, M.D., a man of the greatest skill in these matters, some things in the second Book on the resistance of mediums are somewhat more comprehensively handled than before, and new experiments on the resistance of heavy bodies falling in air are added. In the third Book, the argument to prove that the moon is retained in its orbit by the force of gravity is more fully stated; and there are added new observations made by Mr. *Pound*, concerning the ratio of the diameters of Jupiter to one another. Some observations are also added on the comet which appeared in the year 1680, made in *Germany* in the month of *November* by Mr. *Kirk;* which have lately come to my hands. By the help of these it becomes apparent how nearly parabolic orbits represent the motions of comets. The orbit of that comet is determined somewhat more accurately than before, by the computation of Dr. *Halley,* in an ellipse. And it is shown that, in this elliptic orbit, the comet took its course through the nine signs of the heavens, with as much accuracy as the planets move in the elliptic orbits given in astronomy. The orbit of the comet which appeared in the year 1723 is also added, computed by Mr. *Bradley,* Professor of Astronomy at *Oxford*.[1]

Is. NEWTON

London, Jan. 12, 1725–6.

[1 Appendix, Note 9.]

MATHEMATICAL PRINCIPLES
OF
NATURAL PHILOSOPHY[1]

Definitions

DEFINITION I

The quantity of matter is the measure of the same, arising from its density and bulk conjointly.[2]

THUS AIR of a double density, in a double space, is quadruple in quantity; in a triple space, sextuple in quantity. The same thing is to be understood of snow, and fine dust or powders, that are condensed by compression or liquefaction, and of all bodies that are by any causes whatever differently condensed. I have no regard in this place to a medium, if any such there is, that freely pervades the interstices between the parts of bodies. It is this quantity that I mean hereafter everywhere under the name of body or mass. And the same is known by the weight of each body, for it is proportional to the weight, as I have found by experiments on pendulums, very accurately made, which shall be shown hereafter.

DEFINITION II[3]

The quantity of motion is the measure of the same, arising from the velocity and quantity of matter conjointly.

The motion of the whole is the sum of the motions of all the parts; and therefore in a body double in quantity, with equal velocity, the motion is double; with twice the velocity, it is quadruple.

[1 Appendix, Note 10.]   [2 Appendix, Note 11.]   [3 Appendix, Note 12.]

DEFINITION III

The vis insita, *or innate force of matter, is a power of resisting, by which every body, as much as in it lies, continues in its present state, whether it be of rest, or of moving uniformly forwards in a right line.*

This force is always proportional to the body whose force it is and differs nothing from the inactivity of the mass, but in our manner of conceiving it. A body, from the inert nature of matter, is not without difficulty put out of its state of rest or motion. Upon which account, this *vis insita* may, by a most significant name, be called inertia (*vis inertiæ*) or force of inactivity. But a body only exerts this force when another force, impressed upon it, endeavors to change its condition; and the exercise of this force may be considered as both resistance and impulse; it is resistance so far as the body, for maintaining its present state, opposes the force impressed; it is impulse so far as the body, by not easily giving way to the impressed force of another, endeavors to change the state of that other. Resistance is usually ascribed to bodies at rest, and impulse to those in motion; but motion and rest, as commonly conceived, are only relatively distinguished; nor are those bodies always truly at rest, which commonly are taken to be so.

DEFINITION IV

An impressed force is an action exerted upon a body, in order to change its state, either of rest, or of uniform motion in a right line.

This force consists in the action only, and remains no longer in the body when the action is over. For a body maintains every new state it acquires, by its inertia only. But impressed forces are of different origins, as from percussion, from pressure, from centripetal force.

DEFINITION V

A centripetal force is that by which bodies are drawn or impelled, or any way tend, towards a point as to a centre.

Of this sort is gravity, by which bodies tend to the centre of the earth; magnetism, by which iron tends to the loadstone; and that force, whatever it is, by which the planets are continually drawn aside from the rectilinear motions, which otherwise they would pursue, and made to revolve in curvilinear orbits. A stone, whirled about in a sling, endeavors to recede from the hand that turns it; and by that endeavor, distends the sling, and that with so much the greater force, as it is revolved with the greater velocity, and as soon as it is let go, flies away. That force which opposes itself to this endeavor, and by which the sling continually draws back the stone towards the hand, and retains it in its orbit, because it is directed to the hand as the centre of the orbit, I call the centripetal force. And the same thing is to be understood of all bodies, revolved in any orbits. They all endeavor to recede from the centres of their orbits; and were it not for the opposition of a contrary force which restrains them to, and detains them in their orbits, which I therefore call centripetal, would fly off in right lines, with an uniform motion. A projectile, if it was not for the force of gravity, would not deviate towards the earth, but would go off from it in a right line, and that with an uniform motion, if the resistance of the air was taken away. It is by its gravity that it is drawn aside continually from its rectilinear course, and made to deviate towards the earth, more or less, according to the force of its gravity, and the velocity of its motion. The less its gravity is, or the quantity of its matter, or the greater the velocity with which it is projected, the less will it deviate from a rectilinear course, and the farther it will go. If a leaden ball, projected from the top of a mountain by the force of gunpowder, with a given velocity, and in a direction parallel to the horizon, is carried in a curved line to the distance of two miles before it falls to the ground; the same, if the resistance of the air were taken away, with a double or decuple velocity, would fly twice or ten times as far. And by increasing the velocity, we may at pleasure increase the distance to which it might be projected, and diminish the curvature of the line which it might describe, till at last it should fall at the distance of 10, 30, or 90 degrees, or even might go quite round the whole earth before it falls; or lastly, so that it might never fall to the earth, but go forwards into the celestial spaces, and proceed in its motion *in infinitum*. And after the same manner that a projectile, by the force of gravity, may be made to revolve in an orbit, and go round the whole earth, the moon also, either by the force of gravity, if it is endued with gravity, or by any other force, that impels it towards the earth, may be continually drawn aside towards the earth, out of the rectilinear way which by its innate force it would pursue; and would be made to revolve in the orbit which it now describes; nor could the moon without some such force be retained in its orbit. If this force was too small, it would not sufficiently turn the moon out of a rectilinear course; if it was too great, it would turn it too much, and draw down the moon from its orbit towards the earth. It is necessary that the force be of a just quantity, and it belongs to the mathematicians to find the force that may serve exactly to retain a body in a given orbit with a given velocity; and *vice versa*, to determine the curvilinear way into which a body projected from a given place, with a given velocity, may be made to deviate from its natural rectilinear way, by means of a given force.

The quantity of any centripetal force may be considered as of three kinds: absolute, accelerative, and motive.

DEFINITION VI

The absolute quantity of a centripetal force is the measure of the same, proportional to the efficacy of the cause that propagates it from the centre, through the spaces round about.

Thus the magnetic force is greater in one loadstone and less in another, according to their sizes and strength of intensity.

DEFINITION VII

The accelerative quantity of a centripetal force is the measure of the same, proportional to the velocity which it generates in a given time.

Thus the force of the same loadstone is greater at a less distance, and less at a greater: also the force of gravity is greater in valleys, less on tops of exceeding high mountains; and yet less (as shall hereafter be shown), at greater distances from the body of the earth; but at equal distances, it is the same everywhere; because (taking away, or allowing for, the resistance of the air), it equally accelerates all falling bodies, whether heavy or light, great or small.

DEFINITION VIII

The motive quantity of a centripetal force is the measure of the same, proportional to the motion which it generates in a given time.

Thus the weight is greater in a greater body, less in a less body; and, in the same body, it is greater near to the earth, and less at remoter distances. This sort of quantity is the centripetency, or propension of the whole body towards the centre, or, as I may say, its weight; and it is always known by the quantity of an equal and contrary force just sufficient to hinder the descent of the body.

These quantities of forces, we may, for the sake of brevity, call by the names of motive, accelerative, and absolute forces; and, for the sake of distinction, consider them with respect to the bodies that tend to the centre, to the places of those bodies, and to the centre of force towards which they tend; that is to say, I refer the motive force to the body as an endeavor and propensity of the whole towards a centre, arising from the propensities of the several parts taken together; the accelerative force to the place of the body, as a certain power diffused from the centre to all places around to move the bodies that are in them; and the absolute force to the centre, as endued with some cause, without which those motive forces would not be propagated through the spaces round about; whether that cause be some central body (such as is the magnet in the centre of the magnetic force, or the earth in the centre of the gravitating force), or anything else that does not yet appear. For I here design only to give a mathematical notion of those forces, without considering their physical causes and seats.

Wherefore the accelerative force will stand in the same relation to the motive, as celerity does to motion. For the quantity of motion arises from the celerity multiplied by the quantity of matter; and the motive force arises from the accelerative force multiplied by the same quantity of matter. For the sum of the actions of the accelerative force, upon the several particles of the body, is the motive force of the whole. Hence it is, that near the surface of the earth, where the accelerative gravity, or force productive of gravity, in all bodies is the same, the motive gravity or the weight is as the body; but if we should ascend to higher regions, where the accelerative gravity is less, the weight would be equally diminished, and would always be as the product of the body, by the accelerative gravity. So in those regions, where the accelerative gravity is diminished into one-half, the weight of a body two or three times less, will be four or six times less.

I likewise call attractions and impulses, in the same sense, accelerative, and motive; and use the words attraction, impulse, or propensity of any sort towards a centre, promiscuously, and indifferently, one for another; considering those forces not physically, but mathematically: wherefore the reader is not to imagine that by those words I anywhere take upon me to define the kind, or the manner of any action, the causes or the physical reason thereof, or that I attribute forces, in a true and physical sense, to certain centres (which are only mathematical points); when at any time I happen to speak of centres as attracting, or as endued with attractive powers.

SCHOLIUM[1]

Hitherto I have laid down the definitions of such words as are less known, and explained the sense in which I would have them to be understood in the following discourse. I do not define time, space, place, and motion, as being well known to all. Only I must observe, that the common people conceive those quantities under no other notions but from the relation they bear to sensible objects. And thence arise certain prejudices, for the removing of which it will be convenient to distinguish them into absolute and relative, true and apparent, mathematical and common.

I. Absolute, true, and mathematical time, of itself, and from its own nature, flows equably without relation to anything external, and by another name is called duration: relative, apparent, and common time, is some sensible and external (whether accurate or unequable) measure of duration by the means of motion, which is commonly used instead of true time; such as an hour, a day, a month, a year.

II. Absolute space, in its own nature, without relation to anything external, remains always similar and immovable. Relative space is some movable dimension or measure of the absolute spaces; which our senses determine by its position to bodies; and which is commonly taken for immovable space; such is the dimension of a subterraneous, an aerial, or celestial space, determined by its position in respect of the earth. Absolute and relative space are the same in figure and magnitude; but they do not remain always numerically the same. For if the earth, for instance, moves, a space of our air, which relatively and in respect of the earth remains always the same, will at one time be one part of the absolute space into which the air passes; at another time it will be another part of the same, and so, absolutely understood, it will be continually changed.

III. Place is a part of space which a body takes up, and is according to the space, either absolute or relative. I say, a part of space; not the situation, nor the external surface of the body. For the places of equal solids are always

[1 Appendix, Note 13.]

equal; but their surfaces, by reason of their dissimilar figures, are often unequal. Positions properly have no quantity, nor are they so much the places themselves, as the properties of places. The motion of the whole is the same with the sum of the motions of the parts; that is, the translation of the whole, out of its place, is the same thing with the sum of the translations of the parts out of their places; and therefore the place of the whole is the same as the sum of the places of the parts, and for that reason, it is internal, and in the whole body.

IV. Absolute motion is the translation of a body from one absolute place into another; and relative motion, the translation from one relative place into another. Thus in a ship under sail, the relative place of a body is that part of the ship which the body possesses; or that part of the cavity which the body fills, and which therefore moves together with the ship: and relative rest is the continuance of the body in the same part of the ship, or of its cavity. But real, absolute rest, is the continuance of the body in the same part of that immovable space, in which the ship itself, its cavity, and all that it contains, is moved. Wherefore, if the earth is really at rest, the body, which relatively rests in the ship, will really and absolutely move with the same velocity which the ship has on the earth. But if the earth also moves, the true and absolute motion of the body will arise, partly from the true motion of the earth, in immovable space, partly from the relative motion of the ship on the earth; and if the body moves also relatively in the ship, its true motion will arise, partly from the true motion of the earth, in immovable space, and partly from the relative motions as well of the ship on the earth, as of the body in the ship; and from these relative motions will arise the relative motion of the body on the earth. As if that part of the earth, where the ship is, was truly moved towards the east, with a velocity of 10010 parts; while the ship itself, with a fresh gale, and full sails, is carried towards the west, with a velocity expressed by 10 of those parts; but a sailor walks in the ship towards the east, with 1 part of the said velocity; then the sailor will be moved truly in immovable space towards the east, with a velocity of 10001 parts, and relatively on the earth towards the west, with a velocity of 9 of those parts.

Absolute time, in astronomy, is distinguished from relative, by the equation or correction of the apparent time. For the natural days are truly unequal, though they are commonly considered as equal, and used for a measure of time; astronomers correct this inequality that they may measure the celestial motions by a more accurate time. It may be, that there is no such thing as an equable motion, whereby time may be accurately measured. All motions may be accelerated and retarded, but the flowing of absolute time is not liable to any change. The duration or perseverance of the existence of things remains the same, whether the motions are swift or slow, or none at all: and therefore this duration ought to be distinguished from what are only sensible measures thereof; and from which we deduce it, by means of the astronomical equation. The necessity of this equation, for determining the times of a phenomenon, is evinced as well from the experiments of the pendulum clock, as by eclipses of the satellites of Jupiter.

As the order of the parts of time is immutable, so also is the order of the parts of space. Suppose those parts to be moved out of their places, and they will be moved (if the expression may be allowed) out of themselves. For times and spaces are, as it were, the places as well of themselves as of all other things. All things are placed in time as to order of succession; and in space as to order of situation. It is from their essence or nature that they are places; and that the primary places of things should be movable, is absurd. These are therefore the absolute places; and translations out of those places, are the only absolute motions.

But because the parts of space cannot be seen, or distinguished from one another by our senses, therefore in their stead we use sensible measures of them. For from the positions and distances of things from any body considered as immovable, we define all places; and then with respect to such places, we estimate all motions, considering bodies as transferred from some of those places into others. And so, instead of absolute places and motions, we use relative ones; and that without any inconvenience in common affairs; but in philosophical disquisitions, we ought to abstract from our senses, and consider things themselves, distinct from what are only sensible measures of them. For it may be that there is no body really at rest, to which the places and motions of others may be referred.

But we may distinguish rest and motion, absolute and relative, one from the other by their properties, causes, and effects. It is a property of rest, that bodies really at rest do rest in respect to one another. And therefore as it is possible, that in the remote regions of the fixed stars, or perhaps far beyond them, there may be some body absolutely at rest; but impossible to know, from the position of bodies to one another in our regions, whether any of these do keep the same position to that remote body, it follows that absolute rest cannot be determined from the position of bodies in our regions.

It is a property of motion, that the parts, which retain given positions to their wholes, do partake of the motions of those wholes. For all the parts of revolving bodies endeavor to recede from the axis of motion; and the impetus of bodies moving forwards arises from the joint impetus of all the parts. Therefore, if surrounding bodies are moved, those that are relatively at rest within them will partake of their motion. Upon which account, the true and absolute motion of a body cannot be determined by the translation of it from those which only seem to rest; for the external bodies ought not only to appear at rest, but to be really at rest. For otherwise, all included bodies, besides their translation from near the surrounding ones, partake likewise of their true motions; and though that translation were not made, they would not be really at rest, but only seem to be so. For the surrounding bodies stand in the like relation to the surrounded as the exterior part of a whole does to the interior, or as the shell does to the kernel; but if the shell moves, the kernel will also move, as being part of the whole, without any removal from near the shell.

A property, near akin to the preceding, is this, that if a place is moved, whatever is placed therein moves along with it; and therefore a body, which is moved from a place in motion, partakes also of the motion of its place. Upon which account, all motions, from places in motion, are no other than parts of entire and absolute motions; and every entire motion is composed of the motion of the body out of its first place, and the motion of this place out of its place; and so on, until we come to some immovable place, as in the before-mentioned example of the sailor. Wherefore, entire and absolute motions can be no otherwise determined than by immovable places; and for that reason I did before refer those absolute motions to immovable places, but relative ones to movable places. Now no other places are immovable but those that, from infinity to infinity, do all retain the same given position one to another; and upon this account must ever remain unmoved; and do thereby constitute immovable space.

The causes by which true and relative motions are distinguished, one from the other, are the forces impressed upon bodies to generate motion. True motion is neither generated nor altered, but by some force impressed upon the body moved; but relative motion may be generated or altered without any force impressed upon the body. For it is sufficient only to impress some force on other bodies with which the former is compared, that by their giving way, that relation may be changed, in which the relative rest or motion of this other body did consist. Again, true motion suffers always some change from any force impressed upon the moving body; but relative motion does not necessarily undergo any change by such forces. For if the same forces are likewise impressed on those other bodies, with which the comparison is made, that the relative position may be preserved, then that condition will be preserved in which the relative motion consists. And therefore any relative motion may be changed when the true motion remains unaltered, and the relative may be preserved when the true suffers some change. Thus, true motion by no means consists in such relations.

The effects which distinguish absolute from relative motion are, the forces of receding from the axis of circular motion. For there are no such forces in a circular motion purely relative, but in a true and absolute circular motion, they are greater or less, according to the quantity of the motion. If a vessel, hung by a long cord, is so often turned about that the cord is strongly twisted, then filled with water, and held at rest together with the water; thereupon, by the sudden action of another force, it is whirled about the contrary way, and while the cord is untwisting itself, the vessel continues for some time in this motion; the surface of the water will at first be plain, as before the vessel began to move; but after that, the vessel, by gradually communicating its motion to the water, will make it begin sensibly to revolve, and recede by little and little from the middle, and ascend to the sides of the vessel, forming itself into a concave figure (as I have experienced), and the swifter the motion becomes, the higher will the water rise, till at last, performing its revolutions in the same times with the vessel, it becomes relatively at rest in it. This ascent of the water shows its endeavor to recede from the axis of its motion; and the true and absolute circular motion of the water, which is here directly contrary to the relative, becomes known, and may be measured by this endeavor. At first, when the relative motion of the water in the vessel was greatest, it produced no endeavor to recede from the axis; the water showed no tendency to the circumference, nor any ascent towards the sides of the vessel, but remained of a plain surface, and therefore its true circular motion had not yet begun. But afterwards, when the relative motion of the water had decreased, the ascent thereof towards the sides of the vessel proved its endeavor to recede from the axis; and this endeavor showed the real circular motion of the water continually increasing, till it had acquired its greatest quantity, when the water rested relatively in the vessel. And therefore this endeavor does not depend upon any translation of the water in respect of the ambient bodies, nor can true circular motion be defined by such translation. There is only one real circular motion of any one revolving body, corresponding to only one power of endeavoring to recede from its axis of motion, as its proper and adequate effect; but relative motions, in one and the same body, are innumerable, according to the various relations it bears to external bodies, and, like other relations, are altogether destitute of any real effect, any otherwise than they may perhaps partake of that one only true motion. And therefore in their system who suppose that our heavens, revolving below the sphere of the fixed stars, carry the planets along with them; the several parts of those heavens, and the planets, which are indeed relatively at rest in their heavens, do yet really move. For they change their position one to another (which never happens to bodies truly at rest), and being carried together with their heavens, partake of their motions, and as parts of revolving wholes, endeavor to recede from the axis of their motions.

Wherefore relative quantities are not the quantities themselves, whose names they bear, but those sensible measures of them (either accurate or inaccurate), which are commonly used instead of the measured quantities themselves. And if the meaning of words is to be determined by their use, then by the names time, space, place, and motion, their [sensible] measures are properly to be understood; and the expression will be unusual, and purely mathematical, if the measured quantities themselves are meant. On this account, those violate the accuracy of language, which ought to be kept precise, who interpret these words for the measured quantities. Nor do those less defile the purity of mathematical and philosophical truths, who confound real quantities with their relations and sensible measures.

It is indeed a matter of great difficulty to discover, and effectually to distinguish, the true motions of particular bodies from the apparent; because the parts of that immovable space, in which those motions are performed, do by no means come under the observation of our senses. Yet the thing is not altogether desperate; for we have some arguments to guide us, partly from the apparent motions, which are the differences of the true motions; partly from the forces, which are the causes and effects of the true motions. For instance, if two globes, kept at a given distance one from the other by means of a cord that connects them, were revolved about their common centre of gravity, we might, from the tension of the cord, discover the endeavor of the globes to recede from the axis of their motion, and from thence we might compute the quantity of their circular motions. And then if any equal forces should be impressed at once on the alternate faces of the globes to augment or diminish their circular motions, from the increase or decrease of the tension of the cord, we might infer the increment or decrement of their motions; and thence would be found on what faces those forces ought to be impressed, that the motions of the globes might be most augmented; that is, we might discover their hindmost faces, or those which, in the circular motion, do follow. But the faces which follow being known, and consequently the opposite ones that precede, we should likewise know the determination of their motions. And thus we might find both the quantity and the determination of this circular motion, even in an immense vacuum, where there was nothing external or sensible with which the globes could be compared. But now, if in that space some remote bodies were placed that kept always a given position one to another, as the fixed stars do in our regions, we could not indeed determine from the relative translation of the globes among those bodies, whether the motion did belong to the globes or to the bodies. But if we observed the cord, and found that its tension was that very tension which the motions of the globes required, we might conclude the motion to be in the globes, and the bodies to be at rest; and then, lastly, from the translation of the globes among the bodies, we should find the determination of their motions. But how we are to obtain the true motions from their causes, effects, and apparent differences, and the converse, shall be explained more at large in the following treatise. For to this end it was that I composed it.

AXIOMS, or
LAWS OF MOTION[1]

LAW I

Every body continues in its state of rest, or of uniform motion in a right line, unless it is compelled to change that state by forces impressed upon it.

PROJECTILES continue in their motions, so far as they are not retarded by the resistance of the air, or impelled downwards by the force of gravity. A top, whose parts by their cohesion are continually drawn aside from rectilinear motions, does not cease its rotation, otherwise than as it is retarded by the air. The greater bodies of the planets and comets, meeting with less resistance in freer spaces, preserve their motions both progressive and circular for a much longer time.

LAW II[2]

The change of motion is proportional to the motive force impressed; and is made in the direction of the right line in which that force is impressed.

If any force generates a motion, a double force will generate double the motion, a triple force triple the motion, whether that force be impressed altogether and at once, or gradually and successively. And this motion (being always directed the same way with the generating force), if the body moved before, is added to or subtracted from the former motion, according as they directly conspire with or are directly contrary to each other; or obliquely joined, when they are oblique, so as to produce a new motion compounded from the determination of both.

LAW III

To every action there is always opposed an equal reaction: or, the mutual actions of two bodies upon each other are always equal, and directed to contrary parts.

Whatever draws or presses another is as much drawn or pressed by that other. If you press a stone with your finger, the finger is also pressed by the

[¹ Appendix, Note 14.]   [² Appendix, Note 15.]

stone. If a horse draws a stone tied to a rope, the horse (if I may so say) will be equally drawn back towards the stone; for the distended rope, by the same endeavor to relax or unbend itself, will draw the horse as much towards the stone as it does the stone towards the horse, and will obstruct the progress of the one as much as it advances that of the other. If a body impinge upon another, and by its force change the motion of the other, that body also (because of the equality of the mutual pressure) will undergo an equal change, in its own motion, towards the contrary part. The changes made by these actions are equal, not in the velocities but in the motions of bodies; that is to say, if the bodies are not hindered by any other impediments. For, because the motions are equally changed, the changes of the velocities made towards contrary parts are inversely proportional to the bodies. This law takes place also in attractions, as will be proved in the next Scholium.

COROLLARY I

A body, acted on by two forces simultaneously, will describe the diagonal of a parallelogram in the same time as it would describe the sides by those forces separately.

If a body in a given time, by the force M impressed apart in the place A, should with an uniform motion be carried from A to B, and by the force N impressed apart in the same place, should be carried from A to C, let the

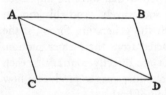

parallelogram ABCD be completed, and, by both forces acting together, it will in the same time be carried in the diagonal from A to D. For since the force N acts in the direction of the line AC, parallel to BD, this force (by the second Law) will not at all alter the velocity generated by the other force M, by which the body is carried towards the line BD. The body therefore will arrive at the line BD in the same time, whether the force N be impressed or not; and therefore at the end of that time it will be found somewhere in the line BD. By the same argument, at the end of the same time it will be found somewhere in the line CD. Therefore it will be found in the point D, where both lines meet. But it will move in a right line from A to D, by Law I.

COROLLARY II

And hence is explained the composition of any one direct force AD, out of any two oblique forces AC and CD; and, on the contrary, the resolution of any one direct force AD into two oblique forces AC and CD: which composition and resolution are abundantly confirmed from mechanics.

As if the unequal radii OM and ON drawn from the centre O of any wheel, should sustain the weights A and P by the cords MA and NP; and the forces of those weights to move the wheel were required. Through the centre O draw the right line KOL, meeting the cords perpendicularly in K and L; and from the centre O, with OL the greater of the distances OK and OL, describe a circle, meeting the cord MA in D; and drawing OD, make AC parallel and DC perpendicular thereto. Now, it being in-

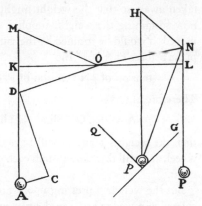

different whether the points K, L, D, of the cords be fixed to the plane of the wheel or not, the weights will have the same effect whether they are suspended from the points K and L, or from D and L. Let the whole force of the weight A be represented by the line AD, and let it be resolved into the forces AC and CD, of which the force AC, drawing the radius OD directly from the centre, will have no effect to move the wheel; but the other force DC, drawing the radius DO perpendicularly, will have the same effect as if it drew perpendicularly the radius OL equal to OD; that is, it will have the same effect as the weight P, if

$$P : A = DC : DA,$$

but because the triangles ADC and DOK are similar,

$$DC : DA = OK : OD = OK : OL.$$

Therefore,

$$P : A = \text{radius } OK : \text{radius } OL.$$

As these radii lie in the same right line they will be equipollent, and so remain in equilibrium; which is the well-known property of the balance, the lever, and the wheel. If either weight is greater than in this ratio, its force to move the wheel will be so much greater.

If the weight $p = P$, is partly suspended by the cord Np, partly sustained by the oblique plane pG; draw pH, NH, the former perpendicular to the horizon, the latter to the plane pG; and if the force of the weight p tending downwards is represented by the line pH, it may be resolved into the forces pN, HN. If there was any plane pQ, perpendicular to the cord pN, cutting the other plane pG in a line parallel to the horizon, and the weight p was supported only by those planes pQ, pG, it would press those planes perpendicularly with the forces pN, HN; to wit, the plane pQ with the force pN, and the plane pG with the force HN. And therefore if the plane pQ was taken away, so that the weight might stretch the cord, because the cord, now sustaining the weight, supplied the place of the plane that was removed, it would be strained by the same force pN which pressed upon the plane before. Therefore, the

tension of pN : tension of PN = line pN : line pH.

Therefore, if

p : A = OK : OL = line pH : line pN,

then the weights p and A will have the same effect towards moving the wheel, and will therefore sustain each other; as anyone may find by experiment.

But the weight p pressing upon those two oblique planes, may be considered as a wedge between the two internal surfaces of a body split by it; and hence the forces of the wedge and the mallet may be determined: because the force with which the weight p presses the plane pQ is to the force with which the same, whether by its own gravity, or by the blow of a mallet, is impelled in the direction of the line pH towards both the planes, as

pN : pH;

and to the force with which it presses the other plane pG, as

pN : NH.

And thus the force of the screw may be deduced from a like resolution of forces; it being no other than a wedge impelled with the force of a lever. Therefore the use of this Corollary spreads far and wide, and by that diffusive extent the truth thereof is further confirmed. For on what has been said depends the whole doctrine of mechanics variously demonstrated by different authors. For from hence are easily deduced the forces of machines, which are compounded of wheels, pullies, levers, cords, and weights, ascending directly or obliquely, and other mechanical powers; as also the force of the tendons to move the bones of animals.

COROLLARY III

The quantity of motion, which is obtained by taking the sum of the motions directed towards the same parts, and the difference of those that are directed to contrary parts, suffers no change from the action of bodies among themselves.

For action and its opposite reaction are equal, by Law III, and therefore, by Law II, they produce in the motions equal changes towards opposite parts. Therefore if the motions are directed towards the same parts, whatever is added to the motion of the preceding body will be subtracted from the motion of that which follows; so that the sum will be the same as before. If the bodies meet, with contrary motions, there will be an equal deduction from the motions of both; and therefore the difference of the motions directed towards opposite parts will remain the same.

Thus, if a spherical body A is 3 times greater than the spherical body B, and has a velocity = 2, and B follows in the same direction with a velocity = 10, then the

$$\text{motion of A : motion of B} = 6 : 10.$$

Suppose, then, their motions to be of 6 parts and of 10 parts, and the sum will be 16 parts. Therefore, upon the meeting of the bodies, if A acquire 3, 4, or 5 parts of motion, B will lose as many; and therefore after reflection A will proceed with 9, 10, or 11 parts, and B with 7, 6, or 5 parts; the sum remaining always of 16 parts as before. If the body A acquire 9, 10, 11, or 12 parts of motion, and therefore after meeting proceed with 15, 16, 17, or

18 parts, the body B, losing so many parts as A has got, will either proceed with 1 part, having lost 9, or stop and remain at rest, as having lost its whole progressive motion of 10 parts; or it will go back with 1 part, having not only lost its whole motion, but (if I may so say) one part more; or it will go back with 2 parts, because a progressive motion of 12 parts is taken off. And so the sums of the conspiring motions,

$$15 + 1 \quad \text{or} \quad 16 + 0,$$

and the differences of the contrary motions,

$$17 - 1 \quad \text{and} \quad 18 - 2,$$

will always be equal to 16 parts, as they were before the meeting and reflection of the bodies. But the motions being known with which the bodies proceed after reflection, the velocity of either will be also known, by taking the velocity after to the velocity before reflection, as the motion after is to the motion before. As in the last case, where the

motion of A before reflection (6) : motion of A after (18)
= velocity of A before (2) : velocity of A after (x);

that is,

$$6 : 18 = 2 : x, \ x = 6.$$

But if the bodies are either not spherical, or, moving in different right lines, impinge obliquely one upon the other, and their motions after reflection are required, in those cases we are first to determine the position of the plane that touches the bodies in the point of impact, then the motion of each body (by Cor. II) is to be resolved into two, one perpendicular to that plane, and the other parallel to it. This done, because the bodies act upon each other in the direction of a line perpendicular to this plane, the parallel motions are to be retained the same after reflection as before; and to the perpendicular motions we are to assign equal changes towards the contrary parts; in such manner that the sum of the conspiring and the difference of the contrary motions may remain the same as before. From such kind of reflections sometimes arise also the circular motions of bodies about their own centres. But these are cases which I do not consider in what follows; and it would be too tedious to demonstrate every particular case that relates to this subject.

COROLLARY IV

The common centre of gravity of two or more bodies does not alter its state of motion or rest by the actions of the bodies among themselves; and therefore the common centre of gravity of all bodies acting upon each other (excluding external actions and impediments) is either at rest, or moves uniformly in a right line.

For if two points proceed with an uniform motion in right lines, and their distance be divided in a given ratio, the dividing point will be either at rest, or proceed uniformly in a right line. This is demonstrated hereafter in Lem. xxiii and Corollary, when the points are moved in the same plane; and by a like way of arguing, it may be demonstrated when the points are not moved in the same plane. Therefore if any number of bodies move uniformly in right lines, the common centre of gravity of any two of them is either at rest, or proceeds uniformly in a right line; because the line which connects the centres of those two bodies so moving is divided at that common centre in a given ratio. In like manner the common centre of those two and that of a third body will be either at rest or moving uniformly in a right line; because at that centre the distance between the common centre of the two bodies, and the centre of this last, is divided in a given ratio. In like manner the common centre of these three, and of a fourth body, is either at rest, or moves uniformly in a right line; because the distance between the common centre of the three bodies, and the centre of the fourth, is there also divided in a given ratio, and so on *in infinitum*. Therefore, in a system of bodies where there is neither any mutual action among themselves, nor any foreign force impressed upon them from without, and which consequently move uniformly in right lines, the common centre of gravity of them all is either at rest or moves uniformly forwards in a right line.

Moreover, in a system of two bodies acting upon each other, since the distances between their centres and the common centre of gravity of both are reciprocally as the bodies, the relative motions of those bodies, whether of approaching to or of receding from that centre, will be equal among themselves. Therefore since the changes which happen to motions are equal and directed to contrary parts, the common centre of those bodies, by their mutual action between themselves, is neither accelerated nor retarded, nor suffers any change as to its state of motion or rest. But in a system of several bodies, because the common centre of gravity of any two acting upon each other suffers no change in its state by that action; and much less the common centre of gravity of the others with which that action does not intervene; but the distance between those two centres is divided by the common centre of gravity of all the bodies into parts inversely proportional to the total sums of those bodies whose centres they are; and therefore while those two centres retain their state of motion or rest, the common centre of all does also retain its state: it is manifest that the common centre of all never suffers any change in the state of its motion or rest from the actions of any two bodies between themselves. But in such a system all the actions of the bodies among themselves either happen between two bodies, or are composed of actions interchanged between some two bodies; and therefore they do never produce any alteration in the common centre of all as to its state of motion or rest. Wherefore since that centre, when the bodies do not act one upon another, either is at rest or moves uniformly forwards in some right line, it will, notwithstanding the mutual actions of the bodies among themselves, always continue in its state, either of rest, or of proceeding uniformly in a right line, unless it is forced out of this state by the action of some power impressed from without upon the whole system. And therefore the same law takes place in a system consisting of many bodies as in one single body, with regard to their persevering in their state of motion or of rest. For the progressive motion, whether of one single body, or of a whole system of bodies, is always to be estimated from the motion of the centre of gravity.

COROLLARY V

The motions of bodies included in a given space are the same among themselves, whether that space is at rest, or moves uniformly forwards in a right line without any circular motion.

For the differences of the motions tending towards the same parts, and the sums of those that tend towards contrary parts, are, at first (by supposition), in both cases the same; and it is from those sums and differences that the collisions and impulses do arise with which the bodies impinge one upon another. Wherefore (by Law II), the effects of those collisions will be equal in both cases; and therefore the mutual motions of the bodies among themselves in the one case will remain equal to the motions of the bodies among themselves in the other. A clear proof of this we have from the experiment of a ship; where all motions happen after the same manner, whether the ship is at rest, or is carried uniformly forwards in a right line.

COROLLARY VI

If bodies, moved in any manner among themselves, are urged in the direction of parallel lines by equal accelerative forces, they will all continue to move among themselves, after the same manner as if they had not been urged by those forces.

For these forces acting equally (with respect to the quantities of the bodies to be moved), and in the direction of parallel lines, will (by Law II) move all the bodies equally (as to velocity), and therefore will never produce any change in the positions or motions of the bodies among themselves.

SCHOLIUM[1]

Hitherto I have laid down such principles as have been received by mathematicians, and are confirmed by abundance of experiments. By the first two Laws and the first two Corollaries, *Galileo* discovered that the descent of bodies varied as the square of the time (*in duplicata ratione temporis*) and that the motion of projectiles was in the curve of a parabola; experience agreeing with both, unless so far as these motions are a little retarded by the resistance of the air. When a body is falling, the uniform force of its gravity acting equally, impresses, in equal intervals of time, equal forces upon that body, and therefore generates equal velocities; and in the whole time impresses a whole force, and generates a whole velocity proportional to the time. And the spaces described in proportional times are as the product of the velocities and the times; that is, as the squares of the times. And when a body is thrown upwards, its uniform gravity impresses forces and reduces velocities proportional to the times; and the times of ascending to the greatest heights are as the velocities to be taken away, and those heights are as the product of the velocities and the times, or as the squares of the velocities. And if a body be projected in any direction, the motion arising from its projection is compounded with the motion arising from its gravity. Thus,

[1 Appendix, Note 16.]

if the body A by its motion of projection alone could describe in a given time the right line AB, and with its motion of falling alone could describe in the same time the altitude AC; complete the parallelogram ABCD, and the body by that compounded motion will at the end of the time be found

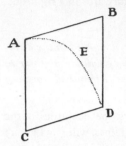

in the place D; and the curved line AED, which that body describes, will be a parabola, to which the right line AB will be a tangent at A; and whose ordinate BD will be as the square of the line AB. On the same Laws and Corollaries depend those things which have been demonstrated concerning the times of the vibration of pendulums, and are confirmed by the daily experiments of pendulum clocks. By the same, together with Law III, Sir *Christopher Wren,* Dr. *Wallis,* and Mr. *Huygens,* the greatest geometers of our times, did severally determine the rules of the impact and reflection of hard bodies, and about the same time communicated their discoveries to the *Royal Society,* exactly agreeing among themselves as to those rules. Dr. *Wallis,* indeed, was somewhat earlier in the publication; then followed Sir *Christopher Wren,* and, lastly, Mr. *Huygens.* But Sir *Christopher Wren* confirmed the truth of the thing before the *Royal Society* by the experiments on pendulums, which M. *Mariotte* soon after thought fit to explain in a treatise entirely upon that subject.[1] But to bring this experiment to an accurate agreement with the theory, we are to have due regard as well to the resistance of the air as to the elastic force of the concurring bodies.

Let the spherical bodies A, B be suspended by the parallel and equal strings AC, BD, from the centres C, D. About these centres, with those lengths as radii, describe the semicircles EAF, GBH, bisected respectively by the radii CA, DB. Bring the

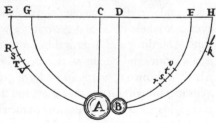

body A to any point R of the arc EAF, and (withdrawing the body B) let it go from thence, and after one oscillation suppose it to return to the point

[1 Appendix, Note 17.]

V: then RV will be the retardation arising from the resistance of the air. Of this RV let ST be a fourth part, situated in the middle, namely, so that
$$RS = TV,$$
and
$$RS : ST = 3 : 2,$$
then will ST represent very nearly the retardation during the descent from S to A. Restore the body B to its place: and, supposing the body A to be let fall from the point S, the velocity thereof in the place of reflection A, without sensible error, will be the same as if it had descended *in vacuo* from the point T. Upon which account this velocity may be represented by the chord of the arc TA. For it is a proposition well known to geometers, that the velocity of a pendulous body in the lowest point is as the chord of the arc which it has described in its descent. After reflection, suppose the body A comes to the place *s*, and the body B to the place *k*. Withdraw the body B, and find the place *v*, from which if the body A, being let go, should after one oscillation return to the place *r*, *st* may be a fourth part of *rv*, so placed in the middle thereof as to leave *rs* equal to *tv*, and let the chord of the arc *t*A represent the velocity which the body A had in the place A immediately after reflection. For *t* will be the true and correct place to which the body A should have ascended, if the resistance of the air had been taken off. In the same way we are to correct the place *k* to which the body B ascends, by finding the place *l* to which it should have ascended *in vacuo*. And thus everything may be subjected to experiment, in the same manner as if we were really placed *in vacuo*. These things being done, we are to take the product (if I may so say) of the body A, by the chord of the arc TA (which represents its velocity), that we may have its motion in the place A immediately before reflection; and then by the chord of the arc *t*A, that we may have its motion in the place A immediately after reflection. And so we are to take the product of the body B by the chord of the arc B*l*, that we may have the motion of the same immediately after reflection. And in like manner, when two bodies are let go together from different places, we are to find the motion of each, as well before as after reflection; and then we may compare the motions between themselves, and collect the effects of the reflection. Thus trying the thing with pendulums of 10 feet, in unequal as well as equal bodies, and making the bodies to concur after a descent through large

spaces, as of 8, 12, or 16 feet, I found always, without an error of 3 inches, that when the bodies concurred together directly, equal changes towards the contrary parts were produced in their motions, and, of consequence, that the action and reaction were always equal. As if the body A impinged upon the body B at rest with 9 parts of motion, and losing 7, proceeded after reflection with 2, the body B was carried backwards with those 7 parts. If the bodies concurred with contrary motions, A with 12 parts of motion, and B with 6, then if A receded with 2, B receded with 8; namely, with a deduction of 14 parts of motion on each side. For from the motion of A subtracting 12 parts, nothing will remain; but subtracting 2 parts more, a motion will be generated of 2 parts towards the contrary way; and so, from the motion of the body B of 6 parts, subtracting 14 parts, a motion is generated of 8 parts towards the contrary way. But if the bodies were made both to move towards the same way, A, the swifter, with 14 parts of motion, B, the slower, with 5, and after reflection A went on with 5, B likewise went on with 14 parts; 9 parts being transferred from A to B. And so in other cases. By the meeting and collision of bodies, the quantity of motion, obtained from the sum of the motions directed towards the same way, or from the difference of those that were directed towards contrary ways, was never changed. For the error of an inch or two in measures may be easily ascribed to the difficulty of executing everything with accuracy. It was not easy to let go the two pendulums so exactly together that the bodies should impinge one upon the other in the lowermost place AB; nor to mark the places s, and k, to which the bodies ascended after impact. Nay, and some errors, too, might have happened from the unequal density of the parts of the pendulous bodies themselves, and from the irregularity of the texture proceeding from other causes.

But to prevent an objection that may perhaps be alleged against the rule, for the proof of which this experiment was made, as if this rule did suppose that the bodies were either absolutely hard, or at least perfectly elastic (whereas no such bodies are to be found in Nature), I must add, that the experiments we have been describing, by no means depending upon that quality of hardness, do succeed as well in soft as in hard bodies. For if the rule is to be tried in bodies not perfectly hard, we are only to diminish the reflection in such a certain proportion as the quantity of the elastic force requires. By the theory of *Wren* and *Huygens,* bodies absolutely hard return one from another with the same velocity with which they meet. But this may be affirmed with more certainty of bodies perfectly elastic. In bodies imperfectly elastic the velocity of the return is to be diminished together with the elastic force; because that force (except when the parts of bodies are bruised by their impact, or suffer some such extension as happens under the strokes of a hammer) is (as far as I can perceive) certain and determined, and makes the bodies to return one from the other with a relative velocity, which is in a given ratio to that relative velocity with which they met. This I tried in balls of wool, made up tightly, and strongly compressed. For, first, by letting go the pendulous bodies, and measuring their reflection, I determined the quantity of their elastic force; and then, according to this force, estimated the reflections that ought to happen in other cases of impact. And with this computation other experiments made afterwards did accordingly agree; the balls always receding one from the other with a relative velocity, which was to the relative velocity with which they met as about 5 to 9. Balls of steel returned with almost the same velocity; those of cork with a velocity something less; but in balls of glass the proportion was as about 15 to 16. And thus the third Law, so far as it regards percussions and reflections, is proved by a theory exactly agreeing with experience.

In attractions, I briefly demonstrate the thing after this manner. Suppose an obstacle is interposed to hinder the meeting of any two bodies A, B, attracting one the other: then if either body, as A, is more attracted towards the other body B, than that other body B is towards the first body A, the obstacle will be more strongly urged by the pressure of the body A than by the pressure of the body B, and therefore will not remain in equilibrium: but the stronger pressure will prevail, and will make the system of the two bodies, together with the obstacle, to move directly towards the parts on which B lies; and in free spaces, to go forwards *in infinitum* with a motion continually accelerated; which is absurd and contrary to the first Law. For, by the first Law, the system ought to continue in its state of rest, or of moving uniformly forwards in a right line; and therefore the bodies must equally press the obstacle, and be equally attracted one by the other. I made the experiment on the loadstone and iron. If these, placed apart in proper vessels, are made to float by one another in standing water, neither of them will propel the other; but, by being equally attracted, they will sustain each other's pressure, and rest at last in an equilibrium.

So the gravitation between the earth and its parts is mutual. Let the earth FI be cut by any plane EG into two parts EGF and EGI, and their weights one towards the other will be mutually equal. For if by another plane HK, parallel to the former EG, the greater part EGI is cut into two parts EGKH

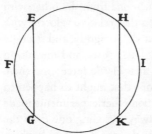

and HKI, whereof HKI is equal to the part EFG, first cut off, it is evident that the middle part EGKH will have no propension by its proper weight towards either side, but will hang as it were, and rest in an equilibrium between both. But the one extreme part HKI will with its whole weight bear upon and press the middle part towards the other extreme part EGF; and therefore the force with which EGI, the sum of the parts HKI and EGKH, tends towards the third part EGF, is equal to the weight of the part HKI, that is, to the weight of the third part EGF. And therefore the weights of the two parts EGI and EGF, one towards the other, are equal, as I was to prove. And indeed if those weights were not equal, the whole earth floating in the nonresisting ether would give way to the greater weight, and, retiring from it, would be carried off *in infinitum*.

And as those bodies are equipollent in the impact and reflection, whose velocities are inversely as their innate forces, so in the use of mechanic instruments those agents are equipollent, and mutually sustain each the contrary pressure of the other, whose velocities, estimated according to the determination of the forces, are inversely as the forces.

So those weights are of equal force to move the arms of a balance, which during the play of the balance are inversely as their velocities upwards and downwards; that is, if the ascent or descent is direct, those weights are of equal force, which are inversely as the distances of the points at which they are suspended from the axis of the balance; but if they are turned aside by the interposition of oblique planes, or other obstacles, and made to ascend or descend obliquely, those bodies will be equipollent, which are inversely as the heights of their ascent and descent taken according to the perpendicular; and that on account of the determination of gravity downwards.

And in like manner in the pulley, or in a combination of pulleys, the force of a hand drawing the rope directly, which is to the weight, whether ascending directly or obliquely, as the velocity of the perpendicular ascent of the weight to the velocity of the hand that draws the rope, will sustain the weight.

In clocks and such like instruments, made up from a combination of wheels, the contrary forces that promote and impede the motion of the wheels, if they are inversely as the velocities of the parts of the wheel on which they are impressed, will mutually sustain each other.

The force of the screw to press a body is to the force of the hand that turns the handles by which it is moved as the circular velocity of the handle in that part where it is impelled by the hand is to the progressive velocity of the screw towards the pressed body.

The forces by which the wedge presses or drives the two parts of the wood it cleaves are to the force of the mallet upon the wedge as the progress of the wedge in the direction of the force impressed upon it by the mallet is to the velocity with which the parts of the wood yield to the wedge, in the direction of lines perpendicular to the sides of the wedge. And the like account is to be given of all machines.

The power and use of machines consist only in this, that by diminishing the velocity we may augment the force, and the contrary; from whence, in all sorts of proper machines, we have the solution of this problem: *To move a given weight with a given power,* or with a given force to overcome any other given resistance. For if machines are so contrived that the velocities of the agent and resistant are inversely as their forces, the agent will just sustain the resistant, but with a greater disparity of velocity will overcome it. So that if the disparity of velocities is so great as to overcome all that resistance which commonly arises either from the friction of contiguous bodies as they slide by one another, or from the cohesion of continuous bodies that are to be separated, or from the weights of bodies to be raised, the excess of the force remaining, after all those resistances are overcome, will produce an acceleration of motion proportional thereto, as well in the parts of the machine as in the resisting body. But to treat of mechanics is not my present business. I was aiming only to show by those examples the great extent and certainty of the third Law of Motion. For if we estimate the action of the agent from the product of its force and velocity, and likewise the reaction of the impediment from the product of the velocities of its several parts, and the forces of resistance arising from the friction, cohesion, weight, and acceleration of those parts, the action and reaction in the use of all sorts of machines will be found always equal to one another. And so far as the action is propagated by the intervening instruments, and at last impressed upon the resisting body, the ultimate action will be always contrary to the reaction.

Book One
THE MOTION OF BODIES

SECTION I

The method of first and last ratios of quantities, by the help of which we demonstrate the propositions that follow.

LEMMA I

Quantities, and the ratios of quantities, which in any finite time converge continually to equality, and before the end of that time approach nearer to each other than by any given difference, become ultimately equal.

If you deny it, suppose them to be ultimately unequal, and let D be their ultimate difference. Therefore they cannot approach nearer to equality than by that given difference D; which is contrary to the supposition.

LEMMA II

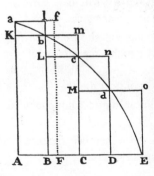

If in any figure AacE, *terminated by the right lines* Aa, AE, *and the curve* acE, *there be inscribed any number of parallelograms* Ab, Bc, Cd, &c., *comprehended under equal bases* AB, BC, CD, &c., *and the sides,* Bb, Cc, Dd, &c., *parallel to one side* Aa *of the figure; and the parallelograms* aKbl, bLcm, cMdn, &c., *are completed: then if the breadth of those parallelograms be supposed to be diminished, and their number to be augmented in infinitum, I say, that the ultimate ratios which the*

[29]

inscribed figure AKbLcMdD, *the circumscribed figure* Aalbmcndo E, *and curvilinear figure* AabcdE, *will have to one another, are ratios of equality.*

For the difference of the inscribed and circumscribed figures is the sum of the parallelograms K*l*, L*m*, M*n*, D*o*, that is (from the equality of all their bases), the rectangle under one of their bases K*b* and the sum of their altitudes A*a*, that is, the rectangle AB*la*. But this rectangle, because its breadth AB is supposed diminished *in infinitum*, becomes less than any given space. And therefore (by Lem. 1) the figures inscribed and circumscribed become ultimately equal one to the other; and much more will the intermediate curvilinear figure be ultimately equal to either. Q.E.D.

LEMMA III

The same ultimate ratios are also ratios of equality, when the breadths AB, BC, DC, &c., *of the parallelograms are unequal, and are all diminished in* infinitum.

For suppose AF equal to the greatest breadth, and complete the parallelogram FA*af*. This parallelogram will be greater than the difference of the

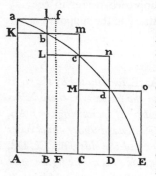

inscribed and circumscribed figures; but, because its breadth AF is diminished *in infinitum*, it will become less than any given rectangle. Q.E.D.

Cor. I. Hence the ultimate sum of those evanescent parallelograms will in all parts coincide with the curvilinear figure.

Cor. II. Much more will the rectilinear figure comprehended under the chords of the evanescent arcs *ab*, *bc*, *cd*, &c., ultimately coincide with the curvilinear figure.

Cor. III. And also the circumscribed rectilinear figure comprehended under the tangents of the same arcs.

Cor. IV. And therefore these ultimate figures (as to their perimeters *ac*E) are not rectilinear, but curvilinear limits of rectilinear figures.

LEMMA IV

If in two figures AacE, PprT, *there are inscribed (as before) two series of parallelograms, an equal number in each series, and, their breadths being diminished* in infinitum, *if the ultimate ratios of the parallelograms in one figure to those in the other, each to each respectively, are the same: I say, that those two figures,* AacE, PprT, *are to each other in that same ratio.*

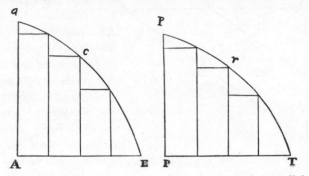

For as the parallelograms in the one are severally to the parallelograms in the other, so (by composition) is the sum of all in the one to the sum of all in the other; and so is the one figure to the other; because (by Lem. III) the former figure to the former sum, and the latter figure to the latter sum, are both in the ratio of equality. Q.E.D.

Cor. Hence if two quantities of any kind are divided in any manner into an equal number of parts, and those parts, when their number is augmented, and their magnitude diminished *in infinitum,* have a given ratio to each other, the first to the first, the second to the second, and so on in order, all of them taken together will be to each other in that same given ratio. For if, in the figures of this Lemma, the parallelograms are taken to each other in the ratio of the parts, the sum of the parts will always be as the sum of the parallelograms; and therefore supposing the number of the parallelograms and parts to be augmented, and their magnitudes diminished *in infinitum,* those sums will be in the ultimate ratio of the parallelogram in the one figure to the correspondent parallelogram in the other; that is (by the supposition), in the ultimate ratio of any part of the one quantity to the correspondent part of the other.

LEMMA V

All homologous sides of similar figures, whether curvilinear or rectilinear, are proportional; and the areas are as the squares of the homologous sides.

LEMMA VI

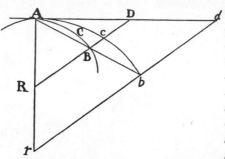

If any arc ACB, given in position, is subtended by its chord AB, and in any point A, in the middle of the continued curvature, is touched by a right line AD, produced both ways; then if the points A and B approach one another and meet, I say, the angle BAD, contained between the chord and the tangent, will be diminished in infinitum, *and ultimately will vanish.*

For if that angle does not vanish, the arc ACB will contain with the tangent AD an angle equal to a rectilinear angle; and therefore the curvature at the point A will not be continued, which is against the supposition.

LEMMA VII

The same things being supposed, I say that the ultimate ratio of the arc, chord, and tangent, any one to any other, is the ratio of equality.

For while the point B approaches towards the point A, consider always AB and AD as produced to the remote points *b* and *d;* and parallel to the secant BD draw *bd;* and let the arc A*cb* be always similar to the arc ACB. Then, supposing the points A and B to coincide, the angle *d*A*b* will vanish, by the preceding Lemma; and therefore the right lines A*b*, A*d* (which are always finite), and the intermediate arc A*cb*, will coincide, and become equal among themselves. Wherefore, the right lines AB, AD, and the intermediate arc ACB (which are always proportional to the former), will vanish, and ultimately acquire the ratio of equality. Q.E.D.

COR. I. Whence if through B we draw BF parallel to the tangent, always cutting any right line AF passing through A in F, this line BF will be

ultimately in the ratio of equality
with the evanescent arc ACB; because, completing the parallelogram
AFBD, it is always in a ratio of
equality with AD.

Cor. II. And if through B and A more right lines are drawn, as BE, BD,
AF, AG, cutting the tangent AD and its parallel BF; the ultimate ratio of
all the abscissas AD, AE, BF, BG, and of the chord and arc AB, any one to
any other, will be the ratio of equality.

Cor. III. And therefore in all our reasoning about ultimate ratios, we may
freely use any one of those lines for any other.

LEMMA VIII

If the right lines AR, BR, *with the arc* ACB, *the chord* AB, *and the tangent*
AD, *constitute three triangles* RAB, RACB, RAD, *and the points* A *and* B
approach and meet: I say, that the ultimate form of these evanescent tri-
angles is that of similitude, and their ultimate ratio that of equality.

For while the point B approaches towards the point A, consider always
AB, AD, AR, as produced to the remote points *b, d,* and *r,* and *rbd* as drawn

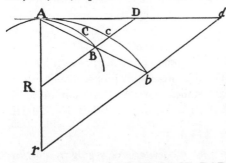

parallel to RD, and let the arc
A*cb* be always similar to the
arc ACB. Then supposing the
points A and B to coincide, the
angle *b*A*d* will vanish; and
therefore the three triangles
*r*A*b, r*A*cb, r*A*d* (which are
always finite), will coincide,
and on that account become
both similar and equal. And
therefore the triangles RAB, RACB, RAD, which are always similar and
proportional to these, will ultimately become both similar and equal
among themselves. Q.E.D.

Cor. And hence in all reasonings about ultimate ratios, we may use any
one of those triangles for any other.

LEMMA IX

If a right line AE, *and a curved line* ABC, *both given by position, cut each other in a given angle,* A; *and to that right line, in another given angle,* BD, CE *are ordinately applied, meeting the curve in* B, C; *and the points* B *and* C *together approach towards and meet in the point* A: *I say, that the areas of the triangles* ABD, ACE, *will ultimately be to each other as the squares of homologous sides.*

For while the points B, C, approach towards the point A, suppose always AD to be produced to the remote points *d* and *e,* so as A*d,* A*e* may be proportional to AD, AE; and the ordinates *db, ec,* to be drawn parallel to the

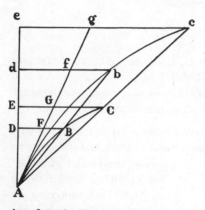

ordinates DB and EC, and meeting AB and AC produced in *b* and *c.* Let the curve A*bc* be similar to the curve ABC, and draw the right line A*g* so as to touch both curves in A, and cut the ordinates DB, EC, *db, ec,* in F, G, *f, g.* Then, supposing the length A*e* to remain the same, let the points B and C meet in the point A; and the angle *c*A*g* vanishing, the curvilinear areas A*bd,* A*ce* will coincide with the rectilinear areas A*fd,* A*ge*; and

therefore (by Lem. v) will be one to the other in the duplicate ratio of the sides A*d,* A*e.* But the areas ABD, ACE are always proportional to these areas; and so the sides AD, AE are to these sides. And therefore the areas ABD, ACE are ultimately to each other as the squares of the sides AD, AE. Q.E.D.

LEMMA X

The spaces which a body describes by any finite force urging it, whether that force is determined and immutable, or is continually augmented or continually diminished, are in the very beginning of the motion to each other as the squares of the times.

Let the times be represented by the lines AD, AE, and the velocities generated in those times by the ordinates DB, EC. The spaces described with these velocities will be as the areas ABD, ACE, described by those ordinates, that is, at the very beginning of the motion (by Lem. ix), in the duplicate ratio of the times AD, AE. Q.E.D.

Cor. i. And hence one may easily infer, that the errors of bodies describing similar parts of similar figures in proportional times, the errors being generated by any equal forces similarly applied to the bodies, and measured by the distances of the bodies from those places of the similar figures, at which, without the action of those forces, the bodies would have arrived in those proportional times—are nearly as the squares of the times in which they are generated.

Cor. ii. But the errors that are generated by proportional forces, similarly applied to the bodies at similar parts of the similar figures, are as the product of the forces and the squares of the times.

Cor. iii. The same thing is to be understood of any spaces whatsoever described by bodies urged with different forces; all which, in the very beginning of the motion, are as the product of the forces and the squares of the times.

Cor. iv. And therefore the forces are directly as the spaces described in the very beginning of the motion, and inversely as the squares of the times.

Cor. v. And the squares of the times are directly as the spaces described, and inversely as the forces.

SCHOLIUM

If in comparing with each other indeterminate quantities of different sorts, any one is said to be directly or inversely as any other, the meaning is, that the former is augmented or diminished in the same ratio as the latter, or as its reciprocal. And if any one is said to be as any other two or more, directly or inversely, the meaning is, that the first is augmented or diminished in the ratio compounded of the ratios in which the others, or the reciprocals of the others, are augmented or diminished. Thus, if A is said to be as B directly, and C directly, and D inversely, the meaning is, that A is augmented or diminished in the same ratio as $B \cdot C \cdot \frac{1}{D}$, that is to say, that A and $\frac{BC}{D}$ are to each other in a given ratio.

LEMMA XI[1]

The evanescent subtense of the angle of contact, in all curves which at the point of contact have a finite curvature, is ultimately as the square of the subtense of the conterminous arc.

CASE 1. Let AB be that arc, AD its tangent, BD the subtense of the angle of contact perpendicular on the tangent, AB the subtense of the arc. Draw BG perpendicular to the subtense AB, and AG perpendicular to the tangent

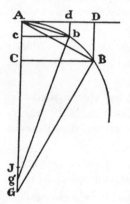

AD, meeting in G; then let the points D, B, and G approach to the points d, b, and g, and suppose J to be the ultimate intersection of the lines BG, AG, when the points D, B have come to A. It is evident that the distance GJ may be less than any assignable distance. But (from the nature of the circles passing through the points A, B, G, and through A, b, g),

$$AB^2 = AG \cdot BD, \text{ and}$$
$$Ab^2 = Ag \cdot bd.$$

But because GJ may be assumed of less length than any assignable, the ratio of AG to Ag may be such as to differ from unity by less than any assignable difference; and therefore the ratio of AB^2 to Ab^2 may be such as to differ from the ratio of BD to bd by less than any assignable difference. Therefore, by Lem. 1, ultimately,

$$AB^2 : Ab^2 = BD : bd. \hspace{3em} \text{Q.E.D.}$$

CASE 2. Now let BD be inclined to AD in any given angle, and the ultimate ratio of BD to bd will always be the same as before, and therefore the same with the ratio of AB^2 to Ab^2. Q.E.D.

CASE 3. And if we suppose the angle D not to be given, but that the right line BD converges to a given point, or is determined by any other condition whatever; nevertheless the angles D, d, being determined by the same law, will always draw nearer to equality, and approach nearer to each other than by any assigned difference, and therefore, by Lem. 1, will at last be equal; and therefore the lines BD, bd are in the same ratio to each other as before. Q.E.D.

[1 Appendix, Note 17.]

Cor. i. Therefore since the tangents AD, Ad, the arcs AB, Ab, and their sines, BC, bc, become ultimately equal to the chords AB, Ab, their squares will ultimately become as the subtenses BD, bd.

Cor. ii. Their squares are also ultimately as the versed sines of the arcs, bisecting the chords, and converging to a given point. For those versed sines are as the subtenses BD, bd.

Cor. iii. And therefore the versed sine is as the square of the time in which a body will describe the arc with a given velocity.

Cor. iv. The ultimate proportion,

$$\triangle \text{ADB} : \triangle \text{A}db = \text{AD}^3 : \text{A}d^3 = \text{DB}^{3/2} : db^{3/2},$$

is derived from

$$\triangle \text{ADB} : \triangle \text{A}db = \text{AD} \cdot \text{DB} : \text{A}d \cdot db$$

and from the ultimate proportion

$$\text{AD}^2 : \text{A}d^2 = \text{DB} : db.$$

So also is obtained ultimately

$$\triangle \text{ABC} : \triangle \text{A}bc = \text{BC}^3 : bc^3.$$

Cor. v. And because DB, db are ultimately parallel and as the squares of the lines AD, Ad, the ultimate curvilinear areas ADB, Adb will be (by the nature of the parabola) two-thirds of the rectilinear triangles ADB, Adb, and the segments AB, Ab will be one-third of the same triangles. And thence those areas and those segments will be as the squares of the tangents AD, Ad, and also of the chords and arcs AB, AB.

SCHOLIUM

But we have all along supposed the angle of contact to be neither infinitely greater nor infinitely less than the angles of contact made by circles and their tangents; that is, that the curvature at the point A is neither infinitely small nor infinitely great, and that the interval AJ is of a finite magnitude. For DB may be taken as AD^3: in which case no circle can be drawn through the point A, between the tangent AD and the curve AB, and therefore the angle of contact will be infinitely less than those of circles. And by a like reasoning, if DB be made successfully as AD^4, AD^5, AD^6, AD^7, &c., we shall have a series of angles of contact, proceeding *in infinitum,* wherein every succeeding term is infinitely less than the preceding.

And if DB be made successively as AD^2, $AD^{3/2}$, $AD^{4/3}$, $AD^{5/4}$, $AD^{6/5}$, $AD^{7/6}$, &c., we shall have another infinite series of angles of contact, the first of which is of the same sort with those of circles, the second infinitely greater, and every succeeding one infinitely greater than the preceding. But between any two of these angles another series of intermediate angles of contact may be interposed, proceeding both ways *in infinitum,* wherein every succeeding angle shall be infinitely greater or infinitely less than the preceding. As if between the terms AD^2 and AD^3 there were interposed the series $AD^{13/6}$, $AD^{11/5}$, $AD^{9/4}$, $AD^{7/3}$, $AD^{5/2}$, $AD^{8/3}$, $AD^{11/4}$, $AD^{14/5}$, $AD^{17/6}$, &c. And again, between any two angles of this series, a new series of intermediate angles may be interposed, differing from one another by infinite intervals. Nor is Nature confined to any bounds.

Those things which have been demonstrated of curved lines, and the surfaces which they comprehend, may be easily applied to the curved surfaces and contents of solids. These Lemmas are premised to avoid the tediousness of deducing involved demonstrations *ad absurdum,* according to the method of the ancient geometers. For demonstrations are shorter by the method of indivisibles; but because the hypothesis of indivisibles seems somewhat harsh, and therefore that method is reckoned less geometrical, I chose rather to reduce the demonstrations of the following Propositions to the first and last sums and ratios of nascent and evanescent quantities, that is, to the limits of those sums and ratios, and so to premise, as short as I could, the demonstrations of those limits. For hereby the same thing is performed as by the method of indivisibles; and now those principles being demonstrated, we may use them with greater safety. Therefore if hereafter I should happen to consider quantities as made up of particles, or should use little curved lines for right ones, I would not be understood to mean indivisibles, but evanescent divisible quantities; not the sums and ratios of determinate parts, but always the limits of sums and ratios; and that the force of such demonstrations always depends on the method laid down in the foregoing Lemmas.

Perhaps it may be objected, that there is no ultimate proportion of evanescent quantities; because the proportion, before the quantities have vanished, is not the ultimate, and when they are vanished, is none. But by the same argument it may be alleged that a body arriving at a certain place, and

there stopping, has no ultimate velocity; because the velocity, before the
body comes to the place, is not its ultimate velocity; when it has arrived,
there is none. But the answer is easy; for by the ultimate velocity is meant
that with which the body is moved, neither before it arrives at its last place
and the motion ceases, nor after, but at the very instant it arrives; that is,
that velocity with which the body arrives at its last place, and with which
the motion ceases. And in like manner, by the ultimate ratio of evanescent
quantities is to be understood the ratio of the quantities not before they
vanish, nor afterwards, but with which they vanish. In like manner the first
ratio of nascent quantities is that with which they begin to be. And the first
or last sum is that with which they begin and cease to be (or to be augmented or diminished). There is a limit which the velocity at the end of the
motion may attain, but not exceed. This is the ultimate velocity. And there
is the like limit in all quantities and proportions that begin and cease to be.
And since such limits are certain and definite, to determine the same is a
problem strictly geometrical. But whatever is geometrical we may use in
determining and demonstrating any other thing that is also geometrical.

It may also be objected, that if the ultimate ratios of evanescent quantities
are given, their ultimate magnitudes will be also given: and so all quantities
will consist of indivisibles, which is contrary to what *Euclid* has demonstrated concerning incommensurables, in the tenth Book of his *Elements*.
But this objection is founded on a false supposition. For those ultimate ratios
with which quantities vanish are not truly the ratios of ultimate quantities,
but limits towards which the ratios of quantities decreasing without limit
do always converge; and to which they approach nearer than by any given
difference, but never go beyond, nor in effect attain to, till the quantities
are diminished *in infinitum*. This thing will appear more evident in quantities infinitely great. If two quantities, whose difference is given, be augmented *in infinitum,* the ultimate ratio of these quantities will be given,
namely, the ratio of equality; but it does not from thence follow, that the
ultimate or greatest quantities themselves, whose ratio that is, will be given.
Therefore if in what follows, for the sake of being more easily understood,
I should happen to mention quantities as least, or evanescent, or ultimate,
you are not to suppose that quantities of any determinate magnitude are
meant, but such as are conceived to be always diminished without end.

SECTION II
The determination of centripetal forces.

PROPOSITION I. THEOREM I

The areas which revolving bodies describe by radii drawn to an immovable centre of force do lie in the same immovable planes, and are proportional to the times in which they are described.

For suppose the time to be divided into equal parts, and in the first part of that time let the body by its innate force describe the right line AB. In the second part of that time, the same would (by Law I), if not hindered,

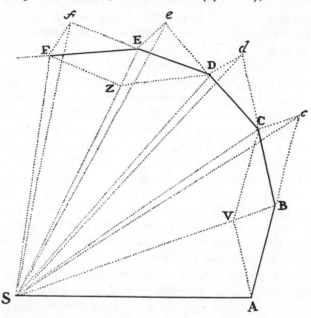

proceed directly to *c,* along the line B*c* equal to AB; so that by the radii AS, BS, *c*S, drawn to the centre, the equal areas ASB, BS*c*, would be described. But when the body is arrived at B, suppose that a centripetal force acts at once with a great impulse, and, turning aside the body from the right line B*c*, compels it afterwards to continue its motion along the right line BC.

Draw cC parallel to BS, meeting BC in C; and at the end of the second part of the time, the body (by Cor. I of the Laws) will be found in C, in the same plane with the triangle ASB. Join SC, and, because SB and Cc are parallel, the triangle SBC will be equal to the triangle SBc, and therefore also to the triangle SAB. By the like argument, if the centripetal force acts successively in C, D, E, &c., and makes the body, in each single particle of time, to describe the right lines CD, DE, EF, &c., they will all lie in the same plane; and the triangle SCD will be equal to the triangle SBC, and SDE to SCD, and SEF to SDE. And therefore, in equal times, equal areas are described in one immovable plane: and, by composition, any sums SADS, SAFS, of those areas, are to each other as the times in which they are described. Now let the number of those triangles be augmented, and their breadth diminished *in infinitum;* and (by Cor. IV, Lem. III) their ultimate perimeter ADF will be a curved line: and therefore the centripetal force, by which the body is continually drawn back from the tangent of this curve, will act continually; and any described areas SADS, SAFS, which are always proportional to the times of description, will, in this case also, be proportional to those times. Q.E.D.

Cor. I. The velocity of a body attracted towards an immovable centre, in spaces void of resistance, is inversely as the perpendicular let fall from that centre on the right line that touches the orbit. For the velocities in those places A, B, C, D, E, are as the bases AB, BC, CD, DE, EF, of equal triangles; and these bases are inversely as the perpendiculars let fall upon them.

Cor. II. If the chords AB, BC of two arcs, successively described in equal times by the same body, in spaces void of resistance, are completed into a parallelogram ABCV, and the diagonal BV of this parallelogram, in the position which it ultimately acquires when those arcs are diminished *in infinitum,* is produced both ways, it will pass through the centre of force.

Cor. III. If the chords AB, BC, and DE, EF, of arcs described in equal times, in spaces void of resistance, are completed into the parallelograms ABCV, DEFZ, the forces in B and E are one to the other in the ultimate ratio of the diagonals BV, EZ, when those arcs are diminished *in infinitum.* For the motions BC and EF of the body (by Cor. I of the Laws) are compounded of the motions Bc, BV, and Ef, EZ; but BV and EZ, which are equal to Cc and Ff, in the demonstration of this Proposition, were generated by the impulses of the centripetal force in B and E, and are therefore proportional to those impulses.

CoR. IV. The forces by which bodies, in spaces void of resistance, are drawn back from rectilinear motions, and turned into curvilinear orbits, are to each other as the versed sines of arcs described in equal times; which versed sines tend to the centre of force, and bisect the chords when those arcs are diminished to infinity. For such versed sines are the halves of the diagonals mentioned in Cor. III.

CoR. V. And therefore those forces are to the force of gravity as the said versed sines to the versed sines perpendicular to the horizon of those parabolic arcs which projectiles describe in the same time.

CoR. VI. And the same things do all hold good (by Cor. v of the Laws) when the planes in which the bodies are moved, together with the centres of force which are placed in those planes, are not at rest, but move uniformly forwards in right lines.

PROPOSITION II. THEOREM II

Every body that moves in any curved line described in a plane, and by a radius drawn to a point either immovable, or moving forwards with an uniform rectilinear motion, describes about that point areas proportional to the times, is urged by a centripetal force directed to that point.

CASE 1. For every body that moves in a curved line is (by Law 1) turned aside from its rectilinear course by the action of some force that impels it. And that force by which the body is turned off from its rectilinear course, and is made to describe, in equal times, the equal least triangles SAB, SBC, SCD, &c., about the immovable point S (by Prop. XL, Book 1, *Elem.* of *Euclid,* and Law II), acts in the place B, according to the direction of a line parallel to cC, that is, in the direction of the line BS; and in the place C, according to the direction of a line parallel to dD, that is, in the direction of the line CS, &c.; and therefore acts always in the direction of lines tending to the immovable point S. Q.E.D.

CASE 2. And (by Cor. v of the Laws) it is indifferent whether the surface in which a body describes a curvilinear figure be at rest, or moves together with the body, the figure described, and its point S, uniformly forwards in a right line.

Cor. i. In nonresisting spaces or mediums, if the areas are not proportional to the times, the forces are not directed to the point in which the radii meet, but deviate therefrom towards the part to which the motion is directed, if the description of the areas is accelerated, and away from that part, if retarded.

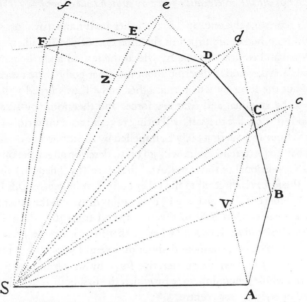

Cor. ii. And even in resisting mediums, if the description of the areas is accelerated, the directions of the forces deviate from the point in which the radii meet, towards the part to which the motion tends.

SCHOLIUM

A body may be urged by a centripetal force compounded of several forces; in which case the meaning of the Proposition is, that the force which results out of all tends to the point S. But if any force acts continually in the direction of lines perpendicular to the described surface, this force will make the body to deviate from the plane of its motion; but will neither augment nor diminish the area of the described surface, and is therefore to be neglected in the composition of forces.

PROPOSITION III. THEOREM III

Every body, that by a radius drawn to the centre of another body, howso-
ever moved, describes areas about that centre proportional to the times, is
urged by a force compounded of the centripetal force tending to that other
body, and of all the accelerative force by which that other body is impelled.

Let L represent the one, and T the other body; and (by Cor. vi of the
Laws) if both bodies are urged in the direction of parallel lines, by a new
force equal and contrary to that by which the second body T is urged, the
first body L will go on to describe about the other body T the same areas as
before: but the force by which that other body T was urged will be now
destroyed by an equal and contrary force; and therefore (by Law i) that
other body T, now left to itself, will either rest, or move uniformly forwards
in a right line: and the first body L, impelled by the difference of the forces,
that is, by the force remaining, will go on to describe about the other body
T areas proportional to the times. And therefore (by Theor. ii) the difference of the forces is directed to the other body T as its centre. Q.E.D.

Cor. i. Hence if the one body L, by a radius drawn to the other body T,
describes areas proportional to the times; and from the whole force, by
which the first body L is urged (whether that force is simple, or, according
to Cor. ii of the Laws, compounded out of several forces), we subtract (by
the same Cor.) that whole accelerative force by which the other body is
urged; the whole remaining force by which the first body is urged will tend
to the other body T, as its centre.

Cor. ii. And, if these areas are proportional to the times nearly, the remaining force will tend to the other body T nearly.

Cor. iii. And *vice versa,* if the remaining force tends nearly to the other
body T, those areas will be nearly proportional to the times.

Cor. iv. If the body L, by a radius drawn to the other body T, describes
areas, which, compared with the times, are very unequal; and that other
body T be either at rest, or moves uniformly forwards in a right line: the
action of the centripetal force tending to that other body T is either none
at all, or it is mixed and compounded with very powerful actions of other
forces: and the whole force compounded of them all, if they are many, is
directed to another (immovable or movable) centre. The same thing

obtains, when the other body is moved by any motion whatsoever; provided that centripetal force is taken, which remains after subtracting that whole force acting upon that other body T.

SCHOLIUM

Since the equable description of areas indicates that there is a centre to which tends that force by which the body is most affected, and by which it is drawn back from its rectilinear motion, and retained in its orbit, why may we not be allowed, in the following discourse, to use the equable description of areas as an indication of a centre, about which all circular motion is performed in free spaces?

PROPOSITION IV. THEOREM IV[1]

The centripetal forces of bodies, which by equable motions describe different circles, tend to the centres of the same circles; and are to each other as the squares of the arcs described in equal times divided respectively by the radii of the circles.

These forces tend to the centres of the circles (by Prop. II, and Cor. II, Prop. I), and are to one another as the versed sines of the least arcs described in equal times (by Cor. IV, Prop. I); that is, as the squares of the same arcs divided by the diameters of the circles (by Lem. VII); and therefore since those arcs are as arcs described in any equal times, and the diameters are as the radii, the forces will be as the squares of any arcs described in the same time divided by the radii of the circles. Q.E.D.

COR. I. Therefore, since those arcs are as the velocities of the bodies, the centripetal forces are as the squares of the velocities divided by the radii.

COR. II. And since the periodic times are as the radii divided by the velocities, the centripetal forces are as the radii divided by the square of the periodic times.

COR. III. Whence if the periodic times are equal, and the velocities therefore as the radii, the centripetal forces will be also as the radii; and conversely.

COR. IV. If the periodic times and the velocities are both as the square roots of the radii, the centripetal forces will be equal among themselves; and conversely.

[1 Appendix, Note 15.]

Cor. v. If the periodic times are as the radii, and therefore the velocities equal, the centripetal forces will be inversely as the radii; and conversely.

Cor. vi. If the periodic times are as the ½th powers of the radii, and therefore the velocities inversely as the square roots of the radii, the centripetal forces will be inversely as the squares of the radii; and conversely.

Cor. vii. And universally, if the periodic time is as any power R^n of the radius R, and therefore the velocity inversely as the power R^{n-1} of the radius, the centripetal force will be inversely as the power R^{2n-1} of the radius; and conversely.

Cor. viii. The same things hold concerning the times, the velocities, and the forces by which bodies describe the similar parts of any similar figures that have their centres in a similar position with those figures; as appears by applying the demonstration of the preceding cases to those. And the application is easy, by only substituting the equable description of areas in the place of equable motion, and using the distances of the bodies from the centres instead of the radii.

Cor. ix. From the same demonstration it likewise follows, that the arc which a body, uniformly revolving in a circle with a given centripetal force, describes in any time, is a mean proportional between the diameter of the circle, and the space which the same body falling by the same given force would describe in the same given time.

SCHOLIUM

The case of the sixth Corollary obtains in the celestial bodies (as Sir *Christopher Wren,* Dr. *Hooke,* and Dr. *Halley* have severally observed); and therefore in what follows, I intend to treat more at large of those things which relate to centripetal force decreasing as the squares of the distances from the centres.

Moreover, by means of the preceding Proposition and its Corollaries, we may discover the proportion of a centripetal force to any other known force, such as that of gravity. For if a body by means of its gravity revolves in a circle concentric to the earth, this gravity is the centripetal force of that body. But from the descent of heavy bodies, the time of one entire revolution, as well as the arc described in any tiven time, is given (by Cor. ix of this Prop.). And by such propositions, Mr. *Huygens,* in his excellent book

De horologio oscillatorio, has compared the force of gravity with the centrifugal forces of revolving bodies.

The preceding Proposition may be likewise demonstrated after this manner. In any circle suppose a polygon to be inscribed of any number of sides.
And if a body, moved with a given velocity along the sides of the polygon,
is reflected from the circle at the several angular points, the force, with
which at every reflection it strikes the circle, will be as its velocity: and therefore the sum of the forces, in a given time, will be as the product of that
velocity and the number of reflections; that is (if the species of the polygon
be given), as the length described in that given time, and increased or
diminished in the ratio of the same length to the radius of the circle; that
is, as the square of that length divided by the radius; and therefore the polygon, by having its sides diminished *in infinitum,* coincides with the circle,
as the square of the arc described in a given time divided by the radius.
This is the centrifugal force, with which the body impels the circle; and to
which the contrary force, wherewith the circle continually repels the body
towards the centre, is equal.

PROPOSITION V. PROBLEM I

*There being given, in any places, the velocity with which a body describes
a given figure, by means of forces directed to some common centre: to find
that centre.*

Let the three right lines PT,
TQV, VR touch the figure described in as many points, P, Q,
R, and meet in T and V. On the
tangents erect the perpendiculars
PA, QB, RC, inversely proportional to the velocities of the body
in the points P, Q, R, from which
the perpendiculars were raised;
that is, so that PA may be to QB

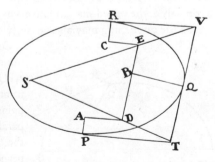

as the velocity in Q to the velocity in P, and QB to RC as the velocity in R
to the velocity in Q. Through the ends A, B, C of the perpendiculars draw

AD, DBE, EC, at right angles, meeting in D and E: and the right lines TD, VE produced, will meet in S, the centre required.

For the perpendiculars let fall from the centre S on the tangents PT, QT, are inversely as the velocities of the bodies in the points P and Q (by Cor. 1, Prop. 1), and therefore, by construction, directly as the perpendiculars AP, BQ; that is, as the perpendiculars let fall from the point D on the tangents. Whence it is easy to infer that the points S, D, T are in one right line. And by the like argument the points S, E, V are also in one right line; and therefore the centre S is in the point where the right lines TD, VE meet. Q.E.D.

PROPOSITION VI. THEOREM V

In a space void of resistance, if a body revolves in any orbit about an immovable centre, and in the least time describes any arc just then nascent; and the versed sine of that arc is supposed to be drawn bisecting the chord, and produced passing through the centre of force: the centripetal force in the middle of the arc will be directly as the versed sine and inversely as the square of the time.

For the versed sine in a given time is as the force (by Cor. IV, Prop. 1); and augmenting the time in any ratio, because the arc will be augmented in the same ratio, the versed sine will be augmented in the square of that ratio (by Cor. II and III, Lem. XI), and therefore is as the force and the square of the time. Divide both sides by the square of the time, and the force will be directly as the versed sine, and inversely as the square of the time. Q.E.D.

And the same thing may also be easily demonstrated by Cor. IV, Lem. X.

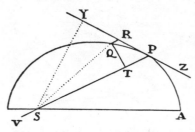

Cor. 1. If a body P revolving about the centre S describes a curved line APQ, which a right line ZPR touches in any point P; and from any other point Q of the curve, QR is drawn parallel to the distance SP, meeting the tangent in R; and QT is drawn perpendicular to the distance SP; the centripetal force will be inversely as the solid $\dfrac{SP^2 \cdot QT^2}{QR}$, if the solid be

taken of that magnitude which it ultimately acquires when the points P and Q coincide. For QR is equal to the versed sine of double the arc QP, whose middle is P: and double the triangle SQP, or SP · QT is proportional to the time in which that double arc is described; and therefore may be used to represent the time.

Cor. ii. By a like reasoning, the centripetal force is inversely as the solid $\dfrac{SY^2 \cdot QP^2}{QR}$; if SY is a perpendicular from the centre of force on PR, the tangent of the orbit. For the rectangles SY · QP and SP · QT are equal.

Cor. iii. If the orbit is either a circle, or touches or cuts a circle concentrically, that is, contains with a circle the least angle of contact or section, having the same curvature and the same radius of curvature at the point P; and if PV be a chord of this circle, drawn from the body through the centre of force; the centripetal force will be inversely as the solid $SY^2 \cdot PV$. For PV is $\dfrac{QP^2}{QR}$.

Cor. iv. The same things being supposed, the centripetal force is as the square of the velocity directly, and that chord inversely. For the velocity is reciprocally as the perpendicular SY, by Cor. i, Prop. i.

Cor. v. Hence if any curvilinear figure APQ is given, and therein a point S is also given, to which a centripetal force is continually directed, that law of centripetal force may be found, by which the body P will be continually drawn back from a rectilinear course, and, being detained in the perimeter of that figure, will describe the same by a continual revolution. That is, we are to find, by computation, either the solid $\dfrac{SP^2 \cdot QT^2}{QR}$ or the solid $SY^2 \cdot PV$, inversely proportional to this force. Examples of this we shall give in the following Problems.

PROPOSITION VII. PROBLEM II

If a body revolves in the circumference of a circle, it is proposed to find the law of centripetal force directed to any given point.

Let VQPA be the circumference of the circle; S the given point to which as to a centre the force tends; P the body moving in the circumference; Q the next place into which it is to move; and PRZ the tangent of the circle at the preceding place. Through the point S draw the chord PV, and the diameter VA of the circle; join AP, and draw QT perpendicular to SP, which produced, may meet the tangent PR in Z; and lastly, through the

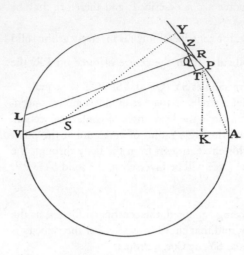

point Q, draw LR parallel to SP, meeting the circle in L, and the tangent PZ in R. And, because of the similar triangles ZQR, ZTP, VPA, we shall have

$$RP^2 : QT^2 = AV^2 : PV^2.$$

Since $RP^2 = RL \cdot QR$,

$$QT^2 = \frac{RL \cdot QR \cdot PV^2}{AV^2}.$$

Multiply those equals by $\dfrac{SP^2}{QR}$, and the points P and Q co-inciding, for RL write PV; then we shall have

$$\frac{SP^2 \cdot PV^3}{AV^2} = \frac{SP^2 \cdot QT^2}{QR}.$$

And therefore (by Cor. I and v, Prop. vi) the centripetal force is inversely as $\dfrac{SP^2 \cdot PV^3}{AV^2}$; that is (because AV^2 is given), inversely as the product of SP^2 and PV^3. Q.E.I.

The same otherwise.

On the tangent PR produced let fall the perpendicular SY; and (because of the similar triangles SYP, VPA) we shall have AV to PV as SP to SY, and therefore $\dfrac{SP \cdot PV}{AV} = SY$, and $\dfrac{SP^2 \cdot PV^3}{AV^2} = SY^2 \cdot PV$. And therefore (by Cor. III and v, Prop. vi) the centripetal force is inversely as $\dfrac{SP^2 \cdot PV^3}{AV^2}$; that is (because AV is given), inversely as $SP^2 \cdot PV^3$. Q.E.I.

Cor. I. Hence if the given point S, to which the centripetal force always tends, is placed in the circumference of the circle, as at V, the centripetal force will be inversely as the fifth power of the altitude SP.

Cor. II. The force by which the body P in the circle APTV revolves about the centre of force S is to the force by which the same body P may revolve in the same circle, and in the same periodic time, about any other centre of force R, as RP² · SP to the cube of the right line SG, which from the first centre of force S is drawn parallel to the distance PR of the body from the second centre of force R, meeting the tangent PG of the orbit in G. For by the construction of this Proposition, the former force is

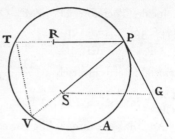

to the latter as RP² · PT³ to SP² · PV³; that is, as SP · RP² to $\dfrac{SP^3 \cdot PV^3}{PT^3}$; or (because of the similar triangles PSG, TPV) to SG³.

Cor. III. The force by which the body P in any orbit revolves about the centre of force S, is to the force by which the same body may revolve in the same orbit, and the same periodic time, about any other centre of force R, as the solid SP · RP², contained under the distance of the body from the first centre of force S, and the square of its distance from the second centre of force R, to the cube of the right line SG, drawn from the first centre of the force S, parallel to the distance RP of the body from the second centre of force R, meeting the tangent PG of the orbit in G. For the force in this orbit at any point P is the same as in a circle of the same curvature.

PROPOSITION VIII. PROBLEM III

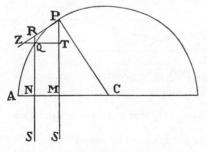

If a body moves in the semicircumference PQA; it is proposed to find the law of the centripetal force tending to a point S, so remote, that all the lines PS, RS drawn thereto, may be taken for parallels.

From C, the centre of the semicircle, let the semidiameter CA be drawn, cutting the parallels at right angles in M and N, and join CP. Because of the similar triangles CPM,

PZT, and RZQ, we shall have $CP^2 : PM^2 = PR^2 : QT^2$. From the nature of the circle, $PR^2 = QR(RN + QN) = QR \cdot 2PM$, when the points P and Q coincide. Therefore $CP^2 : PM^2 = QR \cdot 2PM : QT^2$; and $\dfrac{QT^2}{QR} = \dfrac{2PM^3}{CP^2}$, and $\dfrac{QT^2 \cdot SP^2}{QR} = \dfrac{2PM^3 \cdot SP^2}{CP^2}$. And therefore (by Cor. I and V, Prop. VI) the centripetal force is inversely as $\dfrac{2PM^3 \cdot SP^2}{CP^2}$; that is (neglecting the given ratio $\dfrac{2SP^2}{CP^2}$), inversely as PM^3. Q. E. I.

And the same thing is likewise easily inferred from the preceding Proposition.

SCHOLIUM

And by a like reasoning, a body will be moved in an ellipse, or even in an hyperbola, or parabola, by a centripetal force which is inversely as the cube of the ordinate directed to an infinitely remote centre of force.

PROPOSITION IX. PROBLEM IV

If a body revolves in a spiral PQS, *cutting all the radii* SP, SQ, *&c., in a given angle; it is proposed to find the law of the centripetal force tending to the centre of that spiral.*

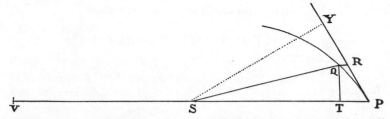

Suppose the indefinitely small angle PSQ to be given; because, then, all the angles are given, the figure SPRQT will be given in kind. Therefore the ratio $\dfrac{QT}{QR}$ is also given, and $\dfrac{QT^2}{QR}$ is as QT, that is (because the figure is given in kind), as SP. But if the angle PSQ is any way changed, the right line QR, subtending the angle of contact QPR (by Lem. XI) will be changed in the ratio of PR^2 or QT^2. Therefore the ratio $\dfrac{QT^2}{QR}$ remains

the same as before, that is, as SP. And $\dfrac{QT^2 \cdot SP^2}{QR}$ is as SP^3, and therefore (by

Cor. I and v, Prop. vi) the centripetal force is inversely as the cube of the distance SP. Q.E.I.

The same otherwise.

The perpendicular SY let fall upon the tangent, and the chord PV of the circle concentrically cutting the spiral, are in given ratios to the height SP; and therefore SP^3 is as $SY^2 \cdot PV$, that is (by Cor. iii and v, Prop. vi) inversely as the centripetal force.

LEMMA XII

All parallelograms circumscribed about any conjugate diameters of a given ellipse or hyperbola are equal among themselves.

This is demonstrated by the writers on the conic sections.

PROPOSITION X. PROBLEM V

If a body revolves in an ellipse; it is proposed to find the law of the centripetal force tending to the centre of the ellipse.

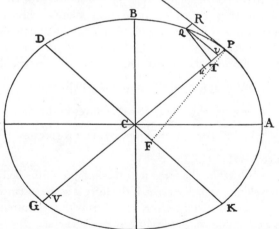

Suppose CA, CB to be semiaxes of the ellipse; GP, DK, conjugate diameters; PF, QT, perpendiculars to those diameters; Q*v*, an ordinate to the diameter GP; and if the parallelogram QvPR be completed, then (by the properties of the conic sections) P$v \cdot v$G : Qv^2 = PC^2 : CD^2, and, because of the similar triangles QvT, PCF, Qv^2 : QT^2 = PC^2 : PF^2; and by eliminating Qv^2, vG : $\dfrac{\text{QT}^2}{\text{P}v}$ = PC^2 : $\dfrac{\text{CD}^2 \cdot \text{PF}^2}{\text{PC}^2}$. Since QR = Pv, and (by Lem. xii) BC $\cdot$ CA = CD $\cdot$ PF, and, when the points P and Q coincide, 2 PC = vG, we shall have, multiplying the extremes and means together,

$$\frac{\text{QT}^2 \cdot \text{PC}^2}{\text{QR}} = \frac{2\,\text{BC}^2 \cdot \text{CA}^2}{\text{PC}}.$$

Therefore (by Cor. v, Prop. vi), the centripetal force is inversely as $\dfrac{2\,\text{BC}^2 \cdot \text{CA}^2}{\text{PC}}$; that is (because 2 BC$^2 \cdot$ CA^2 is given), inversely as $\dfrac{1}{\text{PC}}$; that is, directly as the distance PC. Q.E.I.

The same otherwise.

In the right line PG on the other side of the point T, take the point u so that Tu may be equal to Tv; then take uV, such that uV : vG = DC^2 : PC^2. Since, by the conic sections, Qv^2 : P$v \cdot v$G = DC^2 : PC^2, we have Qv^2 = P$v \cdot u$V. Add P$u \cdot$ Pv to both sides, and the square of the chord of the arc PQ will be equal to the rectangle PV $\cdot$ Pv; and therefore a circle which touches the conic section in P, and passes through the point Q, will pass also through the point V. Now let the points P and Q meet, and the ratio of uV to vG, which is the same with the ratio of DC^2 to PC^2, will become the ratio of PV to PG, or PV to 2PC; and therefore PV will be equal to $\dfrac{2\,\text{DC}^2}{\text{PC}}$. And therefore the force by which the body P revolves in the ellipse will be inversely as $\dfrac{2\,\text{DC}^2}{\text{PC}} \cdot$ PF^2 (by Cor. iii, Prop. vi); that is (because 2 DC$^2 \cdot$ PF^2 is given), directly as PC. Q.E.I.

Cor. i. And therefore the force is as the distance of the body from the centre of the ellipse; and, *vice versa,* if the force is as the distance, the body will move in an ellipse whose centre coincides with the centre of force, or perhaps in a circle into which the ellipse may degenerate.

Cor. ii. And the periodic times of the revolutions made in all ellipses whatsoever about the same centre will be equal. For those times in similar ellipses will be equal (by Cor. III and VIII, Prop. IV); but in ellipses that have their greater axis common, they are to each other as the whole areas of the ellipses directly, and the parts of the areas described in the same time inversely; that is, as the lesser axes directly, and the velocities of the bodies in their principal vertices inversely; that is, as those lesser axes directly, and the ordinates to the same point of the common axes inversely; and therefore (because of the equality of the direct and inverse ratios) in the ratio of equality, $1 : 1$.

SCHOLIUM

If the ellipse, by having its centre removed to an infinite distance, degenerates into a parabola, the body will move in this parabola; and the force, now tending to a centre infinitely remote, will become constant. This is *Galileo's* theorem. And if the parabolic section of the cone (by changing the inclination of the cutting plane to the cone) degenerates into an hyperbola, the body will move in the perimeter of this hyperbola, having its centripetal force changed into a centrifugal force. And in like manner as in the circle, or in the ellipse, if the forces are directed to the centre of the figure placed in the abscissa, those forces by increasing or diminishing the ordinates in any given ratio, or even by changing the angle of the inclination of the ordinates to the abscissa, are always augmented or diminished in the ratio of the distances from the centre; provided the periodic times remain equal; so also in all figures whatsoever, if the ordinates are augmented or diminished in any given ratio, or their inclination is any way changed, the periodic time remaining the same, the forces directed to any centre placed in the abscissa are in the several ordinates augmented or diminished in the ratio of the distances from the centre.

SECTION III

The motion of bodies in eccentric conic sections.

PROPOSITION XI. PROBLEM VI

If a body revolves in an ellipse; it is required to find the law of the centripetal force tending to the focus of the ellipse.

Let S be the focus of the ellipse. Draw SP cutting the diameter DK of the ellipse in E, and the ordinate Qv in x; and complete the parallelogram QxPR. It is evident that EP is equal to the greater semiaxis AC: for drawing HI from the other focus H of the ellipse parallel to EC, because CS, CH are equal, ES, EI will be also equal; so that EP is the half-sum of PS, PI,

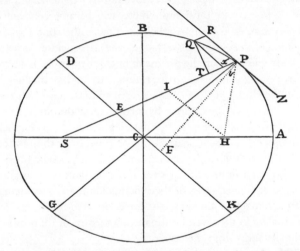

that is (because of the parallels HI, PR, and the equal angles IPR, HPZ), of PS, PH, which taken together are equal to the whole axis 2AC. Draw QT perpendicular to SP, and putting L for the principal latus rectum of the ellipse (or for $\frac{2BC^2}{AC}$), we shall have

$$L \cdot QR : L \cdot Pv = QR : Pv = PE : PC = AC : PC,$$
$$\text{also, } L \cdot Pv : Gv \cdot Pv = L : Gv, \text{ and, } Gv \cdot Pv : Qv^2 = PC^2 : CD^2.$$

By Cor. II, Lem. VII, when the points P and Q coincide, $Qv^2 = Qx^2$, and

Qx^2 or $Qv^2 : QT^2 = EP^2 : PF^2 = CA^2 : PF^2$, and (by Lem. xii) $= CD^2 : CB^2$. Multiplying together corresponding terms of the four proportions, and simplifying, we shall have

$$L \cdot QR : QT^2 = AC \cdot L \cdot PC^2 \cdot CD^2 : PC \cdot Gv \cdot CD^2 \cdot CB^2 = 2PC : Gv,$$

since $AC \cdot L = 2BC^2$. But the points Q and P coinciding, 2PC and Gv are equal. And therefore the quantities $L \cdot QR$ and QT^2, proportional to these, will be also equal. Let those equals be multiplied by $\dfrac{SP^2}{QR}$, and $L \cdot SP^2$ will become equal to $\dfrac{SP^2 \cdot QT^2}{QR}$. And therefore (by Cor. i and v, Prop. vi) the centripetal force is inversely as $L \cdot SP^2$, that is, inversely as the square of the distance SP. Q.E.I.

The same otherwise.

Since the force tending to the centre of the ellipse, by which the body P may revolve in that ellipse, is (by Cor. i, Prop. x) as the distance CP of the body from the centre C of the ellipse, let CE be drawn parallel to the tangent PR of the ellipse; and the force by which the same body P may revolve about any other point S of the ellipse, if CE and PS intersect in E, will be as $\dfrac{PE^3}{SP^2}$ (by Cor. iii, Prop. vii); that is, if the point S is the focus of the ellipse, and therefore PE be given as SP^2 reciprocally. Q.E.I.

With the same brevity with which we reduced the fifth Problem to the parabola, and hyperbola, we might do the like here; but because of the dignity of the Problem and its use in what follows, I shall confirm the other cases by particular demonstrations.

PROPOSITION XII. PROBLEM VII

Suppose a body to move in an hyperbola; it is required to find the law of the centripetal force tending to the focus of that figure.

Let CA, CB be the semiaxes of the hyperbola; PG, KD other conjugate diameters; PF a perpendicular to the diameter KD; and Qv an ordinate to the diameter GP. Draw SP cutting the diameter DK in E, and the ordinate Qv in x, and complete the parallelogram QRPx. It is evident that EP is equal to the semitransverse axis AC; for drawing HI, from the other focus H of the hyperbola, parallel to EC, because CS, CH are equal, ES, EI will be also equal; so that EP is the half difference of PS, PI; that is (because of the parallels IH, PR, and the equal angles IPR, HPZ), of PS, PH, the difference of which is equal to the whole axis 2 AC. Draw QT perpendicular

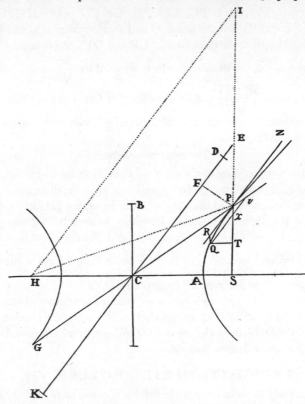

to SP; and putting L for the principal latus rectum of the hyperbola (that is, for $\frac{2\,BC^2}{AC}$), we shall have

$$L \cdot QR : L \cdot Pv = QR : Pv = Px : Pv = PE : PC = AC : PC,$$

also, $L \cdot Pv : Gv \cdot Pv = L : Gv$, and $Gv \cdot Pv : Qv^2 = PC^2 : CD^2$. By Cor. II, Lem. VII, when P and Q coincide, $Qx^2 = Qv^2$, and,

$$Qx^2 \text{ or } Qv^2 : QT^2 = EP^2 : PF^2 = CA^2 : PF^2, \text{ by Lem. XII, } = CD^2 : CB^2.$$

Multiplying together corresponding terms of the four proportions, and simplifying,

$$L \cdot QR : QT^2 = AC \cdot L \cdot PC^2 \cdot CD^2 : PC \cdot Gv \cdot CD^2 \cdot CB^2 = 2PC : Gv,$$

since $AC \cdot L = 2BC^2$. But the points P and Q coinciding, $2PC$ and Gv are equal. And therefore the quantities $L \cdot QR$ and QT^2, proportional to them, will also be equal. Let those equals be drawn into $\dfrac{SP^2}{QR}$, and we shall have $L \cdot SP^2$ equal to $\dfrac{SP^2 \cdot QT^2}{QR}$. And therefore (by Cor. I and V, Prop. VI) the centripetal force is inversely as $L \cdot SP^2$, that is, inversely as the square of the distance SP. Q.E.I.

The same otherwise.

Find out the force tending from the centre C of the hyperbola. This will be proportional to the distance CP. But from thence (by Cor. III, Prop. VII) the force tending to the focus S will be as $\dfrac{PE^3}{SP^2}$, that is, because PE is given reciprocally as SP^2. Q.E.I.

And the same way may it be demonstrated, that the body having its centripetal changed into a centrifugal force, will move in the conjugate hyperbola.

LEMMA XIII

The latus rectum of a parabola belonging to any vertex is four times the distance of that vertex from the focus of the figure.

This is demonstrated by the writers on the conic sections.

LEMMA XIV

The perpendicular, let fall from the focus of a parabola on its tangent, is a mean proportional between the distances of the focus from the point of contact, and from the principal vertex of the figure.

For, let AP be the parabola, S its focus, A its principal vertex, P the point of contact, PO an ordinate to the principal diameter, PM the tangent meeting the principal diameter in M, and SN the perpendicular from the focus on the tangent: join AN, and because of the equal lines MS and SP, MN

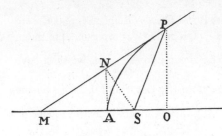

and NP, MA and AO, the right lines AN, OP will be parallel; and thence the triangle SAN will be right-angled at A, and similar to the equal triangles SNM, SNP; therefore PS is to SN as SN is to SA. Q.E.D.

Cor. i. PS² is to SN² as PS is to SA.

Cor. ii. And because SA is given, SN² will vary as PS.

Cor. iii. And the intersection of any tangent PM, with the right line SN, drawn from the focus perpendicular on the tangent, falls in the right line AN that touches the parabola in the principal vertex.

PROPOSITION XIII. PROBLEM VIII

If a body moves in the perimeter of a parabola; it is required to find the law of the centripetal force tending to the focus of that figure.

Retaining the construction of the preceding Lemma, let P be the body in the perimeter of the parabola; and from the place Q, into which it is next

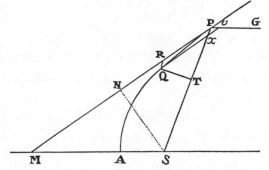

to succeed, draw QR parallel and QT perpendicular to SP, as also Qv parallel to the tangent, and meeting the diameter PG in v, and the distance SP in x. Now, because of the similar triangles Pxv, SPM, and of the equal sides SP, SM of the one, the sides Px or QR and Pv of the other will be also equal. But (by the conic sections) the square of the ordinate Qv is equal to the rectangle under the latus rectum and the segment Pv of the diameter; that is (by Lem. xiii), to the rectangle 4PS · Pv, or 4PS · QR; and the points P and Q coinciding, (by Cor. ii, Lem. vii), Qx = Qv. And therefore Qx^2, in this case, becomes equal to the rectangle 4PS · QR. But (because of the similar triangles QxT, SPN),

$$Qx^2 : QT^2 = PS^2 : SN^2 = PS : SA \text{ (by Cor. i, Lem. xiv),}$$
$$= 4PS · QR : 4SA · QR.$$

Therefore (by Prop. ix, Book v, *Elem.* of *Euclid*), $QT^2 = 4SA · QR$. Multiply these equals by $\dfrac{SP^2}{QR}$, and $\dfrac{SP^2 · QT^2}{QR}$ will become equal to $SP^2 · 4SA$: and therefore (by Cor. i and v, Prop. vi), the centripetal force is inversely as $SP^2 · 4SA$; that is, because 4SA is given, inversely as the square of the distance SP. Q.E.I.

Cor. i. From the three last Propositions it follows, that if any body P goes from the place P with any velocity in the direction of any right line PR, and at the same time is urged by the action of a centripetal force that is inversely proportional to the square of the distance of the places from the centre, the body will move in one of the conic sections, having its focus in the centre of force; and conversely. For the focus, the point of contact, and the position of the tangent, being given, a conic section may be described, which at that point shall have a given curvature. But the curvature is given from the centripetal force and velocity of the body being given; and two orbits, touching one the other, cannot be described by the same centripetal force and the same velocity.

Cor. ii. If the velocity with which the body goes from its place P is such, that in any infinitely small moment of time the small line PR may be thereby described; and the centripetal force such as in the same time to move the same body through the space QR; the body will move in one of the conic sections, whose principal latus rectum is the quantity $\dfrac{QT^2}{QR}$ in its ultimate state, when the small lines PR, QR are diminished *in infinitum*. In these Corollaries I consider the circle as an ellipse; and I except the case where the body descends to the centre in a right line.

PROPOSITION XIV. THEOREM VI

If several bodies revolve about one common centre, and the centripetal force is inversely as the square of the distance of places from the centre: I say, that the principal latera recta of their orbits are as the squares of the areas, which the bodies by radii drawn to the centre describe in the same time.

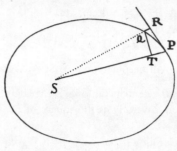

For (by Cor. II, Prop. XIII) the latus rectum L is equal to the quantity $\dfrac{QT^2}{QR}$ in its ultimate state when the points P and Q coincide. But the small line QR in a given time is as the generating centripetal force; that is (by supposition), inversely as SP^2. And therefore $\dfrac{QT^2}{QR}$ is as $QT^2 \cdot SP^2$; that is, the latus rectum L is as the square of the area QT · SP. Q.E.D.

Cor. Hence the whole area of the ellipse, and the rectangle under the axes, which is proportional to it, is as the product of the square root of the latus rectum, and the periodic time. For the whole area is as the area QT · SP, described in a given time, multiplied by the periodic time.

PROPOSITION XV. THEOREM VII

The same things being supposed, I say, that the periodic times in ellipses are as the ⅔th power (in ratione sesquiplicata) *of their greater axes.*

For the lesser axis is a mean proportional between the greater axis and the latus rectum; and, therefore, the product of the axes is equal to the product of the square root of the latus rectum and the ⅔th power of the greater axis. But the product of the axes (by Cor., Prop. XIV) varies as the product of the square root of the latus rectum, and the periodic time. Divide both sides by the square root of the latus rectum and it follows that the ⅔th power of the greater axis varies as the periodic time. Q.E.D.

Cor. Therefore the periodic times in ellipses are the same as in circles whose diameters are equal to the greater axes of the ellipses.

PROPOSITION XVI. THEOREM VIII

The same things being supposed, and right lines being drawn to the bodies that shall touch the orbits, and perpendiculars being let fall on those tangents from the common focus: I say, that the velocities of the bodies vary inversely as the perpendiculars and directly as the square roots of the principal latera recta.

From the focus S draw SY perpendicular to the tangent PR, and the velocity of the body P varies inversely as the square root of the quantity $\frac{SY^2}{L}$. For that velocity is as the infin-

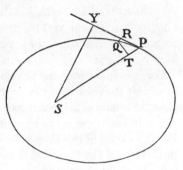

itely small arc PQ described in a given moment of time, that is (by Lem. vii), as the tangent PR; that is (because of the proportion, PR : QT = SP : SY), as $\frac{SP \cdot QT}{SY}$; or inversely as SY, and directly as SP · QT; but SP · QT is as the area described in the given time, that is (by Prop. xiv), as the square root of the latus rectum. Q.E.D.

Cor. i. The principal latera recta vary as the squares of the perpendiculars and the squares of the velocities.

Cor. ii. The velocities of bodies, in their greatest and least distances from the common focus, are inversely as the distances and directly as the square root of the principal latera recta. For those perpendiculars are now the distances.

Cor. iii. And therefore the velocity in a conic section, at its greatest or least distance from the focus, is to the velocity in a circle, at the same distance from the centre, as the square root of the principal latus rectum is to the double of that distance.

Cor. iv. The velocities of the bodies revolving in ellipses, at their mean distances[1] from the common focus, are the same as those of bodies revolving in circles, at the same distances; that is (by Cor. vi, Prop. iv), inversely as the

[1 Appendix, Note 18.]

square root of the distances. For the perpendiculars are now the lesser semi-axes, and these are as mean proportionals between the distances and the latera recta. Let the inverse of this ratio [of the minor semiaxes] be multiplied by the square root of the direct ratio of the latera recta, and we shall have the square root of the inverse ratio of the distances.

Cor. v. In the same figure, or even in different figures, whose principal latera recta are equal, the velocity of a body is inversely as the perpendicular let fall from the focus on the tangent.

Cor. vi. In a parabola, the velocity is inversely as the square root of the ratio of the distance of the body from the focus of the figure; it is more variable in the ellipse, and less in the hyperbola, than according to this ratio. For (by Cor. ii, Lem. xiv) the perpendicular let fall from the focus on the tangent of a parabola is as the square root of the ratio of the distance. In the hyperbola the perpendicular is less variable; in the ellipse, more.

Cor. vii. In a parabola, the velocity of a body at any distance from the focus is to the velocity of a body revolving in a circle, at the same distance from the centre, as the square root of the ratio of the number 2 to 1; in the ellipse it is less, and in the hyperbola greater, than according to this ratio. For (by Cor. ii of this Prop.) the velocity at the vertex of a parabola is in this ratio, and (by Cor. vi of this Prop. and Prop. iv) the same proportion holds in all distances. And hence, also, in a parabola, the velocity is everywhere equal to the velocity of a body revolving in a circle at half the distance; in the ellipse it is less, and in the hyperbola greater.

Cor. viii. The velocity of a body revolving in any conic section is to the velocity of a body revolving in a circle, at the distance of half the principal latus rectum of the section, as that distance to the perpendicular let fall from the focus on the tangent of the section. This appears from Cor. v.

Cor. ix. Wherefore, since (by Cor. vi, Prop. iv) the velocity of a body revolving in this circle is to the velocity of another body revolving in any other circle, inversely as the square root of the ratio of the distances; therefore, likewise, the velocity of a body revolving in a conic section will be to the velocity of a body revolving in a circle at the same distance as a mean proportional between that common distance, and half the principal latus rectum of the section, to the perpendicular let fall from the common focus upon the tangent of the section.

PROPOSITION XVII. PROBLEM IX

Supposing the centripetal force to be inversely proportional to the squares of the distances of places from the centre, and that the absolute value of that force is known; it is required to determine the line which a body will describe that is let go from a given place with a given velocity in the direction of a given right line.

Let the centripetal force tending to the point S be such as will make the body p revolve in any given orbit *pq;* and suppose the velocity of this body in the place p is known. Then from the place P suppose the body P to be let go with a given velocity in the direction of the line PR; but by virtue of a centripetal force to be immediately turned aside from that right line into the conic section PQ. This, the right line PR will therefore touch in P. Suppose likewise that the right line *pr* touches the orbit *pq* in *p;* and if from

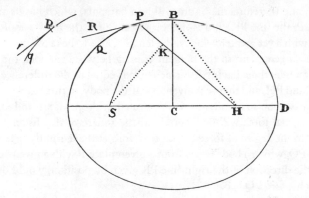

S you suppose perpendiculars let fall on those tangents, the principal latus rectum of the conic section (by Cor. 1, Prop. xvi) will be to the principal latus rectum of that orbit in a ratio compounded of the squared ratio of the perpendiculars, and the squared ratio of the velocities; and is therefore given. Let this latus rectum be L; the focus S of the conic section is also given. Let the angle RPH be the supplement of the angle RPS, and the line PH, in which the other focus H is placed, is given by position. Let fall SK perpendicular on PH, and erect the conjugate semiaxis BC; this done, we shall have

$$SP^2 - 2PH \cdot PK + PH^2 = SH^2 = 4CH^2 = 4(BH^2 - BC^2) =$$
$$(SP + PH)^2 - L(SP + PH) = SP^2 + 2PS \cdot PH + PH^2 - L(SP + PH).$$

Add on both sides

$$2PK \cdot PH - SP^2 - PH^2 + L(SP + PH),$$

and we shall have

$$L(SP + PH) = 2PS \cdot PH + 2PK \cdot PH, \text{ or}$$
$$(SP + PH) : PH = 2(SP + KP) : L.$$

Hence PH is given both in length and position. That is, if the velocity of the body in P is such that the latus rectum L is less than $2SP + 2KP$, PH will lie on the same side of the tangent PR with the line SP; and therefore the figure will be an ellipse, which from the given foci S, H, and the principal axis SP + PH, is given also. But if the velocity of the body is so great, that the latus rectum L becomes equal to $2SP + 2KP$, the length PH will be infinite; and therefore, the figure will be a parabola, which has its axis SH parallel to the line PK, and is thence given. But if the body goes from its place P with a yet greater velocity, the length PH is to be taken on the other side the tangent; and so the tangent passing between the foci, the figure will be an hyperbola having its principal axis equal to the difference of the lines SP and PH, and thence is given. For if the body, in these cases, revolves in a conic section so found, it is demonstrated in Prop. XI, XII, and XIII, that the centripetal force will be inversely as the square of the distance of the body from the centre of force S; and therefore we have rightly determined the line PQ, which a body let go from a given place P with a given velocity, and in the direction of the right line PR given by position, would describe with such a force. Q.E.F.

COR. I. Hence in every conic section, from the principal vertex D, the latus rectum L, and the focus S given, the other focus H is given, by taking DH to DS as the latus rectum to the difference between the latus rectum and 4DS. For the proportion

$$SP + PH : PH = 2SP + 2KP : L$$

becomes, in the case of this Corollary,

$$DS + DH : DH = 4DS : L,$$
$$\text{and } DS : DH = 4DS - L : L.$$

Cor. ii. Whence if the velocity of a body in the principal vertex D is given, the orbit may be readily found; namely, by taking its latus rectum to twice the distance DS, in the squared ratio of this given velocity to the velocity of a body revolving in a circle at the distance DS (by Cor. iii, Prop. xvi), and then taking DH to DS as the latus rectum to the difference between the latus rectum and 4DS.

Cor. iii. Hence also if a body move in any conic section, and is forced out of its orbit by any impulse, you may discover the orbit in which it will afterwards pursue its course. For by compounding the proper motion of the body with that motion, which the impulse alone would generate, you will have the motion with which the body will go off from a given place of impulse in the direction of a right line given in position.

Cor. iv. And if that body is continually disturbed by the action of some foreign force, we may nearly know its course, by collecting the changes which that force introduces in some points, and estimating the continual changes it will undergo in the intermediate places, from the analogy that appears in the progress of the series.

SCHOLIUM

If a body P, by means of a centripetal force tending to any given point R, move in the perimeter of any given conic section whose centre is C; and the law of the centripetal force is required: draw CG parallel to the radius RP, and meeting the tangent PG of the orbit in G; and the force required (by Cor. i and Schol., Prop. x, and Cor. iii, Prop. vii) will be as $\frac{CG^3}{RP^2}$.

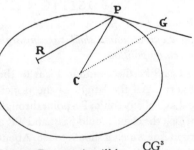

SECTION IV

The finding of elliptic, parabolic, and hyperbolic orbits, from the
focus given.

LEMMA XV

If from the two foci S, H, of any ellipse or hyperbola, we draw to any third
point V the right lines SV, HV, whereof one HV is equal to the principal
axis of the figure, that is, to the axis in which the foci are situated, the other,
SV, is bisected in T by the perpendicular TR let fall upon it; that perpen-
dicular TR will somewhere touch the conic section: and, vice versa, *if it*
does touch it, HV *will be equal to the principal axis of the figure.*

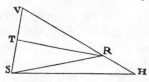

For, let the perpendicular TR cut the right
line HV, produced, if need be, in R; and
join SR. Because TS, TV are equal, therefore the right lines SR, VR, as well as the
angles TRS, TRV, will be also equal.
Whence the point R will be in the conic section, and the perpendicular TR
will touch the same; and the contrary. Q.E.D.

PROPOSITION XVIII. PROBLEM X

From a focus and the principal axes given, to describe elliptic and hyper-
bolic curves which shall pass through given points, and touch right lines
given by position.

Let S be the common focus of the
figures; AB the length of the principal axis of any conic; P a point through
which the conic should pass; and TR a
right line which it should touch. About
the centre P, with the radius AB − SP,
if the orbit is an ellipse, or AB + SP, if

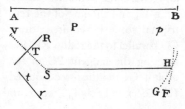

the orbit is an hyperbola, describe the circle HG. On the tangent TR let fall
the perpendicular ST, and produce the same to V, so that TV may be equal
to ST; and about V as a centre with the interval AB describe the circle FH.
In this manner, whether two points P, *p,* are given, or two tangents TR, *tr,*

or a point P and a tangent TR, we are to describe two circles. Let H be their common intersection, and from the foci S, H, with the given axis describe the conic: I say, the thing is done. For (because PH + SP in the ellipse, and PH − SP in the hyperbola, is equal to the axis) the described conic will pass through the point P, and (by the preceding Lemma) will touch the right line TR. And by the same argument it will either pass through the two points P, *p,* or touch the two right lines TR, *tr.* Q.E.F.

PROPOSITION XIX. PROBLEM XI

About a given focus, to describe a parabola which shall pass through given
points and touch right lines given by position.

Let S be the focus, P a point, and TR a tangent of the curve to be described. About P as a centre, with the radius PS, describe the circle FG. From the focus let fall ST perpendicular on the tangent, and produce the same to V, so as TV may be equal to ST. After

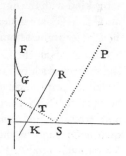

the same manner another circle *fg* is to be described, if another point *p* is given; or another point *v* is to be found, if another tangent *tr* is given; then draw the right line IF, which shall touch the two circles FG, *fg,* if two points P, *p* are given; or pass through the two points V, *v,* if two tangents TR, *tr,* are given; or touch the circle FG, and pass through the point V, if the point P and the tangent TR are given. On FI let fall the perpendicular SI, and bisect the same in K; and with the axis SK and principal vertex K describe a parabola: I say, the thing is done. For this parabola (because SK is equal to IK, and SP to FP) will pass through the point P; and (by Cor. III, Lem. XIV) because ST is equal to TV, and STR a right angle, it will touch the right line TR. Q.E.F.

PROPOSITION XX. PROBLEM XII

About a given focus, to describe any given conic which shall pass through
given points and touch right lines given by position.

CASE I. About the focus S it is required to describe a conic ABC, passing through two points B, C. Because the conic is given in kind, the ratio of the principal axis to the distance of the foci will be given. In that ratio take KB to BS, and LC to CS. About the centres B, C, with the intervals BK, CL, describe two circles; and on the right line KL, that touches the same in K and L, let fall the perpendicular SG; which cut in A and *a*, so that GA may

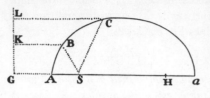

be to AS, and G*a* to *a*S, as KB to BS; and with the axis A*a*, and vertices A, *a*, describe a conic: I say, the thing is done. For let H be the other focus of the described figure, and seeing that GA : AS = G*a* : *a*S, we shall have

G*a* – GA : *a*S – AS = GA : AS, or A*a* : SH = GA : AS, and therefore GA and AS are in the ratio which the principal axis of the figure to be described has to the distance of its foci; and therefore the described figure is of the same kind with the figure which was to be described. And since KB to BS, and LC to CS, are in the same ratio, this figure will pass through the points B, C, as is manifest from the conic sections.

CASE 2. About the focus S it is required to describe a conic which shall somewhere touch two right lines TR, *tr*. From the focus on those tangents let fall the perpendiculars ST, S*t*, which produce to V, *v*, so that TV, *tv* may be equal to TS, *t*S. Bisect V*v* in O,

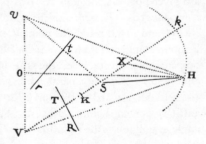

and erect the indefinite perpendicular OH, and cut the right line VS infinitely produced in K and *k*, so that VK be to KS, and V*k* to *k*S, as the principal axis of the conic to be described is to the distance of its foci. On the diameter K*k* describe a circle cutting OH in H; and with the foci S, H, and principal axis equal to VH, describe a conic: I say, the thing is done. For bisecting K*k* in X, and joining HX, HS, HV, H*v*, because VK is to KS as V*k* to *k*S; and by composition, as VK + V*k* to KS + *k*S; and by subtraction, as V*k* – VK to *k*S – KS, that is, as 2VX to 2KX, and 2KX to 2SX, and therefore as VX to HX and HX to SX, the triangles VXH, HXS will be similar; therefore VH will be to SH as VX to XH; and therefore as VK to KS. Wherefore VH, the principal axis of the described conic, has the same ratio to SH, the distance of the foci, as the principal axis of the conic which was to be described has to the distance of its foci; and is therefore of the same kind. And seeing VH, *v*H are equal to the principal axis, and VS, *v*S are perpendicularly bisected by the right lines TR, *tr*, it is evident (by Lem. xv) that those right lines touch the described conic. Q.E.F.

Case 3. About the focus S it is required to describe a conic which shall touch a right line TR in a given point R. On the right line TR let fall the perpendicular ST, which produce to V, so that TV may be equal to ST; join VR, and cut the right line VS indefinitely produced in K and *k*, so that VK may be to SK, and V*k* to S*k*, as the principal axis of the ellipse to be described to the distance of its foci;

and on the diameter K*k* describing a circle, cut the right line VR produced in H; then with the foci S, H, and principal axis equal to VH, describe a conic: I say, the thing is done. For VH : SH = VK : SK, and therefore as the principal axis of the

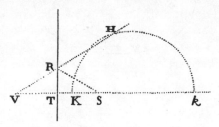

conic which was to be described to the distance of its foci (as appears from what we have demonstrated in Case 2); and therefore the described conic is of the same kind with that which was to be described; but that the right line TR, by which the angle VRS is bisected, touches the conic in the point R, is certain from the properties of the conic sections. Q.E.F.

Case 4. About the focus S it is required to describe a conic APB that shall touch a right line TR, and pass through any given point P without the tangent, and shall be similar to the figure *apb*, described with the principal axis *ab*, and foci *s*, *h*. On the tangent TR let fall the perpendicular ST, which produce to V, so that TV may be equal to ST; and making the angles *hsq*, *shq*, equal to the angles VSP, SVP, about *q* as a centre, and with a radius which shall be to *ab* as SP to VS, describe a circle cutting the figure *apb* in *p*. Join *sp*, and draw SH such that it may be to *sh* as SP is to *sp*, and may make the angle PSH equal to the angle *psh*, and the angle VSH equal to the angle *psq*. Then with the foci S, H, and principal axis AB, equal to the distance VH, describe a conic section: I say, the thing is done; for if *sv* is drawn so that it shall be to *sp* as *sh* is to *sq*, and shall make the angle *vsp* equal to the angle *hsq*, and the angle *vsh* equal to the angle *psq*, the triangles *svh*, *spq*, will be similar, and therefore *vh* will be to *pq* as *sh* is to *sq*; that is (because of the similar triangles VSP, *hsq*), as VS is to SP, or as *ab* to *pq*. Wherefore *vh* and *ab* are equal. But, because of the similar triangles VSH,

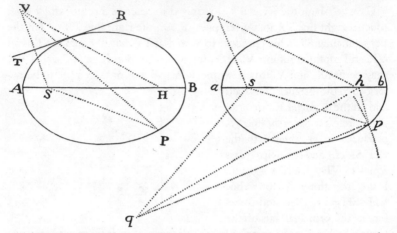

vsh, VH is to SH as *vh* to *sh*; that is, the axis of the conic section now described is to the distance of its foci as the axis *ab* to the distance of the foci *sh*; and therefore the figure now described is similar to the figure *aph*. But, because the triangle PSH is similar to the triangle *psh*, this figure passes through the point P; and because VH is equal to its axis, and VS is perpendicularly bisected by the right line TR, the said figure touches the right line TR. Q.E.F.

LEMMA XVI

From three given points to draw to a fourth point that is not given three right lines whose differences either shall be given or are zero.

CASE I. Let the given points be A, B, C, and Z the fourth point which we are to find; because of the given difference of the lines AZ, BZ, the locus of the point Z will be an hyperbola whose foci are A and B, and whose principal axis is the given difference. Let that axis be MN. Taking PM to MA as MN to AB, erect PR perpendicular to AB, and let fall ZR perpendicular to PR; then from the nature of the hyperbola, ZR : AZ = MN : AB. And by the

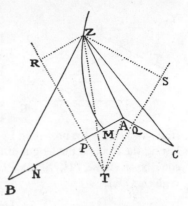

like argument, the locus of the point Z will be another hyperbola, whose foci are A, C, and whose principal axis is the difference between AZ and CZ; and QS a perpendicular on AC may be drawn, to which (QS) if from any point Z of this hyperbola a perpendicular ZS is let fall, (this ZS) shall be to AZ as the difference between AZ and CZ is to AC. Wherefore the ratios of ZR and ZS to AZ are given, and consequently the ratio of ZR to ZS one to the other; and therefore if the right lines RP, SQ, meet in T, and TZ and TA are drawn, the figure TRZS will be given in kind, and the right line TZ, in which the point Z is somewhere placed, will be given in position. There will be given also the right line TA, and the angle ATZ; and because the ratios of AZ and TZ to ZS are given, their ratio to each other is given also; and thence will be given likewise the triangle ATZ, whose vertex is the point Z. Q.E.I.

Case 2. If two of the three lines, for example AZ and BZ, are equal, draw the right line TZ so as to bisect the right line AB; then find the triangle ATZ as above. Q.E.I.

Case 3. If all the three are equal, the point Z will be placed in the centre of a circle that passes through the points A, B, C. Q.E.I.

This problematic Lemma is likewise solved in the *Book of Tactions* of *Apollonius* restored by *Vieta*.

PROPOSITION XXI. PROBLEM XIII

About a given focus, to describe a conic that shall pass through given points and touch right lines given by position.

Let the focus S, the point P, and the tangent TR be given, and suppose that the other focus H is to be found. On the tangent let fall the perpendicular ST, which produce to Y, so that TY may be equal to ST, and YH will be equal to the principal axis. Join SP, HP, and SP will be the difference between HP and the principal axis. After this manner, if more tangents

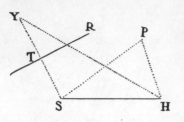

TR are given, or more points P, we shall always determine as many lines YH, or PH, drawn from the said points Y or P, to the focus H, which either shall be equal to the axes, or differ from the axes by given lengths SP; and therefore which shall either be equal among themselves, or shall have given differences; from whence (by the preceding Lemma), that other focus H is given. But having the foci and the length of the axis (which is either YH, or, if the conic be an ellipse, PH + SP; or PH − SP, if it be an hyperbola), the conic is given. Q.E.I.

SCHOLIUM

When the conic is an hyperbola, I do not include its conjugate hyperbola under the name of this conic. For a body going on with a continued motion can never pass out of one hyperbola into its conjugate hyperbola.

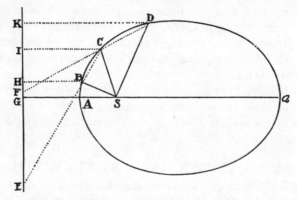

The case when three points are given is more readily solved thus. Let B, C, D be the given points. Join BC, CD, and produce them to E, F, so as EB may be to EC as SB to SC; and FC to FD as SC to SD. On EF drawn and

produced let fall the perpendiculars SG, BH, and in GS produced indefinitely take GA to AS, and Ga to aS, as HB is to BS: then A will be the vertex, and Aa the principal axis of the conic; which, according as GA is greater than, equal to, or less than AS, will be either an ellipse, a parabola, or an hyperbola; the point a in the first case falling on the same side of the line GF as the point A; in the second, going off to an infinite distance; in the third, falling on the other side of the line GF. For if on GF the perpendiculars CI, DK are let fall, IC will be to HB as EC to EB; that is, as SC to SB; and by permutation, IC to SC as HB to SB, or as GA to SA. And, by the like argument, we may prove that KD is to SD in the same ratio. Wherefore the points B, C, D lie in a conic section described about the focus S, in such manner that all the right lines drawn from the focus S to the several points of the section, and the perpendiculars let fall from the same points on the right line GF, are in that given ratio.

That excellent geometer M. *de la Hire* has solved this Problem much after the same way, in his *Conics,* Prop. xxv, Book viii.

SECTION V

How the orbits are to be found when neither focus is given.

LEMMA XVII

If from any point P *of a given conic section, to the four produced sides* AB, CD, AC, DB *of any trapezium*[1] ABDC *inscribed in that section, as many right lines* PQ, PR, PS, PT *are drawn in given angles, each line to each side; the rectangle* PQ · PR *of those on the opposite sides* AB, CD, *will be to the rectangle* PS · PT *of those on the other two opposite sides* AC, BD, *in a given ratio.*

CASE I. Let us suppose, first, that the lines drawn to one pair of opposite sides are parallel to either of the other sides; as PQ and PR to the side AC, and PS and PT to the side AB. And further, that one pair of the opposite

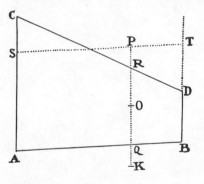

sides, as AC and BD, are parallel between themselves; then the right line which bisects those parallel sides will be one of the diameters of the conic section, and will likewise bisect RQ. Let O be the point in which RQ is bisected, and PO will be an ordinate to that diameter. Produce PO to K, so that OK may be equal to PO, and OK will be an ordinate on the other side of that diameter. Since, therefore, the points A, B, P, and K are placed in the conic section, and PK cuts AB in a given angle, the rectangle PQ · QK (by Prop. XVII, XIX, XXI, and XXIII, Book III, *Conics* of *Apollonius*) will be to the rectangle AQ · QB in a given ratio. But QK and PR are equal, as being the differences of the equal lines OK, OP, and OQ, OR; whence the rectangles PQ · QK and PQ · PR are equal; and therefore the rectangle PQ · PR is to the rectangle AQ · QB, that is, to the rectangle PS · PT, in a given ratio. Q.E.D.

[1 Appendix, Note 19.]

Case 2. Let us next suppose that the opposite sides AC and BD of the trapezium are not parallel. Draw B*d* parallel to AC, and meeting as well the right line ST in *t,* as the conic section in *d.* Join C*d* cutting PQ in *r,* and draw DM parallel to PQ, cutting C*d* in M, and AB in N. Then (because of the similar triangles BT*t,* DBN) B*t* or PQ : T*t* = DN : NB. And so R*r* : AQ or PS = DM : AN. Wherefore, by multiplying the antecedents by the antecedents, and the consequents by the consequents, as the rectangle PQ · R*r* is to the rectangle PS · T*t,* so will the rectangle DN · DM be to the rectangle

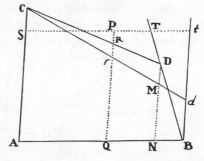

NA · NB; and (by Case 1) so is the rectangle PQ · P*r* to the rectangle PS · P*t,* and, by division, so is the rectangle PQ · PR to the rectangle PS · PT. Q.E.D.

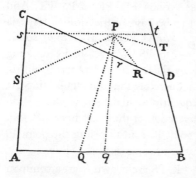

Case 3. Let us suppose, lastly, the four lines PQ, PR, PS, PT not to be parallel to the sides AC, AB, but any way inclined to them. In their place draw P*q,* P*r,* parallel to AC; and P*s,* P*t* parallel to AB; and because the angles of the triangles PQ*q,* PR*r,* PS*s,* PT*t* are given, the ratios of PQ to P*q,* PR to P*r,* PS to P*s,* PT to P*t* will be also given; and therefore the compounded ratios PQ · PR to P*q* · P*r,* and PS · PT to P*s* · P*t* are given. But from what we have demonstrated before, the ratio of P*q* · P*r* to P*s* · P*t* is given; and therefore also the ratio of PQ · PR to PS · PT. Q.E.D.

LEMMA XVIII

The same things supposed, if the rectangle PQ · PR *of the lines drawn to the two opposite sides of the trapezium is to the rectangle* PS · PT *of those*

drawn to the other two sides in a given ratio, the point P, from whence those lines are drawn, will be placed in a conic section described about the trapezium.

Conceive a conic section to be described passing through the points A, B, C, D, and any one of the infinite number of points P, as for example *p*: I say, the point P will be always placed in this section. If you deny the thing,

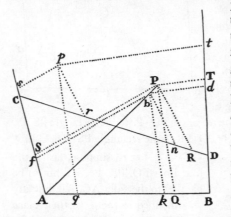

join AP cutting this conic section somewhere else, if possible, than in P, as in *b*. Therefore if from those points *p* and *b,* in the given angles to the sides of the trapezium, we draw the right lines *pq, pr, ps, pt,* and *bk, bn, bf, bd,* we shall have, as *bk · bn* to *bf · bd,* so (by Lem. XVII) *pq · pr* to *ps · pt*; and so (by supposition) PQ · PR to PS · PT. And because of the similar trapezia *bkAf,* PQAS, as *bk* to *bf,* so PQ to PS. Wherefore by dividing the terms of the preceding proportion by the correspondent terms of this, we shall have *bn* to *bd* as PR to PT. And therefore the equiangular trapezia D*nbd,* DRPT, are similar, and consequently their diagonals D*b,* DP do coincide. Wherefore *b* falls in the intersection of the right lines AP, DP, and consequently coincides with the point P. And therefore the point P, wherever it is taken, falls within the assigned conic section. Q.E.D.

Cor. Hence if three right lines PQ, PR, PS are drawn from a common point P, to as many other right lines given in position, AB, CD, AC, each to each, in as many angles respectively given, and the rectangle PQ · PR under any two of the lines drawn be to the square of the third PS in a given ratio; the point P, from which the right lines are drawn, will be placed in a conic section that touches the lines AB, CD in A and C; and the contrary. For the position of the three right lines AB, CD, AC remaining the same, let the line BD approach to and coincide with the line AC; then let the line

PT come likewise to coincide with the line PS; and the rectangle PS · PT
will become PS², and the right lines AB, CD, which before did cut the curve
in the points A and B, C and D, can no longer cut, but only touch, the curve
in those coinciding points.

SCHOLIUM

In this Lemma, the name of conic section is to be understood in a large
sense, comprehending as well the rectilinear section through the vertex of
the cone, as the circular one parallel to the base. For if the point p happens
to be in a right line, by which
the points A and D, or C and B
are joined, the conic section will
be changed into two right lines,
one of which is that right line
upon which the point p falls, and
the other is a right line that joins
the other two of the four points.
If the two opposite angles of the
trapezium taken together are
equal to two right angles, and if
the four lines PQ, PR, PS, PT
are drawn to the sides thereof at
right angles, or any other equal

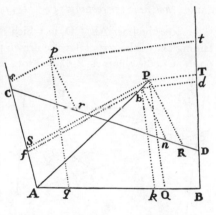

angles, and the rectangle PQ · PR under two of the lines drawn PQ and
PR, is equal to the rectangle PS · PT under the other two PS and PT, the
conic section will become a circle. And the same thing will happen if the
four lines are drawn in any angles, and the rectangle PQ · PR, under one
pair of the lines drawn, is to the rectangle PS · PT under the other pair as
the rectangle under the sines of the angles S, T, in which the two last lines
PS, PT are drawn, to the rectangle under the sines of the angles Q, R, in
which the first two PQ, PR are drawn. In all other cases the locus of the
point P will be one of the three figures which pass commonly by the name
of the conic sections. But in place of the trapezium ABCD, we may substitute a quadrilateral figure whose two opposite sides cross one another
like diagonals. And one or two of the four points A, B, C, D may be sup-

posed to be removed to an infinite distance, by which means the sides of
the figure which converge to those points, will become parallel; and in this
case the conic section will pass through the other points, and will go the
same way as the parallels *in infinitum.*

LEMMA XIX

To find a point P *from which if four right lines* PQ, PR, PS, PT *are drawn
to as many other right lines* AB, CD, AC, BD, *given by position, each to
each, at given angles, the rectangle* PQ · PR, *under any two of the lines
drawn, shall be to the rectangle* PS · PT, *under the other two, in a given ratio.*

Suppose the lines AB, CD, to which the two right lines PQ, PR, containing one of the rectangles, are drawn to meet two other lines, given by
position, in the points A, B, C, D. From one of those, as A, draw any right

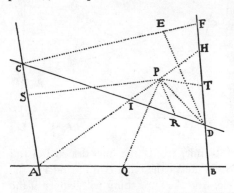

line AH, in which you would
find the point P. Let this cut
the opposite lines BD, CD, in
H and I; and, because all the
angles of the figure are given,
the ratio of PQ to PA, and
PA to PS, and therefore of
PQ to PS, will be also given.
This ratio taken as a divisor
of the given ratio of PQ · PR
to PS · PT, gives the ratio of
PR to PT; and multiplying
the given ratios of PI to PR, and PT to PH, the ratio of PI to PH, and therefore the point P, will be given. Q.E.I.

Cor. 1. Hence also a tangent may be drawn to any point D of the locus
of all the points P. For the chord PD, where the points P and D meet, that
is, where AH is drawn through the point D, becomes a tangent. In which
case the ultimate ratio of the evanescent lines IP and PH will be found as
above. Therefore draw CF parallel to AD, meeting BD in F, and cut it in
E in the same ultimate ratio, then DE will be the tangent; because CF and
the evanescent IH are parallel, and similarly cut in E and P.

Cor. ii. Hence also the locus of all the points P may be determined. Through any of the points A, B, C, D, as A, draw AE touching the locus, and through any other point B, parallel to the tangent, draw BF meeting the locus in F; and find the point F by this Lemma. Bisect BF in G, and, drawing the indefinite line AG, this will be the position of the diameter to which BG and FG are ordinates. Let this AG meet the locus in H, and AH will be its diameter or latus transversum, to which the latus rectum will be as BG² to AG·GH. If AG nowhere meets the locus, the line AH being infinite, the locus will be a parabola; and its latus rectum corresponding to the diameter AG will be $\frac{BG^2}{AG}$. But if it does meet it

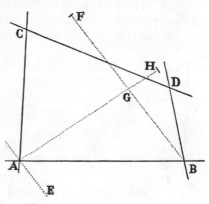

anywhere, the locus will be an hyperbola, when the points A and H are placed on the same side of the point G; and an ellipse, if the point G falls between the points A and H; unless, perhaps, the angle AGB is a right angle, and at the same time BG² equal to the rectangle GA · GH, in which case the locus will be a circle.

And so we have given in this Corollary a solution of that famous Problem of the ancients concerning four lines, begun by *Euclid,* and carried on by *Apollonius;* and this not an analytical calculus but a geometrical composition, such as the ancients required.

LEMMA XX

If the two opposite angular points A *and* P *of any parallelogram* ASPQ *touch any conic section in the points* A *and* P; *and the sides* AQ, AS *of one of those angles, indefinitely produced, meet the same conic section in* B *and* C; *and from the points of meeting* B *and* C *to any fifth point* D *of the conic section, two right lines* BD, CD *are drawn meeting the two other sides* PS, PQ *of the parallelogram, indefinitely produced in* T *and* R; *the parts* PR *and* PT, *cut off from the sides, will always be one to the other in a given ratio. And conversely, if those parts cut off are one to the other in a given ratio, the locus of the point* D *will be a conic section passing through the four points* A, B, C, P.

CASE 1. Join BP, CP, and from the point D draw the two right lines DG, DE, of which the first DG shall be parallel to AB, and meet PB, PQ, CA, in H, I, G; and the other DE shall be parallel to AC, and meet PC, PS, AB, in F, K, E; and (by Lem. xvii)

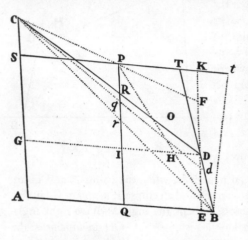

the rectangle DE · DF will be to the rectangle DG · DH in a given ratio. But PQ is to DE (or IQ) as PB to HB, and consequently as PT to DH; and by permutation PQ is to PT as DE to DH. Likewise PR is to DF as RC to DC, and therefore as (IG or) PS to DG; and by permutation PR is to PS as DF to DG; and, by compounding those ratios, the rectangle PQ · PR will be to the rectangle PS · PT as the rectangle DE · DF is to the rectangle DG · DH, and consequently in a given ratio. But PQ and PS are given, and therefore the ratio of PR to PT is given. Q.E.D.

CASE 2. But if PR and PT are supposed to be in a given ratio one to the other, then by going back again, by a like reasoning, it will follow that the rectangle DE · DF is to the rectangle DG · DH in a given ratio; and so the point D (by Lem. xviii) will lie in a conic section passing through the points A, B, C, P, as its locus. Q.E.D.

COR. 1. Hence if we draw BC cutting PQ in *r* and in PT take P*t* to P*r* in the same ratio which PT has to PR; then B*t* will touch the conic section in the point B. For suppose the point D to coalesce with the point B, so that the chord BD vanishing, BT shall become a tangent; and CD and BT will coincide with CB and B*t*.

Cor. II. And, *vice versa,* if B*t* is a tangent, and the lines BD, CD meet in any point D of a conic section, PR will be to PT as P*r* to P*t*. And, on the contrary, if PR is to PT as P*r* to P*t*, then BD and CD will meet in some point D of a conic section.

Cor. III. One conic section cannot cut another conic section in more than four points. For, if it is possible, let two conic sections pass through the five points A, B, C, P, O; and let the right line BD cut them in the points D, *d,* and the right line C*d* cut the right line PQ in *q*. Therefore PR is to PT as P*q* to PT: whence PR and P*q* are equal one to the other, against the supposition.

LEMMA XXI

If two movable and indefinite right lines BM, CM *drawn through given points* B, C, *as poles, do by their point of meeting* M *describe a third right line* MN *given by position; and other two indefinite right lines* BD, CD *are drawn, making with the former two at those given points* B, C, *given angles,* MBD, MCD: *I say, that those two right lines* BD, CD *will by their point of meeting* D *describe a conic section passing through the points* B, C. *And conversely, if the right lines* BD, CD *do by their point of meeting* D *describe a conic section passing through the given points* B, C, A, *and the angle* DBM *is always equal to the given angle* ABC, *as well as the angle* DCM *always equal to the given angle* ACB, *the point* M *will lie in a right line given by position, as its locus.*

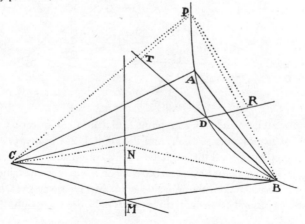

For in the right line MN let a point N be given, and when the movable point M falls on the immovable point N, let the movable point D fall on an immovable point P. Join CN, BN, CP, BP, and from the point P draw the right lines PT, PR meeting BD, CD in T and R, and making the angle BPT equal to the given angle BNM, and the angle CPR equal to the given angle CNM. Wherefore since (by supposition) the angles MBD, NBP are equal, as also the angles MCD, NCP, take away the angles NBD and NCD that are common, and there will remain the angles NBM and PBT, NCM and PCR equal; and therefore the triangles NBM, PBT are similar, as also the triangles NCM, PCR. Wherefore PT is to NM as PB to NB; and PR to NM as PC to NC. But the points B, C, N, P are immovable: wherefore PT and PR have a given ratio to NM, and consequently a given ratio between themselves; and therefore, (by Lem. xx) the point D wherein the movable right lines BT and CR continually concur, will be placed in a conic section passing through the points B, C, P. Q.E.D.

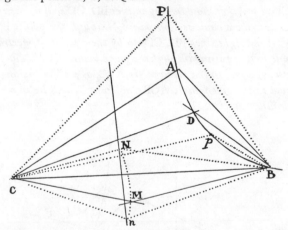

And conversely, if the movable point D lies in a conic section passing through the given points B, C, A; and the angle DBM is always equal to the given angle ABC, and the angle DCM always equal to the given angle ACB, and when the point D falls successively on any two immovable points p, P, of the conic section, the movable point M falls successively on two immovable points n, N. Through these points n, N, draw the right line nN:

this line nN will be the continual locus of that movable point M. For, if possible, let the point M be placed in any curved line. Therefore the point D will be placed in a conic section passing through the five points B, C, A, p, P, when the point M is continually placed in a curved line. But from what was demonstrated before, the point D will be also placed in a conic section passing through the same five points B, C, A, p, P, when the point M is continually placed in a right line. Wherefore the two conic sections will both pass through the same five points, against Cor. III, Lem. XX. It is therefore absurd to suppose that the point M is placed in a curved line. Q.E.D.

PROPOSITION XXII. PROBLEM XIV

To describe a conic that shall pass through five given points.

Let the five given points be A, B, C, P, D. From any one of them, as A, to any other two as B, C, which may be called the poles, draw the right lines AB, AC, and parallel to those the lines TPS, PRQ through the fourth point

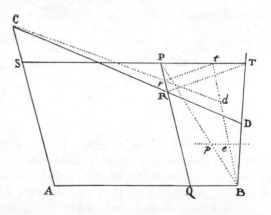

P. Then from the two poles B, C, draw through the fifth point D two indefinite lines BDT, CRD, meeting with the last drawn lines TPS, PRQ (the former with the former, and the latter with the latter) in T and R. And then draw the right line tr parallel to TR, cutting off from the right lines PT, PR, any segments Pt, Pr, proportional to PT, PR; and if through their extremities t, r, and the poles B, C, the right lines Bt, Cr are drawn, meeting in d, that point d will be placed in the conic required. For (by Lem. XX) that point d is placed in a conic section passing through the four points A, B, C, P; and the lines Rr, Tt vanishing, the point d comes to coincide with the point D. Wherefore the conic section passes through the five points A, B, C, P, D. Q.E.D.

The same otherwise.

Of the given points join any three, as A, B, C; and about two of them B, C, as poles, making the angles ABC, ACB of a given magnitude to revolve, apply the legs BA, CA, first to the point D, then to the point P, and mark the points M, N, in which the other legs BL, CL intersect each other in both cases. Draw the indefinite right line MN, and let those movable angles revolve about their poles B, C, in such manner that the intersection, which

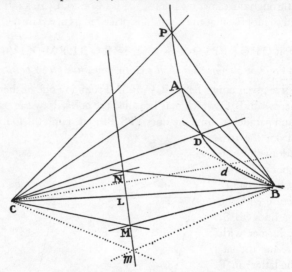

is now supposed to be *m,* of the legs BL, CL, or BM, CM, may always fall in that indefinite right line MN; and the intersection, which is now supposed to be *d,* of the legs BA, CA, or BD, CD, will describe the conic required, PAD*d*B. For (by Lem. xxi) the point *d* will be placed in a conic section passing through the points B, C; and when the point *m* comes to coincide with the points L, M, N, the point *d* will (by construction) come to coincide with the points A, D, P. Wherefore a conic section will be described that shall pass through the five points A, B, C, P, D. Q.E.F.

Cor. i. Hence a right line may be readily drawn which shall be a tangent to the conic in any given point B. Let the point *d* come to coincide with the point B, and the right line B*d* will become the tangent required.

Cor. ii. Hence also may be found the centres, diameters, and latera recta of the conics, as in Cor. ii, Lem. xix.

SCHOLIUM

The former of these constructions will become something more simple by joining B, P, and in that line, produced, if need be, taking B*p* to BP as PR is to PT; and through *p* draw the indefi- 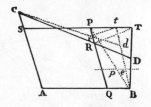 nite right line *pe* parallel to SPT, and in that line *pe* taking always *pe* equal to P*r*; and draw the right lines B*e*, C*r* to meet in *d*. For since P*r* to P*t*, PR to PT, *p*B to PB, *pe* to P*t*, are all in the same ratio, *pe* and P*r* will be always equal. After this manner the points of the conic are most readily found, unless you would rather describe the curve mechanically, as in the second construction.

PROPOSITION XXIII. PROBLEM XV

To describe a conic that shall pass through four given points, and touch a given right line.

Case i. Suppose that HB is the given tangent, B the point of contact, and C, D, P, the three other given points. Join BC, and draw PS parallel to BH, and PQ parallel to BC; complete the parallelogram BSPQ. Draw BD cut-

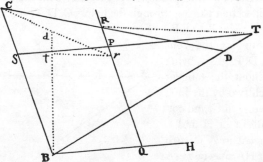

ting SP in T, and CD cutting PQ in R. Lastly, draw any line *tr* parallel to TR, cutting off from PQ, PS, the segments P*r*, P*t* proportional to PR, PT respectively, and draw C*r*, B*t*; their point of intersection *d* will (by Lem. xx) always fall on the conic to be described.

The same otherwise.

Let the angle CBH of a given magnitude revolve about the pole B, as also the rectilinear radius DC, both ways produced, about the pole C. Mark the points M, N, on which the leg BC of the angle cuts that radius when BH,

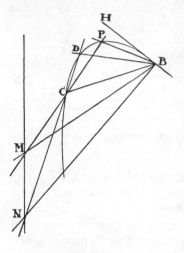

the other leg thereof, meets the same radius in the points P and D. Then drawing the indefinite line MN, let that radius CP or CD and the leg BC of the angle continually meet in this line; and the point of meeting of the other leg BH with the radius will delineate the conic required.

For if in the constructions of the preceding Problem the point A comes to a coincidence with the point B, the lines CA and CB will coincide, and the line AB, in its last situation, will become the tangent BH; and therefore the constructions there set down will become the same with the constructions here described. Wherefore the intersection of the leg BH with the radius will describe a conic section passing through the points C, D, P, and touching the line BH in the point B. Q.E.F.

CASE 2. Suppose the four points B, C, D, P, given, being situated without the tangent HI. Join each two by the lines BD, CP meeting in G, and cutting the tangent in H and I. Cut the tangent in A in such manner that HA may be to IA as the product of the mean proportional between CG and GP, and the mean proportional between BH and HD is to the

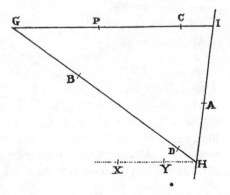

product of the mean proportional between GD and GB, and the mean proportional between PI and IC, and A will be the point of contact. For if HX, a parallel to the right line PI, cuts the conic in any points X and Y, the point A (by the properties of the conic sections) will come to be so placed, that HA^2 will become to AI^2 in a ratio that is compounded out of the ratio of the rectangle $HX \cdot HY$ to the rectangle $BH \cdot HD$, or of the rectangle $CG \cdot GP$ to the rectangle $DG \cdot GB$; and the ratio of the rectangle $BH \cdot HD$ to the rectangle $PI \cdot IC$. But after the point of contact A is found, the conic will be described as in the first Case. Q.E.F. But the point A may be taken either between or without the points H and I, upon which account a twofold conic may be described.

PROPOSITION XXIV. PROBLEM XVI

To describe a conic that shall pass through three given points, and touch two given right lines.

Suppose HI, KL to be the given tangents and B, C, D the given points. Through any two of those points, as B, D, draw the indefinite right line BD meeting the tangents in the points H, K. Then likewise through any other

two of these points, as C, D, draw the indefinite right line CD meeting the tangents in the points I, L. Cut the lines drawn in R and S, so that HR may be to KR as the mean proportional between BH and HD is to the mean proportional between BK and KD, and IS to LS as the mean proportional between CI and ID is to the mean proportional between CL

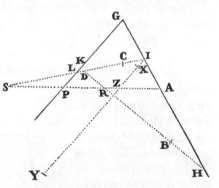

and LD. But you may cut, at pleasure, either within or between the points K and H, I and L, or without them. Then draw RS cutting the tangents in A and P, and A and P will be the points of contact. For if A and P are supposed to be the points of contact, situated anywhere else in the tangents, and through any of the points H, I, K, L, as I, situated in either tangent HI, a right line IY is drawn parallel to the other tangent KL, and meeting the curve in X and Y, and in that right line there be taken IZ equal to a mean proportional between IX and IY, the rectangle XI · IY or IZ² will (by the properties of the conic sections) be to LP² as the rectangle CI · ID is to the rectangle CL · LD; that is (by the construction), as SI is to SL², and therefore IZ : LP = SI : SL. Wherefore the points S, P, Z are in one right line. Moreover, since the tangents meet in G, the rectangle XI · IY or IZ² will (by the properties of the conic sections) be to IA² as GP² is to GA², and consequently IZ : IA = GP : GA. Wherefore the points P, Z, A lie in one right line, and therefore the points S, P, and A are in one right line. And the same argument will prove that the points R, P, and A are in one right line. Wherefore the points of contact A and P lie in the right line RS. But after these points are found, the conic may be described, as in the first Case of the preceding Problem. Q.E.F.

In this Proposition, and Case 2 of the foregoing, the constructions are the same, whether the right line XY cuts the conic in X and Y, or not; neither do they depend upon that section. But the constructions being demonstrated where that right line does cut the conic, the constructions where it does not are also known; and therefore, for brevity's sake, I omit any further demonstration of them.

LEMMA XXII[1]

To transform figures into other figures of the same kind.

Suppose that any figure HGI is to be transformed. Draw, at pleasure, two parallel lines AO, BL, cutting any given third line AB in A and B, and from any point G of the figure, draw out any right line GD, parallel to OA, till it meets the right line AB. Then from any given point O in the line OA, draw to the point

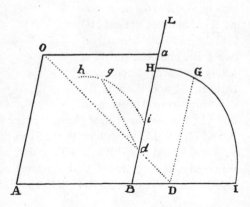

[1 Appendix, Note 20.]

D the right line OD, meeting BL in d; and from the point of intersection raise the right line dg containing any given angle with the right line BL, and having such ratio to Od as DG has to OD; and g will be the point in the new figure hgi, corresponding to the point G. And in like manner the several points of the first figure will give as many correspondent points of the new figure. If we therefore conceive the point G to be carried along by a continual motion through all the points of the first figure, the point g will be likewise carried along by a continual motion through all the points of the new figure, and describe the same. For distinction's sake, let us call DG the first ordinate, dg the new ordinate, AD the first abscissa, ad the new abscissa, O the pole, OD the abscinding radius, OA the first ordinate radius, and Oa (by which the parallelogram OABa is completed) the new ordinate radius.

I say, then, that if the point G is placed in a given right line, the point g will be also placed in a given right line. If the point G is placed in a conic section, the point g will be likewise placed in a conic section. And here I understand the circle as one of the conic sections. But further, if the point G is placed in a line of the third analytical order, the point g will also be placed in a line of the third order, and so on in curved lines of higher orders. The two lines in which the points G, g are placed, will be always of the same analytical order. For as $ad : \mathrm{O}A = \mathrm{O}d : \mathrm{O}D = dg : \mathrm{D}G = \mathrm{A}B : \mathrm{A}D$; and therefore AD is equal to $\dfrac{\mathrm{OA} \cdot \mathrm{AB}}{ad}$, and DG equal to $\dfrac{\mathrm{OA} \cdot dg}{ad}$. Now if the point G is placed in a right line, and therefore, in any equation by which the relation between the abscissa AD and the ordinate GD is expressed, those indetermined lines AD and DG rise no higher than to one dimension, by writing this equation $\dfrac{\mathrm{OA} \cdot \mathrm{AB}}{ad}$ in place of AD, and $\dfrac{\mathrm{OA} \cdot dg}{ad}$ in place of DG, a new equation will be produced, in which the new abscissa ad and new ordinate dg rise only to one dimension; and which therefore must denote a right line. But if AD and DG (or either of them) had risen to two dimensions in the first equation, ad and dg would likewise have risen to two dimensions in the second equation. And so on in three or more dimensions. The indetermined lines, ad, dg in the second equation, and AD, DG in the first, will always rise to the same number of dimensions; and therefore the lines in which the points G, g are placed are of the same analytical order.

I say, further, that if any right line touches the curved line in the first figure, the same right line transferred the same way with the curve into the new figure will touch that curved line in the new figure, and conversely. For if any two points of the curve in the first figure are supposed to approach one the other till they come to coincide, the same points transferred will approach one the other till they come to coincide in the new figure; and therefore the right lines with which those points are joined will become together tangents of the curves in both figures. I might have given demonstrations of these assertions in a more geometrical form; but I study to be brief.

Wherefore if one rectilinear figure is to be transformed into another, we need only transfer the intersections of the right lines of which the first figure consists, and through the transferred intersections to draw right lines in the new figure. But if a curvilinear figure is to be transformed, we must transfer the points, the tangents, and other right lines, by means of which the curved line is defined. This Lemma is of use in the solution of the more difficult Problems; for thereby we may transform the proposed figures, if they are intricate, into others that are more simple. Thus any right lines converging to a point are transformed into parallels, by taking for the first ordinate radius any right line that passes through the point of intersection of the converging lines, and that because their point of intersection is by this means made to go off *in infinitum;* and parallel lines are such as tend to a point infinitely remote. And after the problem is solved in the new figure, if by the inverse operations we transform the new into the first figure, we shall have the solution required.

This Lemma is also of use in the solution of solid problems. For as often as two conic sections occur, by the intersection of which a problem may be solved, any one of them may be transformed, if it is an hyperbola or a parabola, into an ellipse, and then this ellipse may be easily changed into a circle. So also a right line and a conic section, in the construction of plane problems, may be transformed into a right line and a circle.

PROPOSITION XXV. PROBLEM XVII

To describe a conic that shall pass through two given points, and touch three given right lines.

Through the intersection of any two of the tangents one with the other, and the intersection of the third tangent with the right line which passes through the two given points, draw an indefinite right line; and, taking this line for the first ordinate radius, transform the figure by the preceding Lemma into a new figure. In this figure those two tangents will become parallel to each other, and the third tangent will be parallel to the right line that passes through the two given points. Suppose *hi, kl* to be those two parallel tangents, *ik* the third tangent, and *hl* a right line parallel thereto, passing through those points *a, b,* through which the conic section ought to pass in this new figure; and completing the parallelogram *hikl,* let the right lines *hi, ik, kl* be so cut in *c, d, e,* that *hc* may be to the square root of the rectangle *ahb, ic* to *id,* and *ke* to *kd,* as the sum of the right lines *hi* and *kl* is to the sum of the three lines, the first whereof is the right line *ik,*

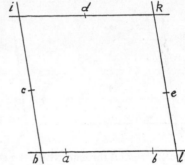

and the other two are the square roots of the rectangles *ahb* and *alb;* and *c, d, e* will be the points of contact. For by the properties of the conic sections,

$$hc^2 : ah \cdot hb = ic^2 : id^2 = ke^2 : kd^2 = el^2 : al \cdot lb.$$

Therefore,

$$hc : \sqrt{(ah \cdot hb)} = ic : id = ke : kd = el : \sqrt{(al \cdot lb)}$$
$$= hc + ic + ke + el : \sqrt{(ah \cdot hb)} + id + kd + \sqrt{al \cdot lb}$$
$$= hi + kl : \sqrt{(ah \cdot hb)} + ik + \sqrt{(al \cdot lb)}.$$

Wherefore from that given ratio we have the points of contact *c, d, e,* in the new figure. By the inverted operations of the last Lemma, let those points be transferred into the first figure, and the conic will be there described by Prob. xiv. Q.E.F. But according as the points *a, b,* fall between the points *h, l,* or without them, the points *c, d, e* must be taken either between the points *h, i, k, l,* or without them. If one of the points *a, b* falls between the points *h, i,* and the other without the points *h, l,* the Problem is impossible.

PROPOSITION XXVI. PROBLEM XVIII

To describe a conic that shall pass through a given point, and touch four given right lines.

From the common intersections of any two of the tangents to the common intersection of the other two, draw an indefinite right line; and taking this line for the first ordinate radius, transform the figure (by Lem. xxii)

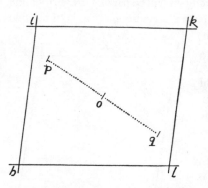

into a new figure, and the two pairs of tangents, each of which before concurred in the first ordinate radius, will now become parallel. Let *hi* and *kl, ik* and *hl,* be those pairs of parallels completing the parallelogram *hikl.* And let *p* be the point in this new figure corresponding to the given point in the first figure. Through O the centre of the figure draw *pq:* and O*q* being equal to O*p, q* will be the other point through which the conic section must pass in this new figure. Let this point be transferred, by the inverse operation of Lem. xxii, into the first figure, and there we shall have the two points through which the conic is to be described. But through those points that conic may be described by Prop. xvii.

LEMMA XXIII

If two given right lines, as AC, BD, *terminating in given points* A, B, *are in a given ratio one to the other, and the right line* CD, *by which the indetermined points* C, D *are joined is cut in* K *in a given ratio: I say, that the point* K *will be placed in a given right line.*

For let the right lines AC, BD meet in E, and in BE take BG to AE as BD is to AC, and let FD be always equal to the given line EG; and, by construction, EC will be to GD, that is, to EF, as AC to BD, and therefore in a given ratio; and therefore the triangle EFC will be given in kind. Let CF be cut in L so as CL may be to CF in the ratio of CK to CD; and because that is a given ratio, the triangle EFL will be given in kind, and therefore the point L will be placed in the given right line EL. Join LK, and the triangles CLK, CFD will be similar; and because FD is a given line, and LK is to FD in a given ratio, LK will be also given. To this let EH be taken equal, and ELKH will be always a parallelogram. And therefore the point K is always placed in the given side HK of that parallelogram. Q.E.D.

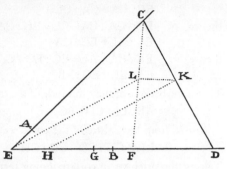

Cor. Because the figure EFLC is given in kind, the three right lines EF, EL, and EC, that is, GD, HK, and EC, will have given ratios to each other.

LEMMA XXIV

If three right lines, two whereof are parallel, and given in position, touch any conic section: I say, that the semidiameter of the section which is parallel to those two is a mean proportional between the segments of those two that are intercepted between the points of contact and the third tangent.

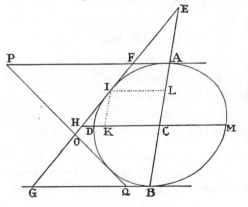

Let AF, GB be the two parallels touching the conic section ADB in A and B; EF the third right line touching the conic section in I, and meeting the two former tangents in F and G, and let CD be the semidiameter of the figure parallel to those tangents: I say, that AF, CD, BG are continually proportional.

For if the conjugate diameters AB, DM meet the tangent FG in E and H, and cut one the other in C, and the parallelogram IKCL be completed; from the nature of the conic sections,

$$EC : CA = CA : CL;$$

thence, $\qquad EC - CA : CA - CL = EC : CA$

or $\qquad EA : AL = EC : CA;$

thence, $\qquad EA : EA + AL = EC : EC + CA$

or $\qquad EA : EL = EC : EB.$

Therefore, because of the similitude of the triangles EAF, ELI, ECH, EBG,

$$AF : LI = CH : BG.$$

Likewise, from the nature of the conic sections,

$$LI \text{ or } CK : CD = CD : CH.$$

Taking the products of corresponding terms in the last two proportions and simplifying,

$$AF : CD = CD : BG. \qquad\qquad \text{Q.E.D.}$$

Cor. i. Hence if two tangents FG, PQ meet two parallel tangents AF, BG in F and G, P and Q, and cut one the other in O; then by the Lemma applied to EG and PQ,

$$AF : CD = CD : BG,$$
$$BQ : CD = CD : AP.$$

Therefore, $\qquad AF : AP = BQ : BG$

and $\qquad AP - AF : AP = BG - BQ : BG$

or $\qquad PF : AP = GQ : BG,$

and $\qquad AP : BG = PF : GQ = FO : GO = AF : BQ.$

Cor. ii. Whence also the two right lines PG, FQ drawn through the points P and G, F and Q, will meet in the right line ACB passing through the centre of the figure and the points of contact A, B.

LEMMA XXV

If four sides of a parallelogram indefinitely produced touch any conic section, and are cut by a fifth tangent: I say, that, taking those segments of any two conterminous sides that terminate in opposite angles of the parallelo-

*gram, either segment is to the side from which it is cut off as that part of
the other conterminous side which is intercepted between the point of contact and the third side is to the other segment.*

Let the four sides ML, IK, KL, MI of the parallelogram MLIK touch
the conic section in A, B, C, D; and let the fifth tangent FQ cut those sides
in F, Q, H, and E; and taking the segments ME, KQ of the sides MI, KI,
or the segments KH, MF of the sides KL, ML: I say, that

$$ME : MI = BK : KQ,$$
and $\qquad$ KH : KL = AM : MF.
For, by Cor. 1 of the preceding Lemma,
$$ME : EI = AM \text{ or } BK : BQ,$$
and by addition,
$$ME : MI = BK : KQ. \qquad\qquad\qquad\qquad\qquad \text{Q.E.D.}$$
Also, $\qquad$ KH : HL = BK or AM : AF,
and by subtraction,
$$KH : KL = AM : MF. \qquad\qquad\qquad\qquad\qquad \text{Q.E.D.}$$

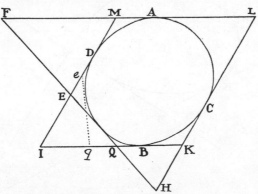

COR. I. Hence if a parallelogram IKLM described about a given conic
section is given, the rectangle KQ · ME, as also the rectangle KH · MF equal
thereto, will be given. For, by reason of the similar triangles KQH, MFE,
those rectangles are equal.

COR. II. And if a sixth tangent *eq* is drawn meeting the tangents KI, MI
in *q* and *e*, the rectangle KQ · ME will be equal to the rectangle K*q* · M*e*,

and   KQ : Me = Kq : ME,

and by subtraction

KQ : Me = Qq : Ee.

Cor. iii. Hence, also, if Eq, eQ are joined and bisected, and a right line is drawn through the points of bisection, this right line will pass through the centre of the conic section. For since Qq : Ee = KQ : Me, the same right line will pass through the middle of all the lines Eq, eQ, MK (by Lem. xxiii), and the middle point of the right line MK is the centre of the section.

PROPOSITION XXVII. PROBLEM XIX

To describe a conic that may touch five right lines given in position.

Supposing ABG, BCF, GCD, FDE, EA to be the tangents given in position. Bisect in M and N, AF, BE, the diagonals of the quadrilateral figure ABFE contained under any four of them; and (by Cor. iii, Lem. xxv) the

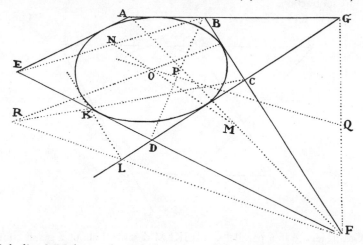

right line MN drawn through the points of bisection will pass through the centre of the conic. Again, bisect in P and Q the diagonals (if I may so call them) BD, GF of the quadrilateral figure BGDF contained under any other four tangents, and the right line PQ drawn through the points of bisection will pass through the centre of the conic; and therefore the centre will be given in the intersection of the bisecting lines. Suppose it to be O. Parallel to any tangent BC draw KL at such distance that the centre O may be placed in the middle between the parallels; this KL will touch the conic to be described. Let this cut any other two tangents GCD, FDE, in L and K. Through the points C and K, F and L, where the tangents not parallel, CL, FK, meet the parallel tangents CF, KL, draw CK, FL meeting in R; and the right line OR, drawn and produced, will cut the parallel tangents CF, KL, in the points of contact. This appears from Cor. II, Lem. XXIV. And by the same method the other points of contact may be found, and then the conic may be described by Prob. XIV. Q.E.F.

SCHOLIUM

Under the preceding Propositions are comprehended those Problems wherein either the centres or asymptotes of the conics are given. For when points and tangents and the centre are given, as many other points and as many other tangents are given at an equal distance on the other side of the centre. And an asymptote is to be considered as a tangent, and its infinitely remote extremity (if we may say so) is a point of contact. Conceive the point of contact of any tangent removed *in infinitum,* and the tangent will degenerate into an asymptote, and the constructions of the preceding Problems will be changed into the constructions of those Problems wherein the asymptote is given.

After the conic is described, we may find its axes and foci in this manner. In the construction and figure of Lem. XXI, let those legs BP, CP, of the movable angles PBN, PCN, by the intersection of which the conic was described, be made parallel one to the other; and retaining that position, let them revolve about their poles B, C, in that figure. In the meanwhile let the other legs CN, BN, of those angles, by their intersection K or

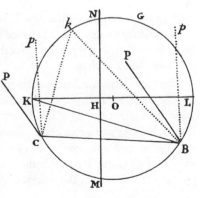

k, describe the circle BKGC. Let O be the centre of this circle; and from this centre upon the ruler MN, wherein those legs CN, BN did concur while the conic was described, let fall the perpendicular OH meeting the circle in K and L. And when those other legs CK, BK meet in the point K that is nearest to the ruler, the first legs CP, BP will be parallel to the greater axis, and perpendicular on the lesser; and the contrary will happen if those legs meet in the remotest point L. Whence if the centre of the conic is given, the axes will be given; and those being given, the foci will be readily found.

But the squares of the axes are one to the other as KH to LH, and thence it is easy to describe a conic given in kind through four given points. For if two of the given points are made the poles C, B, the third will give the

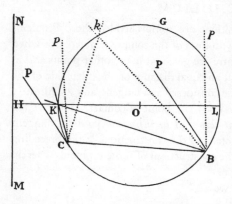

movable angles PCK, PBK; but those being given, the circle BGKC may be described. Then, because the conic is given in kind, the ratio of OH to OK, and therefore OH itself, will be given. About the centre O, with the interval OH, describe another circle, and the right line that touches this circle, and passes through the intersection of the legs CK, BK, when the first legs CP, BP meet in the fourth given point, will be the ruler MN, by means of which the conic may be described. Whence also on the other hand a trapezium given in kind (excepting a few cases that are impossible) may be inscribed in a given conic section.

There are also other Lemmas, by the help of which conics given in kind may be described through given points, and touching given lines. Of such a sort is this, that if a right line is drawn through any point given in position, that may cut a given conic section in two points, and the distance of the intersections is bisected, the point of bisection will touch another conic section of the same kind with the former, and having its axes parallel to the axes of the former. But I hasten to things of greater use.

LEMMA XXVI

To place the three angles of a triangle, given both in kind and in magnitude, in respect to as many right lines given in position, provided they are not all parallel among themselves, in such manner that the several angles may touch the several lines.

Three indefinite right lines AB, AC, BC are given in position, and it is required so to place the triangle DEF that its angle D may touch the line AB, its angle E the line AC, and its angle F the line BC. Upon DE, DF, and

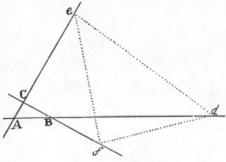

EF describe three segments of circles DRE, DGF, EMF, capable of angles equal to the angles BAC, ABC, ACB respectively. But those segments are to be described towards such sides of the lines DE, DF, EF, that the letters DRED may turn round about in the same order with the letters BACB; the letters DGFD in the same order with the letters ABCA; and the letters EMFE in the same order with the letters ACBA; then, completing those segments into entire circles, let the two former circles cut each other in G, and suppose P and Q to be their centres. Then joining GP, PQ, take

$$Ga : AB = GP : PQ;$$

and about the centre G, with the interval Ga, describe a circle that may cut the first circle DGE in *a*. Join *a*D cutting the second circle DFG in *b*, as well as *a*E cutting the third circle EMF in *c*. Complete the figure ABC*def* similar and equal to the figure *abc*DEF: I say, the thing is done.

For drawing F*c* meeting *a*D in *n*, and joining *a*G, *b*G, QG, QD, PD, by construction the angle E*a*D is equal to the angle CAB, and the angle *ac*F equal to the angle ACB; and therefore the triangle *anc* equiangular to the triangle ABC. Wherefore the angle *anc* or F*n*D is equal to the angle ABC, and consequently to the angle F*b*D; and therefore the point *n* falls on the point *b*. Moreover the angle GPQ, which is half the angle GPD at the centre, is equal to the angle GaD at the circumference; and the angle GQP, which is half the angle GQD at the centre, is equal to the supplement of the angle GbD at the circumference, and therefore equal to the angle Gba. Upon which account the triangles GPQ, Gab are similar, and

$$Ga : ab = GP : PQ$$

and, by construction,

$$GP : PQ = Ga : AB.$$

Wherefore ab and AB are equal; and consequently the triangles abc, ABC, which we have now proved to be similar, are also equal. And therefore since

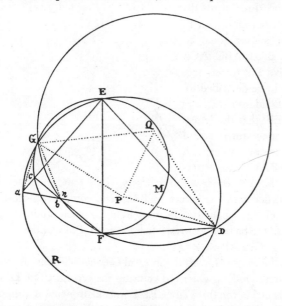

the angles D, E, F of the triangle DEF do respectively touch the sides ab, ac, bc of the triangle abc, the figure ABCdef may be completed similar and equal to the figure abcDEF, and by completing it the Problem will be solved. Q.E.F.

Cor. Hence a right line may be drawn whose parts given in length may be intercepted between three right lines given in position. Suppose the triangle DEF, by the approach of its point D to the side EF, and by having

the sides DE, DF placed into the same straight line, to be itself changed
into a right line whose given part DE is to be placed between the right lines
AB, AC given in position; and its given part DF is to be placed between
the right lines AB, BC given in position; then, by applying the preceding
construction to this case, the Problem will be solved.

PROPOSITION XXVIII. PROBLEM XX

*To describe a conic given both in kind and in magnitude, given parts of
which shall be placed between three right lines given in position.*

Suppose a conic is to be described that may be similar and equal to the
curved line DEF, and may be cut by three right lines AB, AC, BC, given in
position, into parts DE and EF, similar and equal to the given parts of this
curved line.

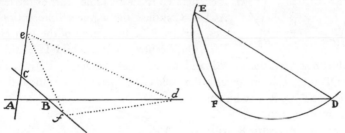

Draw the right lines DE, EF, DF; and place the angles D, E, F, of this
triangle DEF, so as to touch those right lines given in position (by Lem.
xxvi). Then about the triangle describe the conic, similar and equal to the
curve DEF. Q.E.F.

LEMMA XXVII

*To describe a trapezium given in kind, the angles whereof may respectively
touch four right lines given in position, that are neither all parallel among
themselves, nor converge to one common point.*

Let the four right lines ABC, AD, BD, CE be given in position; the first
cutting the second in A, the third in B, and the fourth in C; and suppose a
trapezium *fghi* is to be described that may be similar to the trapezium
FGHI, and whose angle *f,* equal to the given angle F, may touch the right

line ABC; and the other angles *g, h, i*, equal to the other given angles G, H, I, may touch the other lines AD, BD, CE respectively. Join FH, and

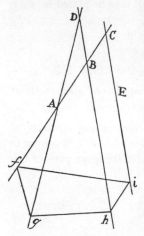

upon FG, FH, FI describe as many segments of circles FSG, FTH, FVI, the first of which FSG may be capable of an angle equal to the angle BAD; the second FTH capable of an angle equal to the angle CBD; and the third FVI of an angle equal to the angle ACE. But the segments are to be described towards those sides of the lines FG, FH, FI, that the circular order of the letters FSGF may be the same as of the letters BADB, and that the letters FTHF may turn about in the same order as the letters CBDC, and the letters FVIF in the same order as the letters ACEA. Complete the segments into entire circles, and let P be the centre of the first circle FSG, Q the centre of the second FTH. Join and produce both ways the line PQ, and in it take QR so that QR : PQ=BC : AB. But QR is to be taken towards that side of the point Q, that the order of

the letters P, Q, R may be the same as of the letters A, B, C; and about the centre R with the radius RF describe a fourth circle FN*c* cutting the third circle FVI in *c*. Join F*c* cutting the first circle in *a*, and the second in *b*. Draw *a*G, *b*H, *c*I, and let the figure ABC*fghi* be made similar to the figure *abc*FGHI; and the trapezium *fghi* will be that which was required to be described.

For let the two first circles FSG, FTH cut one the other in K; join PK, QK, RK, *a*K,

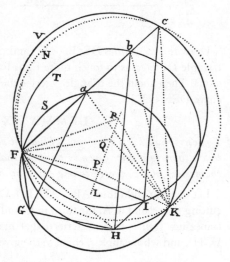

*b*K, *c*K, and produce QP to L. The angles F*a*K, F*b*K, F*c*K at the circumferences are the halves of the angles FPK, FQK, FRK at the centres, and therefore equal to LPK, LQK, LRK, the halves of those angles. Therefore the figure PQRK is equiangular and similar to the figure *abc*K, and consequently *ab* is to *bc* as PQ to QR, that is, as AB to BC. But by construction the angles *f*A*g*, *f*B*h*, *f*C*i* are equal to the angles F*a*G, F*b*H, F*c*I. And therefore the figure ABC*fghi* may be completed similar to the figure *abc*FGHI. This done, a trapezium *fghi* will be constructed similar to the trapezium FGHI, and by its angles *f*, *g*, *h*, *i* will touch the right lines ABC, AD, BD, CE. Q.E.F.

Cor. Hence a right line may be drawn whose parts intercepted in a given order, between four right lines given by position, shall have a given proportion among themselves. Let the angles FGH, GHI be so far increased that the right lines FG, GH, HI may lie in the same line; and by constructing the Problem in this case, a right line *fghi* will be drawn, whose parts *fg*, *gh*, *hi*, intercepted between the four right lines given in position, AB and AD, AD and BD, BD and CE, will be to each other as the lines FG, GH, HI, and will observe the same order among themselves. But the same thing may be more readily done in this manner.

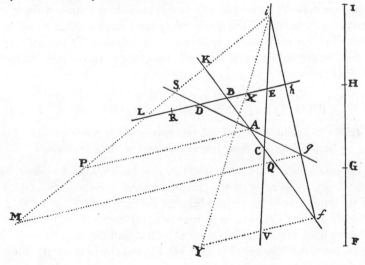

Produce AB to K and BD to L, so as BK may be to AB as HI to GH; and DL to BD as GI to FG; and join KL meeting the right line CE in i. Produce iL to M, so as LM may be to iL as GH to HI; then draw MQ parallel to LB, and meeting the right line AD in g, and join gi cutting AB, BD in f, h: I say, the thing is done.

For let Mg cut the right line AB in Q, and AD the right line KL in S, and draw AP parallel to BD and meeting iL in P, and gM to Lh (gi to hi, Mi to Li, GI to HI, AK to BK) and AP to BL will be in the same ratio. Cut DL in R, so as DL to RL may be in that same ratio; and because gS to gM, AS to AP, and DS to DL are proportional; therefore, as gS to Lh, so will AS be to BL, and DS to RL; and mixtly, BL – RL to Lh – BL, as AS – DS to gS – AS. That is, BR is to Bh as AD is to Ag, and therefore as BD to gQ. And alternately BR is to BD as Bh to gQ, or as fh to fg. But by construction the line BL was cut in D and R in the same ratio as the line FI in G and H; and therefore BR is to BD as FH to FG. Therefore fh is to fg as FH to FG. Since, therefore, gi to hi likewise is as Mi to Li, that is, as GI to HI, it is manifest that the lines FI, fi are similarly cut in G and H, g and h. Q.E.F.

In the construction of this Corollary, after the line LK is drawn cutting CE in i, we may produce iE to V, so as EV may be to Ei as FH to HI, and then draw Vf parallel to BD. It will come to the same, if about the centre i with an interval IH, we describe a circle cutting BD in X, and produce iX to Y so as iY may be equal to IF, and then draw Yf parallel to BD.

Sir *Christopher Wren* and Dr. *Wallis* have long ago given other solutions of this Problem.

PROPOSITION XXIX. PROBLEM XXI

To describe a conic given in kind, that may be cut by four right lines given in position, into parts given in order, kind, and proportion.

Suppose a conic is to be described that may be similar to the curved line FGHI, and whose parts, similar and proportional to the parts FG, GH, HI of the other, may be intercepted between the right lines AB and AD, AD and BD, BD and CE given in position, viz., the first between the first pair of those lines, the second between the second, and the third between the third. Draw the right lines FG, GH, HI, FI; and (by Lem. xxvii) describe a trapezium *fghi* that may be similar to the trapezium FGHI, and whose angles *f, g, h, i* may touch the right lines given in position AB, AD, BD, CE, severally according to their order. And then about this trapezium describe a conic, that conic will be similar to the curved line FGHI.

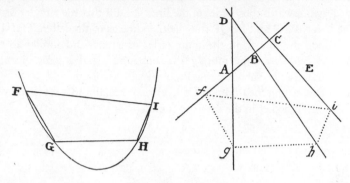

SCHOLIUM

This problem may be likewise constructed in the following manner. Joining FG, GH, HI, FI, produce GF to V, and join FH, IG, and make the angles CAK, DAL equal to the angles FGH, VFH. Let AK, AL meet the right line BD in K and L, and thence draw KM, LN, of which let KM make the angle AKM equal to the angle GHI, and be itself to AK as HI is to GH; and let LN make the angle ALN equal to the angle FHI, and be

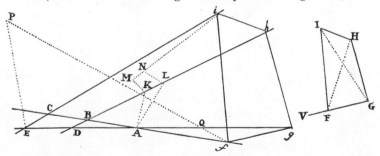

itself to AL as HI to FH. But AK, KM, AL, LN are to be drawn towards those sides of the lines AD, AK, AL, that the letters CAKMC, ALKA, DALND may be carried round in the same order as the letters FGHIF;

and draw MN meeting the right line CE in i. Make the angle iEP equal to the angle IGF, and let PE be to Ei as FG to GI; and through P draw PQf that may with the right line ADE contain an angle PQE equal to the angle FIG, and may meet the right line AB in f, and join $f i$. But PE and PQ are to be drawn towards those sides of the lines CE, PE that the circular order of the letters PEiP and PEQP may be the same as of the letters FGHIF; and if upon the line $f i$, in the same order of letters, and similar to the trapezium FGHI, a trapezium $fghi$ is constructed, and a conic given in kind is circumscribed about it, the Problem will be solved.

So far concerning the finding of the orbits. It remains that we determine the motions of bodies in the orbits so found.

SECTION VI

How the motions are to be found in given orbits.

PROPOSITION XXX. PROBLEM XXII

To find at any assigned time the place of a body moving in a given parabola.

Let S be the focus, and A the principal vertex of the parabola; and suppose $4AS \cdot M$ equal to the parabolic area to be cut off APS, which either was described by the radius SP, since the body's departure from the vertex,

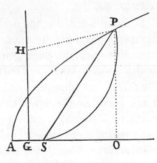

or is to be described thereby before its arrival there. Now the quantity of that area to be cut off is known from the time which is proportional to it. Bisect AS in G, and erect the perpendicular GH equal to $3M$, and a circle described about the centre H, with the radius HS, will cut the parabola in the place P required. For letting fall PO perpendicular on the axis, and drawing PH, there will be $AG^2 + GH^2 \ (= HP^2 = (AO - AG)^2 + (PO - GH)^2) = AO^2 + PO^2 - 2AO \cdot AG - 2GH \cdot PO + AG^2 + GH^2$. Whence $2GH \cdot PO \ (= AO^2 + PO^2 - 2AO \cdot AG) = AO^2 + \tfrac{3}{4} PO^2$. For AO^2 write $AO \cdot \dfrac{PO^2}{4AS}$; then dividing all the terms by $3PO$, and multiplying them by $2AS$, we shall have $\tfrac{2}{3} GH \cdot AS \ (= \tfrac{1}{6} AO \cdot PO + \tfrac{1}{2} AS \cdot PO = \dfrac{AO + 3AS}{6} \cdot PO = \dfrac{4AO - 3SO}{6} \cdot PO =$ to the area, $APO - SPO) =$ to the area APS. But GH was $3M$, and therefore $\tfrac{2}{3} GH \cdot AS$ is $4AS \cdot M$. Therefore the area cut off APS is equal to the area that was to be cut off $4AS \cdot M$. Q.E.D.

Cor. i. Hence GH is to AS as the time in which the body described the arc AP to the time in which the body described the arc between the vertex A and the perpendicular erected from the focus S upon the axis.

Cor. ii. And supposing a circle ASP continually to pass through the moving body P, the velocity of the point H is to the velocity which the body had in the vertex A as 3 to 8; and therefore in the same ratio is the line GH to the right line which the body, in the time of its moving from A to P, would describe with that velocity which it had in the vertex A.

Cor. iii. Hence, also, on the other hand, the time may be found in which the body has described any assigned arc AP. Join AP, and on its middle point erect a perpendicular meeting the right line GH in H.

LEMMA XXVIII[1]

There is no oval figure whose area, cut off by right lines at pleasure, can be universally found by means of equations of any number of finite terms and dimensions.

Suppose that within the oval any point is given, about which as a pole a right line is continually revolving with an uniform motion, while in that right line a movable point going out from the pole moves always forwards with a velocity proportional to the square of that right line within the oval. By this motion that point will describe a spiral with infinite circumgyrations. Now if a portion of the area of the oval cut off by that right line could be found by a finite equation, the distance of the point from the pole, which is proportional to this area, might be found by the same equation, and therefore all the points of the spiral might be found by a finite equation also; and therefore the intersection of a right line given in position with the spiral might also be found by a finite equation. But every right line infinitely produced cuts a spiral in an infinite number of points; and the equation by which any one intersection of two lines is found at the same time exhibits all their intersections by as many roots, and therefore rises to as many dimensions as there are intersections. Because two circles cut one another in two points, one of those intersections is not to be found but by an equation of two dimensions, by which the other intersection may be also found. Because there may be four intersections of two conic sections, any one of them is not to be found universally, but by an equation of four dimensions, by which they may be all found together. For if those intersections are severally sought, because the law and condition of all is the same, the calculus will be the same in every case, and therefore the conclu-

[1 Appendix, Note 21.]

sion always the same, which must therefore comprehend all those intersections at once within itself, and exhibit them all indifferently. Hence it
is that the intersections of the conic sections with the curves of the third
order, because they may amount to six, come out together by equations
of six dimensions; and the intersections of two curves of the third order,
because they may amount to nine, come out together by equations of nine
dimensions. If this did not necessarily happen, we might reduce all solid
to plane Problems, and those higher than solid to solid Problems. But here
I speak of curves irreducible in power. For if the equation by which the
curve is defined may be reduced to a lower power, the curve will not be
one single curve, but composed of two, or more, whose intersections may
be severally found by different calculi. After the same manner the two
intersections of right lines with the conic sections come out always by
equations of two dimensions; the three intersections of right lines with the
irreducible curves of the third order, by equations of three dimensions; the
four intersections of right lines with the irreducible curves of the fourth
order, by equations of four dimensions; and so on *in infinitum*. Wherefore the innumerable intersections of a right line with a spiral, since this is
but one simple curve, and not reducible to more curves, require equations
infinite in number of dimensions and roots, by which they may be all
exhibited together. For the law and calculus of all is the same. For if a perpendicular is let fall from the pole upon that intersecting right line, and that
perpendicular together with the intersecting line revolves about the pole,
the intersections of the spiral will mutually pass the one into the other; and
that which was first or nearest, after one revolution, will be the second;
after two, the third; and so on: nor will the equation in the meantime be
changed but as the magnitudes of those quantities are changed, by which
the position of the intersecting line is determined. Therefore since those
quantities after every revolution return to their first magnitudes, the equation will return to its first form; and consequently one and the same equation will exhibit all the intersections, and will therefore have an infinite
number of roots, by which they may be all exhibited. Therefore the intersection of a right line with a spiral cannot be universally found by any finite
equation; and hence there is no oval figure whose area, cut off by right lines
at pleasure, can be universally exhibited by any such equation.

By the same argument, if the interval of the pole and point by which the spiral is described is taken proportional to that part of the perimeter of the oval which is cut off, it may be proved that the length of the perimeter cannot be universally exhibited by any finite equation. But here I speak of ovals that are not touched by conjugate figures running out *in infinitum*.

Cor. Hence the area of an ellipse, described by a radius drawn from the focus to the moving body, is not to be found from the time given by a finite equation; and therefore cannot be determined by the description of curves geometrically rational. Those curves I call geometrically rational, all the points whereof may be determined by lengths that are definable by equations; that is, by the complicated ratios of lengths. Other curves (such as spirals, quadratrixes, and cycloids) I call geometrically irrational. For the lengths which are or are not as number to number (according to Book x, *Elem.* of *Euclid*) are arithmetically rational or irrational. And therefore I cut off an area of an ellipse proportional to the time in which it is described by a curve geometrically irrational, in the following manner:

PROPOSITION XXXI.   PROBLEM XXIII
To find the place of a body moving in a given ellipse at any assigned time.

Suppose A to be the principal vertex, S the focus, and O the centre of the ellipse APB; and let P be the place of the body to be found. Produce OA to G so that OG : OA = OA : OS. Erect the perpendicular GH; and about the centre O, with the radius OG, describe the circle GEF; and on the ruler

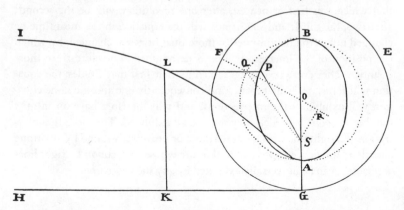

GH, as a base, suppose the wheel GEF to move forwards, revolving about its axis, and in the meantime by its point A describing the cycloid ALI. This done, take GK to the perimeter GEFG of the wheel, in the ratio of the time in which the body proceeding from A described the arc AP, to the time of a whole revolution in the ellipse. Erect the perpendicular KL meeting the cycloid in L; then LP drawn parallel to KG will meet the ellipse in P, the required place of the body.

For about the centre O with the radius OA describe the semicircle AQB, and let LP, produced, if need be, meet the arc AQ in Q, and join SQ, OQ. Let OQ meet the arc EFG in F, and upon OQ let fall the perpendicular SR. The area APS varies as the area AQS, that is, as the difference between the sector OQA and the triangle OQS, or as the difference of the rectangles ½ OQ · AQ, and ½ OQ · SR, that is, because ½ OQ is given, as the difference between the arc AQ and the right line SR; and therefore (because of the equality of the given ratios SR to the sine of the arc AQ, OS to OA, OA to OG, AQ to GF; and by division, AQ−SR to GF−sine of the arc AQ) as GK, the difference between the arc GF and the sine of the arc AQ. Q.E.D.

SCHOLIUM

But since the description of this curve is difficult, a solution by approximation will be preferable.[1] First, then, let there be found a certain angle B which may be to an angle of 57.29578 degrees, which an arc equal to the radius

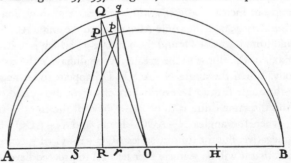

subtends, as SH, the distance of the foci, to AB, the diameter of the ellipse. Secondly, a certain length L, which may be to the radius in the same ratio inversely. And these being found, the Problem may be solved by the follow-

[1 Appendix, Note 22.]

ing analysis. By any construction (or even by conjecture), suppose we know P the place of the body near its true place p. Then letting fall on the axis of the ellipse the ordinate PR from the proportion of the diameters of the ellipse, the ordinate RQ of the circumscribed circle AQB will be given; which ordinate is the sine of the angle AOQ, supposing AO to be the radius, and also cuts the ellipse in P. It will be sufficient if that angle is found by a rude calculus in numbers near the truth. Suppose we also know the angle proportional to the time, that is, which is to four right angles as the time in which the body described the arc Ap to the time of one revolution in the ellipse. Let this angle be N. Then take an angle D, which may be to the angle B as the sine of the angle AOQ to the radius; and an angle E which may be to the angle N − AOQ + D as the length L to the same length L diminished by the cosine of the angle AOQ, when that angle is less than a right angle, or increased thereby when greater. In the next place, take an angle F that may be to the angle B as the sine of the angle AOQ + E to the radius, and an angle G, that may be to the angle N − AOQ − E + F as the length L to the same length L diminished by the cosine of the angle AOQ + E, when that angle is less than a right angle, or increased thereby when greater. For the third time take an angle H, that may be to the angle B as the sine of the angle AOQ + E + G to the radius; and an angle I to the angle N − AOQ − E − G + H, as the length L is to the same length L diminished by the cosine of the angle AOQ + E + G, when that angle is less than a right angle, or increased thereby when greater. And so we may proceed *in infinitum*. Lastly, take the angle AOq equal to the angle AOQ + E + G + I +, &c., and from its cosine Or and the ordinate pr, which is to its sine qr as the lesser axis of the ellipse to the greater, we shall have p the correct place of the body. When the angle N − AOQ + D happens to be negative, the sign + of the angle E must be everywhere changed into −, and the sign − into +. And the same thing is to be understood of the signs of the angles G and I, when the angles N − AOQ − E + F, and N − AOQ − E − G + H come out negative. But the infinite series AOQ + E + G + I +, &c., converges so very fast, that it will be scarcely ever needful to proceed beyond the second term E. And the calculus is founded upon this Theorem, that the area APS varies as the difference between the arc AQ and the right line let fall from the focus S perpendicularly upon the radius OQ.

And by a calculus not unlike, the Problem is solved in the hyperbola. Let its centre be O, its vertex A, its focus S, and asymptote OK; and suppose the amount of the area to be cut off is known, as being proportional to the time. Let that be A, and by conjecture suppose we know the position of a right line SP, that cuts off an area APS near the truth. Join OP, and from A and P to the asymptote draw AI, PK, parallel to the other asymptote; and by the table of logarithms the area AIKP will be given, and equal thereto the area OPA, which, subtracted from the triangle OPS, will leave the

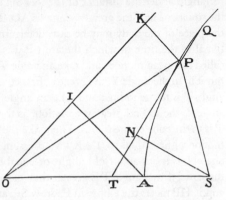

area cut off APS. And by applying $2APS - 2A$, or $2A - 2APS$, the double difference of the area A that was to be cut off, and the area APS that is cut off, to the line SN that is let fall from the focus S, perpendicular upon the tangent TP, we shall have the length of the chord PQ. Which chord PQ is to be inscribed between A and P, if the area APS that is cut off be greater than the area A that was to be cut off, but towards the contrary side of the point P, if otherwise: and the point Q will be the place of the body more accurately. And by repeating the computation the place may be found continually to greater and greater accuracy.

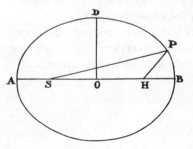

And by such computations we have a general analytical resolution of the Problem. But the particular calculus that follows is better fitted for astronomical purposes. Supposing AO, OB, OD to be the semiaxes of the ellipse, and L its latus rectum, and D the difference between the lesser semi-axis OD, and ½L the half of the latus rectum: let an angle Y be found, whose sine may be to the radius as the rectangle under that difference D, and AO + OD the half sum of the axes, to the square of the greater axis AB. Find also an angle Z, whose sine may be to the radius as the double rectangle under the distance of the foci SH and that difference D, to triple the square of half the greater semiaxis AO. Those angles being once found, the place of the body may be thus determined. Take the angle T proportional to the time in which the arc BP was described, or equal to what is called the mean motion; and take an angle V, the first equation of the mean motion, to the angle Y, the greatest first equation, as the sine of double the angle T is to the radius; and take an angle X, the second equation, to the angle Z, the second greatest equation, as the cube of the sine of the angle T is to the cube of the radius. Then take the angle BHP, the mean equated motion either equal to T + X + V, the sum of the angles T, V, X, if the angle T is less than a right angle, or equal to T + X − V, the difference of the same, if that angle T is greater than one and less than two right angles; and if HP meets the ellipse in P, draw SP, and it will cut off the area BSP, nearly proportional to the time.

This practice seems to be expeditious enough, because the angles V and X, taken in fractions of seconds, if you please, being very small, it will be sufficient to find two or three of their first figures. But it is likewise sufficiently accurate to answer to the theory of the planets' motions. For even in the orbit of Mars, where the greatest equation of the centre amounts to ten degrees, the error will scarcely exceed one second. But when the angle of the mean motion equated BHP is found, the angle of the true motion BSP, and the distance SP, are readily had by the known methods.

And so far concerning the motion of bodies in curved lines. But it may also come to pass that a moving body shall ascend or descend in a right line; and I shall now go on to explain what belongs to such kind of motions.

SECTION VII

The rectilinear ascent and descent of bodies.

PROPOSITION XXXII. PROBLEM XXIV

Supposing that the centripetal force is inversely proportional to the square of the distance of the places from the centre; it is required to define the spaces which a body, falling directly, describes in given times.

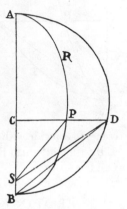

CASE 1. If the body does not fall perpendicularly, it will (by Cor. 1, Prop. XIII) describe some conic section whose focus is placed in the centre of force. Suppose that conic section to be ARPB and its focus S. And, first, if the figure be an ellipse, upon the greater axis thereof AB describe the semicircle ADB, and let the right line DPC pass through the falling body, making right angles with the axis; and drawing DS, PS, the area ASD will be proportional to the area ASP, and therefore also to the time. The axis AB still remaining the same, let the breadth of the ellipse be continually diminished, and the area ASD will always remain proportional to the time. Suppose that breadth to be diminished *in infinitum;* and the orbit APB in that case coinciding with the axis AB, and the focus S with the extreme point of the axis B, the body will descend in the right line AC, and the area ABD will become proportional to the time. Therefore the space AC will be given which the body describes in a given time by its perpendicular fall from the place A, if the area ABD is taken proportional to the time, and from the point D the right line DC is let fall perpendicularly on the right line AB. Q.E.I.

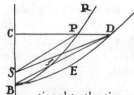

CASE 2. If the figure RPB is an hyperbola, on the same principal diameter AB describe the rectangular hyperbola BED; and because there exist between the several areas and the heights CP and CD

relations, CSP : CSD = CB*f*P : CBED = SP*f*B : SDEB = CP : CD, and since the area SP*f*B varies as the time in which the body P will move through the arc P*f*B, the area SDEB will also vary as that time. Let the latus rectum of the hyperbola RPB be diminished *in infinitum,* the transverse axis remaining the same; and the arc PB will come to coincide with the right line CB, and the focus S with the vertex B, and the right line SD with the right line

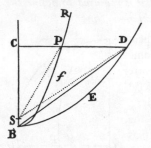

BD. And therefore the area BDEB will vary as the time in which the body C, by its perpendicular descent, describes the line CB. Q.E.I.

CASE 3. And by the like argument, if the figure RPB is a parabola, and to the same principal vertex B another parabola BED is described, that may always remain given while the former parabola in whose perimeter the body P moves, by having its latus rectum diminished and reduced to nothing, comes to coincide with the line CB, the parabolic segment BDEB will vary as the time in which that body P or C will descend to the centre S or B. Q.E.I.

PROPOSITION XXXIII. THEOREM IX

The things above found being supposed, I say, that the velocity of a falling body in any place C is to the velocity of a body, describing a circle about the centre B at the distance BC, as the square root of the ratio of AC, the distance of the body from the remoter vertex A of the circle or rectangular hyperbola, to ½AB, the principal semidiameter of the figure.

Let AB, the common diameter of both figures RPB, DEB, be bisected in O; and draw the right line PT that may touch the figure RPB in P, and likewise cut that common diameter AB (produced, if need be) in T; and let SY be perpendicular to this line, and BQ perpendicular to this diameter, and suppose the latus rectum of the figure RPB to be L. From Cor. IX, Prop. XVI, it is manifest that the velocity of a body, moving in the line RPB about the centre S, in any place P, is to the velocity of a body describing a circle about the same centre, at the distance SP, as the square root of the ratio of the rectangle ½ L · SP to SY². For by the properties of the conic sections AC · CB is to CP² as 2AO to L, and therefore $\frac{2CP^2 \cdot AO}{AC \cdot CB}$ is equal to L. Therefore those velocities are to each other as the square root of the ratio of $\frac{CP^2 \cdot AO \cdot SP}{AC \cdot CB}$ to SY². Moreover, by the properties of the conic sec-

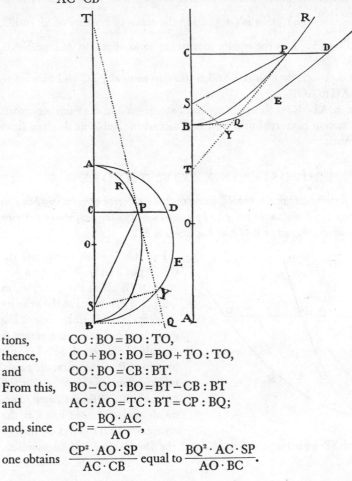

tions,   CO : BO = BO : TO,

thence,   CO + BO : BO = BO + TO : TO,

and   CO : BO = CB : BT.

From this,   BO − CO : BO = BT − CB : BT

and   AC : AO = TC : BT = CP : BQ;

and, since $CP = \frac{BQ \cdot AC}{AO}$,

one obtains $\frac{CP^2 \cdot AO \cdot SP}{AC \cdot CB}$ equal to $\frac{BQ^2 \cdot AC \cdot SP}{AO \cdot BC}$.

Now suppose CP, the breadth of the figure RPB, to be diminished *in infinitum,* so that the point P may come to coincide with the point C, and the point S with the point B, and the line SP with the line BC, and the line SY with the line BQ; and the velocity of the body now descending perpendicularly in the line CB will be to the velocity of a body describing a circle about the centre B, at the distance BC, as the square root of the ratio of $\frac{BQ^2 \cdot AC \cdot SP}{AO \cdot BC}$ to SY^2, that is (neglecting the ratios of equality of SP to BC, and BQ^2 to SY^2), as the square root of the ratio of AC to AO, or ½AB. Q.E.D.

Cor. i. When the points B and S come to coincide, TC will become to TS as AC to AO.

Cor. ii. A body revolving in any circle at a given distance from the centre, by its motion converted upwards, will ascend to double its distance from the centre.

PROPOSITION XXXIV. THEOREM X

If the figure BED is a parabola, I say, that the velocity of a falling body in any place C is equal to the velocity by which a body may uniformly describe a circle about the centre B at half the interval BC.

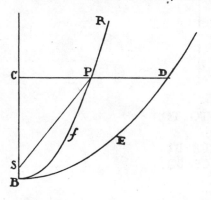

For (by Cor. vii, Prop. xvi) the velocity of a body describing a parabola RPB about the centre S, in any place P, is equal to the velocity of a body uniformly describing a circle about the same centre S at half the interval SP. Let the breadth CP of the parabola be diminished *in infinitum,* so that the parabolic arc PfB may come to coincide with the right line CB, the centre S with the vertex B, and the interval SP with the interval BC, and the Proposition will be manifest. Q.E.D.

PROPOSITION XXXV. THEOREM XI

The same things supposed, I say, that the area of the figure DES, *described by the indefinite radius* SD, *is equal to the area which a body with a radius equal to half the latus rectum of the figure* DES *describes in the same time, by uniformly revolving about the centre* S.

For suppose a body C in the smallest moment of time describes in falling the infinitely little line Cc, while another body K, uniformly revolving about the centre S in the circle OKk, describes the arc Kk. Erect the perpendiculars CD, cd, meeting the figure DES in D, d. Join SD, Sd, SK, Sk, and draw Dd meeting the axis AS in T, and thereon let fall the perpendicular SY.

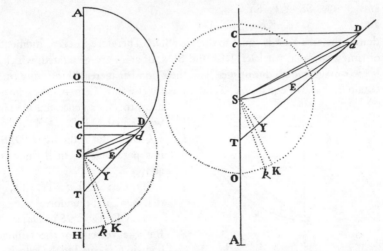

CASE I. If the figure DES is a circle, or a rectangular hyperbola, bisect its transverse diameter AS in O, and SO will be half the latus rectum. And
because   TC : TD = Cc : Dd
and   TD : TS = CD : Sy,
there follows TC : TS = CD · Cc : SY · Dd.
But (by Cor. I, Prop. xxxiii)
                TC : TS = AC : AO,

namely, if in the coalescence of the points D, *d* the ultimate ratios of the lines are taken. Therefore,

$$AC : AO \text{ or } SK = CD \cdot Cc : SY \cdot Dd.$$

Further, the velocity of the descending body in C is to the velocity of a body describing a circle about the centre S, at the interval SC, as the square root of the ratio of AC to AO or SK (by Prop. xxxiii); and this velocity is to the velocity of a body describing the circle OK*k* as the square root of the ratio of SK to SC (by Cor. vi, Prop. iv); and, consequently, the first velocity is to the last, that is, the little line C*c* to the arc K*k*, as the square root of the ratio of AC to SC, that is, in the ratio of AC to CD. Therefore,

$$CD \cdot Cc = AC \cdot Kk,$$

hence, $\quad AC : SK = AC \cdot Kk : SY \cdot Dd,$

and $\quad SK \cdot Kk = SY \cdot Dd,$

and $\quad \tfrac{1}{2}SK \cdot Kk = \tfrac{1}{2}SY \cdot Dd,$

that is, the area KS*k* is equal to the area SD*d*. Therefore in every moment of time two equal particles, KS*k* and SD*d*, of areas are generated, which, if their magnitude is diminished, and their number increased *in infinitum,* obtain the ratio of equality, and consequently (by Cor., Lem. iv) the whole areas together generated are always equal. Q.E.D.

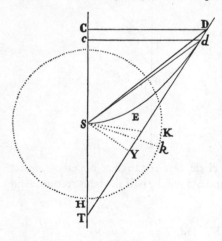

CASE 2. But if the figure DES is a parabola, we shall find, as above,

$$CD \cdot Cc : SY \cdot Dd = TC : TS,$$

that is, $= 2 : 1$; therefore,

$$\tfrac{1}{4} CD \cdot Cc = \tfrac{1}{2} SY \cdot Dd.$$

But the velocity of the falling body in C is equal to the velocity with which a circle may be uniformly described at the interval ½ SC (by Prop. xxxiv). And this velocity to the velocity with which a circle may be described with the radius SK, that is, the little line C*c* to the arc K*k*, is (by Cor. vi, Prop. iv) as the square root of the ratio of SK to ½ SC; that is, in the ratio

of SK to ½ CD. Therefore ½ SK · K*k* is equal to ¼ CD · C*c*, and therefore
equal to ½ SY · D*d*; that is, the area KS*k* is equal to the area SD*d*, as above.
Q.E.D.

PROPOSITION XXXVI. PROBLEM XXV

To determine the times of the descent of a body falling
from a given place A.

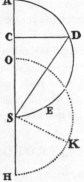

Upon the diameter AS, the distance of the body from
the centre at the beginning, describe the semicircle
ADS, as likewise the semicircle OKH equal thereto,
about the centre S. From any place C of the body erect
the ordinate CD. Join SD, and make the sector OSK
equal to the area ASD. It is evident (by Prop. xxxv) that
the body in falling will describe the space AC in the
same time in which another body, uniformly revolving
about the centre S, may describe the arc OK. Q.E.F.

PROPOSITION XXXVII. PROBLEM XXVI

To define the times of the ascent or descent of a body projected upwards
or downwards from a given place.

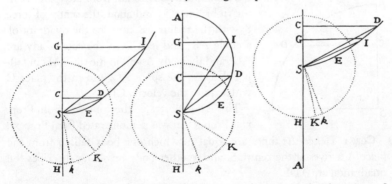

Suppose the body to go off from the given place G, in the direction of
the line GS, with any velocity. Take GA to ½ AS as the square of the ratio
of this velocity to the uniform velocity in a circle, with which the body may

revolve about the centre S at the given interval SG. If that ratio is the same as of the number 2 to 1, the point A is infinitely remote; in which case a parabola is to be described with any latus rectum to the vertex S, and axis SG; as appears by Prop. xxxiv. But if that ratio is less or greater than the ratio of 2 to 1, in the former case a circle, in the latter a rectangular hyperbola, is to be described on the diameter SA; as appears by Prop. xxxiii. Then about the centre S, with a radius equal to half the latus rectum, describe the circle H*k*K; and at the place G of the ascending or descending body, and at any other place C, erect the perpendiculars GI, CD, meeting the conic section or circle in I and D. Then joining SI, SD, let the sectors HSK, HS*k* be made equal to the segments SEIS, SEDS, and (by Prop. xxxv) the body G will describe the space GC in the same time in which the body K may describe the arc K*k*. Q.E.F.

PROPOSITION XXXVIII. THEOREM XII

Supposing that the centripetal force is proportional to the altitude or distance of places from the centre, I say, that the times and velocities of falling bodies, and the spaces which they describe, are respectively proportional to the arcs, and the sines and versed sines of the arcs.

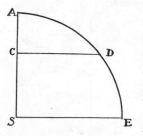

Suppose the body to fall from any place A in the right line AS; and about the centre of force S, with the radius AS, describe the quadrant of a circle AE; and let CD be the sine of any arc AD; and the body A will in the time AD in falling describe the space AC, and in the place C will acquire the velocity CD.

This is demonstrated the same way from Prop. x, as Prop. xxxii was demonstrated from Prop. xi.

Cor. i. Hence the times are equal in which one body falling from the place A arrives at the centre S, and another body revolving describes the quadrantal arc ADE.

Cor. ii. Therefore all the times are equal in which bodies falling from whatsoever places arrive at the centre. For all the periodic times of revolving bodies are equal (by Cor. iii, Prop. iv).

PROPOSITION XXXIX. PROBLEM XXVII

Supposing a centripetal force of any kind, and granting the quadratures of curvilinear figures; it is required to find the velocity of a body, ascending or descending in a right line, in the several places through which it passes, as also the time in which it will arrive at any place; and conversely.

Suppose the body E to fall from any place A in the right line ADEC; and from its place E imagine a perpendicular EG always erected proportional to the centripetal force in that place tending to the centre C; and let BFG

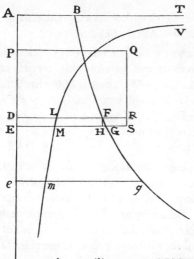

be a curved line, the locus of the point G. And in the beginning of the motion suppose EG to coincide with the perpendicular AB; and the velocity of the body in any place E will be as a right line whose square is equal to the curvilinear area ABGE. Q.E.I.

In EG take EM inversely proportional to a right line whose square is equal to the area ABGE, and let VLM be a curved line wherein the point M is always placed, and to which the right line AB produced is an asymptote; and the time in which the body in falling describes the line AE, will be as the curvilinear area ABTVME. Q.E.I.

For in the right line AE let there be taken the very small line DE of a given length, and let DLF be the place of the line EMG, when the body was in D; and if the centripetal force be such, that a right line, whose square is equal to the area ABGE, is as the velocity of the descending body, the area itself will be as the square of that velocity; that is, if for the velocities in D and E we write V and V + I, the area ABFD will be as VV, and the area ABGE as $VV + 2VI + II$; and by subtraction, the area DFGE as $2VI + II$, and therefore $\dfrac{DFGE}{DE}$ will be as $\dfrac{2VI + II}{DE}$; that is, if we take the first ratios of those quantities when just nascent, the length DF is as the quantity $\frac{2VI}{DE}$, and therefore also as half that quantity $\frac{I \cdot V}{DE}$. But the time in which the body in falling describes the very small line DE, is directly as that line and inversely as the velocity V; and the force will be directly as the increment I of the velocity and inversely as the time; and therefore if we take the first ratios when those quantities are just nascent, as $\frac{I \cdot V}{DE}$, that is, as the length DF. Therefore a force proportional to DF or EG will cause the body to descend with a velocity that is as the right line whose square is equal to the area ABGE. Q.E.D.

Moreover, since the time in which a very small line DE of a given length may be described is inversely as the velocity and therefore also inversely as a right line whose square is equal to the area ABFD; and since the line DL, and by consequence the nascent area DLME, will be inversely as the same right line, the time will be as the area DLME, and the sum of all the times will be as the sum of all the areas; that is (by Cor., Lem. IV), the whole time in which the line AE is described will be as the whole area ATVME. Q.E.D.

Cor. I. Let P be the place from whence a body ought to fall, so as that, when urged by any known uniform centripetal force (such as gravity is commonly supposed to be), it may acquire in the place D a velocity equal to the velocity which another body, falling by any force whatever, hath acquired in that place D. In the perpendicular DF let there be taken DR, which may be to DF as that uniform force to the other force in the place D. Complete the rectangle PDRQ, and cut off the area ABFD equal to that rectangle. Then A will be the place from whence the other body fell. For completing the rectangle DRSE, since the area ABFD is to the area DFGE as VV to 2VI, and therefore as ½ V to I, that is, as half the whole velocity to the increment of the velocity of the body falling by the variable force; and in like manner the area PQRD to the area DRSE as half the whole velocity to the increment of the velocity of the body falling by the uniform force; and since those increments (by reason of the equality of the nascent times) are as the generating forces, that is, as the ordinates DF, DR, and consequently as the nascent areas DFGE, DRSE; therefore, the whole areas

ABFD, PQRD will be to each other as the halves of the whole velocities; and therefore, because the velocities are equal, they become equal also.

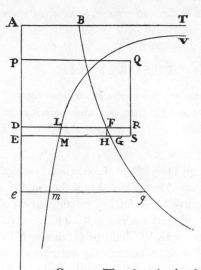

Cor. II. Whence if any body be projected either upwards or downwards with a given velocity from any place D, and there be given the law of centripetal force acting on it, its velocity will be found in any other place, as *e*, by erecting the ordinate *eg*, and taking that velocity to the velocity in the place D as a right line whose square is equal to the rectangle PQRD, either increased by the curvilinear area DF*ge*, if the place *e* is below the place D, or diminished by the same area DF*ge*, if it be higher, is to the right line whose square is equal to the rectangle PQRD alone.

Cor. III. The time is also known by erecting the ordinate *em* inversely proportional to the square root of PQRD + or − DF*ge*, and taking the time in which the body has described the line D*e* to the time in which another body has fallen with an uniform force from P, and in falling arrived at D in the proportion of the curvilinear area DL*me* to the rectangle 2PD · DL. For the time in which a body falling with an uniform force hath described the line PD, is to the time in which the same body hath described the line PE as the square root of the ratio of PD to PE; that is (the very small line DE being just nascent), in the ratio of PD to PD + ½DE or 2PD to 2PD + DE, and, by subtraction, to the time in which the body hath described the small line DE, as 2PD to DE, and therefore as the rectangle 2PD · DL to the area DLME; and the time in which both the bodies described the very small line DE is to the time in which the body with the variable motion described the line D*e* as the area DLME to the area DL*me*; and therefore the first mentioned of these times is to the last as the rectangle 2PD · DL to the area DL*me*.

SECTION VIII

The determination of orbits in which bodies will revolve, being acted upon by any sort of centripetal force.

PROPOSITION XL. THEOREM XIII

If a body, acted upon by any centripetal force, is moved in any manner, and another body ascends or descends in a right line, and their velocities be equal in any one case of equal altitudes, their velocities will be also equal at all equal altitudes.

Let a body descend from A through D and E, to the centre C; and let another body move from V in the curved line VIKk. From the centre C, with any distances, describe the concentric circles DI, EK, meeting the right line AC in D and E, and the curve VIK in I and K. Draw IC meeting KE in N, and on IK let fall the perpendicular NT; and let the interval DE or IN between the circumferences of the circles be very small; and imagine the bodies in D and I to have equal velocities. Then because the distances CD and CI are equal, the centripetal forces in D and I will be also equal. Let those forces be expressed by the equal short lines DE and IN; and let the force IN (by Cor. II of the Laws of Motion) be resolved into two others, NT and IT. Then the force NT acting in the direction of the line NT perpendicular to the path ITK of the body will not at all affect or change the velocity of the body in that path, but only draw it aside from a rectilinear course, and make it deflect continually from the tangent of the orbit, and proceed in the curvilinear path ITKk. That whole force, therefore, will be spent in producing this effect; but the other force IT, acting in the direction of the course of the body, will be all employed in accelerating it, and in the least given time will produce an acceleration proportional to itself. Therefore the accelerations of

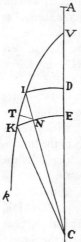

the bodies in D and I, produced in equal times, are as the lines DE, IT (if we take the first ratios of the nascent lines DE, IN, IK, IT, NT); and in unequal times as the product of those lines and the times. But the times in which DE and IK are described, are, by reason of the equal velocities (in D and I), as the spaces described DE and IK, and therefore the accelerations in the course of the bodies through the lines DE and IK are as DE and IT, and DE and IK conjointly; that is, as the square of DE to the rectangle IT·IK. But the rectangle IT·IK is equal to the square of IN, that is, equal to the square of DE; and therefore the accelerations generated in the passage of the bodies from D and I to E and K are equal. Therefore the velocities of the bodies in E and K are also equal: and by the same reasoning they will always be found equal in any subsequent equal distances. Q.E.D.

By the same reasoning, bodies of equal velocities and equal distances from the centre will be equally retarded in their ascent to equal distances. Q.E.D.

Cor. i. Therefore if a body either oscillates by hanging to a string, or by any polished and perfectly smooth impediment is forced to move in a curved line; and another body ascends or descends in a right line, and their velocities be equal at any one equal altitude, their velocities will be also equal at all other equal altitudes. For by the string of the pendulous body, or by the impediment of a vessel perfectly smooth, the same thing will be effected as by the transverse force NT. The body is neither accelerated nor retarded by it, but only is obliged to leave its rectilinear course.

Cor. ii. Suppose the quantity P to be the greatest distance from the centre to which a body can ascend, whether it be oscillating, or revolving in a curve, and so the same projected upwards from any point of a curve with the velocity it has in that point. Let the quantity A be the distance of the body from the centre in any other point of the orbit; and let the centripetal force be always as the power A^{n-1}, of the quantity A, the index of which power $n-1$ is any number n diminished by unity. Then the velocity in every altitude A will be as $\sqrt{(P^n - A^n)}$, and therefore will be given. For by Prop. xxxix, the velocity of a body ascending and descending in a right line is in that very ratio.

PROPOSITION XLI. PROBLEM XXVIII

Supposing a centripetal force of any kind, and granting the quadratures of curvilinear figures; it is required to find as well the curves in which bodies will move, as the times of their motions in the curves found.

Let any centripetal force tend to the centre C, and let it be required to find the curve VIK*k*. Let there be given the circle VR, described from the centre C with any radius CV; and from the same centre describe any other circles ID, KE, cutting the curve in I and K, and the right line CV

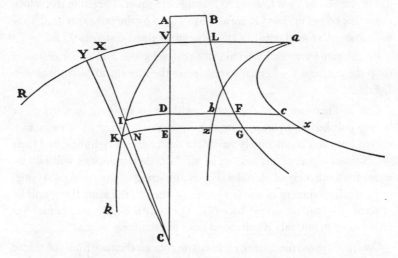

in D and E. Then draw the right line CNIX cutting the circles KE, VR in N and X, and the right line CKY meeting the circle VR in Y. Let the points I and K be indefinitely near; and let the body go on from V through I and K to *k*; and let the point A be the place from which another body is to fall, so as in the place D to acquire a velocity equal to the velocity of the first body in I. And things remaining as in Prop. XXXIX, the short line IK, described in the least given time, will be as the velocity, and therefore as the right line whose square is equal to the area ABFD, and the triangle ICK proportional to the time will be given, and therefore KN will be inversely as the altitude IC; that is (if there be given any quantity Q, and the altitude IC be called A), as $\frac{Q}{A}$. This quantity $\frac{Q}{A}$ call Z, and suppose the magnitude of Q to be such that in some one case

$$\sqrt{ABFD} : Z = IK : KN,$$

and then in all cases

$$\sqrt{ABFD} : Z = IK : KN,$$

and $\quad ABFD : ZZ = IK^2 : KN^2,$

and by subtraction,

$$ABFD - ZZ : ZZ = IN^2 : KN^2,$$

and therefore

$$\sqrt{(ABFD - ZZ)} : Z \text{ or } \frac{Q}{A} = IN : KN,$$

and $\quad A \cdot KN = \dfrac{Q \cdot IN}{\sqrt{(ABFD - ZZ)}}.$

Since $\quad YX \cdot XC : A \cdot KN = CX^2 : AA,$

it follows that

$$YX \cdot XC = \frac{Q \cdot IN \cdot CX^2}{AA\sqrt{(ABFD - ZZ)}}.$$

Therefore in the perpendicular DF let there be taken continually Db, Dc equal to $\dfrac{Q}{2\sqrt{(ABFD - ZZ)}}$, $\dfrac{Q \cdot CX^2}{2AA\sqrt{(ABFD - ZZ)}}$ respectively, and let the curved lines ab, ac, the foci of the points b and c, be described; and from the point V let the perpendicular Va be erected to the line AC, cutting off the curvilinear areas VDba, VDca, and let the ordinates Ez, Ex, be erected also. Then because the rectangle D$b \cdot$ IN or DbzE is equal to half the rectangle A $\cdot$ KN, or to the triangle ICK; and the rectangle D$c \cdot$ IN or DcxE is equal to half the rectangle YX $\cdot$ XC, or to the triangle XCY; that is, because the nascent particles DbzE, ICK of the areas VDba, VIC are always equal; and the nascent particles DcxE, XCY of the areas VDca, VCX are always equal: therefore the generated area VDba will be equal to the generated area VIC, and therefore proportional to the time; and the generated area VDca is equal to the generated sector VCX. If, therefore, any time be given during which the body has been moving from V, there will be also given the area proportional to it VDba; and thence will be given the altitude of the body CD or CI; and the area VD*ca,* and the sector VCX equal thereto, together with its angle VCI. But the angle VCI, and the altitude CI being given, there is also given the place I, in which the body will be found at the end of that time. Q.E.I.

Cor. i. Hence the greatest and least altitudes of the bodies, that is, the apsides of the curves, may be found very readily. For the apsides are those points in which a right line IC drawn through the centre falls perpendicularly upon the curves VIK; which comes to pass when the right lines IK and NK become equal; that is, when the area ABFD is equal to ZZ.

Cor. ii. So also the angle KIN, in which the curve at any place cuts the line IC, may be readily found by the given altitude IC of the body; namely, by making the sine of that angle to the radius as KN to IK, that is, as Z to the square root of the area ABFD.

Cor. iii. If to the centre C, and the principal vertex V, there be described a conic section VRS; and from any point thereof, as R, there be drawn the tangent RT meeting the axis CV indefinitely produced in the point T; and

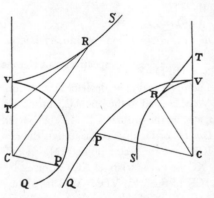

then joining CR there be drawn the right line CP, equal to the abscissa CT, making an angle VCP proportional to the sector VCR; and if a centripetal force inversely proportional to the cubes of the distances of the places from the centre, tends to the centre C; and from the place V there sets out a body with a just velocity in the direction of a line perpendicular to the right line CV; that body will proceed in a curve VPQ, which the point P will always touch; and therefore if the conic section VRS be an hyperbola, the body will descend to the centre; but if it be an ellipse, it will ascend continually, and go farther and farther off *in infinitum.* And, on the contrary, if a body endued with any velocity goes off from the place V, and according as it begins either to descend obliquely to the centre, or to ascend obliquely from it, the figure VRS be either an hyperbola or an ellipse, the curve may be found by increasing or diminishing the angle VCP in a given ratio. And the centripetal force becoming centrifugal, the body will ascend obliquely in the curve VPQ, which is found by taking the angle VCP proportional to the elliptic sector VRC, and the length CP equal to the length CT, as before. All these things follow from the foregoing Proposition, by the quadrature of a certain curve, the invention of which, as being easy enough, for brevity's sake I omit.

PROPOSITION XLII. PROBLEM XXIX

The law of centripetal force being given, it is required to find the motion of a body setting out from a given place, with a given velocity, in the direction of a given right line.

Suppose the same things as in the three preceding Propositions; and let the body go off from the place I in the direction of the little line IK, with

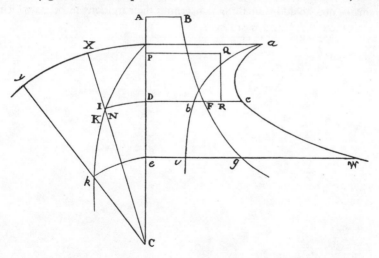

the same velocity as another body, by falling with an uniform centripetal force from the place P, may acquire in D; and let this uniform force be to the force with which the body is at first urged in I, as DR to DF. Let the body go on towards k; and about the centre C, with the radius Ck, describe the circle ke, meeting the right line PD in e, and let there be erected the lines eg, ev, ew, ordinately applied to the curves BFg, abv, acw. From the given rectangle PDRQ and the given law of centripetal force, by which the first body is acted on, the curved line BFg is also given, by the construction of Prop. xxvii, and its Cor. i. Then from the given angle CIK is given the proportion of the nascent lines IK, KN; and thence, by the construction of Prob. xxviii, there is given the quantity Q, with the curved lines abv, acw; and therefore, at the end of any time $Dbve$, there is given both the altitude of the body Ce or Ck, and the area $Dcwe$, with the sector equal to it XCy, the angle ICk, and the place k, in which the body will then be found. Q.E.I.

We suppose in these Propositions the centripetal force to vary in its recess from the centre according to some law, which anyone may imagine at pleasure, but at equal distances from the centre to be everywhere the same.

I have hitherto considered the motions of bodies in immovable orbits. It remains now to add something concerning their motions in orbits which revolve round the centres of force.

SECTION IX

The motion of bodies in movable orbits; and the motion of the apsides.

PROPOSITION XLIII. PROBLEM XXX

It is required to make a body move in a curve that revolves about the centre of force in the same manner as another body in the same curve at rest.

In the fixed orbit VPK, let the body P revolve, proceeding from V towards K. From the centre C let there be continually drawn C*p*, equal to CP, making the angle VC*p* proportional to the angle VCP; and the area which the line C*p* describes will be to the area VCP, which the line CP describes at the same time, as the velocity of the describing line C*p* to the velocity of the describing line CP; that is, as the angle VC*p* to the angle VCP, therefore in a given ratio, and therefore proportional to the time. Since, then, the area described by the line C*p* in a fixed plane is proportional to the time, it is manifest that a body, being acted upon by a suitable centripetal force, may revolve with 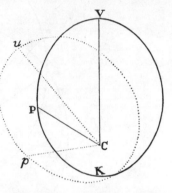 the point *p* in the curved line which the same point *p*, by the method just now explained, may be made to describe in a fixed plane. Make the angle VC*u* equal to the angle PC*p*, and the line C*u* equal to CV, and the figure *u*C*p* equal to the figure VCP, and the body being always in the point *p*, will move in the perimeter of the revolving figure *u*C*p*, and will describe its (revolving) arc *up* in the same time that the other body P describes the similar and equal arc VP in the fixed figure VPK. Find, then, by Cor. v, Prop. vi, the centripetal force by which the body may be made to revolve in the curved line which the point *p* describes in a fixed plane, and the Problem will be solved. Q.E.F.

PROPOSITION XLIV. THEOREM XIV

The difference of the forces, by which two bodies may be made to move equally, one in a fixed, the other in the same orbit revolving, varies inversely as the cube of their common altitudes.

Let the parts of the fixed orbit VP, PK be similar and equal to the parts of the revolving orbit *up*, *pk*; and let the distance of the points P and K be supposed of the utmost smallness. Let fall a perpendicular *kr* from the point *k* to the right line *p*C, and produce it to *m*, so that *mr* may be to *kr*

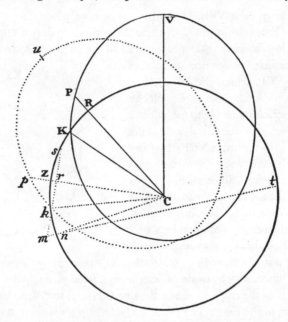

as the angle VC*p* to the angle VCP. Because the altitudes of the bodies PC and *p*C, KC and *k*C, are always equal, it is manifest that the increments or decrements of the lines PC and *p*C are always equal; and therefore if each of the several motions of the bodies in the places P and *p* be resolved into two (by Cor. II of the Laws of Motion), one of which is directed towards the centre, or according to the lines PC, *p*C, and the other, transverse to the former, hath a direction perpendicular to the lines PC and pC; the motions towards the centre will be equal, and the transverse motion of the body p will be to the transverse motion of the body P as the angular motion of the line pC to the angular motion of the line PC; that is, as the angle VCp to the angle VCP. Therefore, at the same time that the body P, by both its motions, comes to the point K, the body p, having an equal motion towards the centre, will be equally moved from p towards C; and therefore that time being expired, it will be found somewhere in the line mkr, which, passing through the point k, is perpendicular to the line pC; and by its transverse motion will acquire a distance from the line pC, that will be to the distance which the other body P acquires from the line PC as the transverse motion of the body p to the transverse motion of the other body P. Therefore since kr is equal to the distance which the body P acquires from the line PC, and mr is to kr as the angle VCp to the angle VCP, that is, as the transverse motion of the body p to the transverse motion of the body P, it is manifest that the body p, at the expiration of that time, will be found in the place m. These things will be so, if the bodies p and P are equally moved in the directions of the lines pC and PC, and are therefore urged with equal forces in those directions. But if we take an angle pCn that is to the angle pCk as the angle VCp to the angle VCP, and nC be equal to kC, in that case the body p at the expiration of the time will really be in n; and is therefore urged with a greater force than the body P, if the angle nCp is greater than the angle kCp, that is, if the orbit upk moves either progressively, or in a retrograde direction, with a velocity greater than the double of that with which the line CP is carried forwards; and with a less force if the retrograde motion of the orbit is slower. And the difference of the forces will be as the interval mn of the places through which the body would be carried by the action of that difference in that given space of time. About the centre C with the interval Cn or Ck suppose a circle described cutting the lines mr, mn produced in s and t, and the rectangle $mn \cdot mt$ will be equal to the rectangle $mk \cdot ms$, and therefore mn will be equal to $\dfrac{mk \cdot ms}{mt}$. But since the triangles pCk, pCn, in a given time, are of a given magnitude, kr and mr, and their difference mk, and their sum ms, are inversely as the altitude pC, and therefore the rectangle $mk \cdot ms$

is inversely as the square of the altitude pC. Moreover, mt is directly as $\frac{1}{2}mt$, that is, as the altitude pC. These are the first ratios of the nascent lines; and hence $\frac{mk \cdot ms}{mt}$, that is, the nascent short line mn, and the difference of the forces proportional thereto, are inversely as the cube of the altitude pC. Q.E.D.

COR. I. Hence the difference of the forces in the places P and p, or K and k, is to the force with which a body may revolve with a circular motion from R to K, in the same time that the body P in a fixed orbit describes the arc PK, as the nascent line mn to the versed sine of the nascent arc RK, that is, as $\frac{mk \cdot ms}{mt}$ to $\frac{rk^2}{2kC}$, or as $mk \cdot ms$ to the square of rk; that is, if we take given quantities F and G in the same ratio to each other as the angle VCP bears to the angle VCp, as GG – FF to FF. And, therefore, if from the centre C, with any distance CP or Cp, there be described a circular sector equal to the whole area VPC, which the body revolving in a fixed orbit hath by a radius drawn to the centre described in any certain time, the difference of the forces, with which the body P revolves in a fixed orbit, and the body p in a movable orbit, will be to the centripetal force, with which another body by a radius drawn to the centre can uniformly describe that sector in the same time as the area VPC is described, as GG – FF to FF. For that sector and the area pCk are to each other as the times in which they are described.

COR. II. If the orbit VPK be an ellipse, having its focus C, and its highest apse V, and we suppose the ellipse upk similar and equal to it, so that pC may be always equal to PC, and the angle VCp be to the angle VCP in the given ratio of G to F; and for the altitude PC or pC we put A, and 2R for the latus rectum of the ellipse, the force with which a body may be made to revolve in a movable ellipse will be as $\frac{FF}{AA} + \frac{RGG - RFF}{A^3}$, and conversely.

Let the force with which a body may revolve in a fixed ellipse be expressed by the quantity $\frac{FF}{AA}$, and the force in V will be $\frac{FF}{CV^2}$. But the force with which a body may revolve in a circle at the distance CV, with the same velocity as a body revolving in an ellipse has in V, is to the force with which a body revolving in an ellipse is acted upon in the apse V, as half the latus rectum of the ellipse to the semidiameter CV of the circle, and therefore is as $\dfrac{RFF}{CV^3}$; and the force which is to this as GG – FF to FF, is as $\dfrac{RGG-RFF}{CV^3}$; and this force (by Cor. 1 of this Prop.) is the difference of the forces in V, with which the body P revolves in the fixed ellipse VPK, and the body p

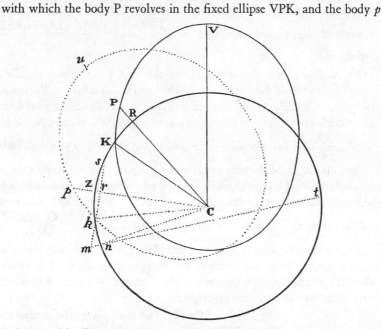

in the movable ellipse upk. Then since by this Proposition that difference at any other altitude A is to itself at the altitude CV as $\dfrac{1}{A^3}$ to $\dfrac{1}{CV^3}$, the same difference in every altitude A will be as $\dfrac{RGG-RFF}{A^3}$. Therefore to the force $\dfrac{FF}{AA}$, by which the body may revolve in a fixed ellipse VPK, add the excess $\dfrac{RGG-RFF}{A^3}$, and the sum will be the whole force $\dfrac{FF}{AA} + \dfrac{RGG-RFF}{A^3}$ by which a body may revolve in the same time in the movable ellipse upk.

Cor. III. In the same manner it will be found, that, if the fixed orbit VPK be an ellipse having its centre in the centre of the forces C, and there be supposed a movable ellipse *upk,* similar, equal, and concentric to it; and 2R be the principal latus rectum of that ellipse, and 2T the latus transversum, or greater axis; and the angle VC*p* be continually to the angle VCP as G to F; the forces with which bodies may revolve in the fixed and movable ellipse, in equal times, will be as $\dfrac{FFA}{T^3}$ and $\dfrac{FFA}{T^3} + \dfrac{RGG-RFF}{A^3}$ respectively.

Cor. IV. And universally, if the greatest altitude CV of the body be called T, and the radius of the curvature which the orbit VPK has in V, that is, the radius of a circle equally curved, be called R, and the centripetal force with which a body may revolve in any fixed curve VPK at the place V be called $\dfrac{VFF}{TT}$, and in other places P be indefinitely styled X; and the altitude CP be called A, and G be taken to F in the given ratio of the angle VC*p* to the angle VCP; the centripetal force with which the same body will perform the same motions in the same time, in the same curve *upk* revolving with a circular motion, will be as the sum of the forces $X + \dfrac{VRGG-VRFF}{A^3}$.

Cor. V. Therefore the motion of a body in a fixed orbit being given, its angular motion round the centre of the forces may be increased or diminished in a given ratio; and thence new fixed orbits may be found in which bodies may revolve with new centripetal forces.

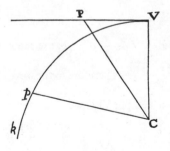

Cor. VI. Therefore if there be erected the line VP of an indeterminate length, perpendicular to the line CV given by position, and CP be drawn, and C*p* equal to it, making the angle VC*p* having a given ratio to the angle VCP, the force with which a body may revolve in the curved line V*pk,* which the point *p* is continually describing, will be inversely as the cube of the altitude C*p*. For the body P, by its inertia alone, no other force impelling it, will proceed uniformly in the right line VP. Add, then, a force tending to the centre C inversely as the cube of the altitude CP or Cp, and (by what was just demonstrated) the body will deflect from the rectilinear motion into the curved line Vpk. But this curve Vpk is the same with the curve VPQ found in Cor. III, Prop. XLI, in which, I said, bodies attracted with such forces would ascend obliquely.

PROPOSITION XLV. PROBLEM XXXI

To find the motion of the apsides in orbits approaching very near to circles.[1]

This problem is solved arithmetically by reducing the orbit, which a body revolving in a movable ellipse (as in Cor. II and III of the above Prop.) describes in a fixed plane, to the figure of the orbit whose apsides are required; and then seeking the apsides of the orbit which that body describes in a fixed plane. But orbits acquire the same figure, if the centripetal forces with which they are described, compared between themselves, are made proportional at equal altitudes. Let the point V be the highest apse, and write T for the greatest altitude CV, A for any other altitude CP or Cp, and X for the difference of the altitudes CV − CP; and the force with which a body moves in an ellipse revolving about its focus C (as in Cor. II), and which in Cor. II was as $\dfrac{FF}{AA} + \dfrac{RGG - RFF}{A^3}$, that is, as $\dfrac{FFA + RGG - RFF}{A^3}$, by substituting T − X for A, will become as $\dfrac{RGG - RFF + TFF - FFX}{A^3}$. In like manner any other centripetal force is to be reduced to a fraction whose denominator is A^3, and the numerators are to be made analogous by collating together the homologous terms. This will be made plainer by Examples.

EXAM. I. Let us suppose the centripetal force to be uniform, and therefore as $\dfrac{A^3}{A^3}$, or, writing T − X for A in the numerator, as

$$\frac{T^3 - 3TTX + 3TXX - X^3}{A^3}.$$

Then collating together the correspondent terms of the numerators, that is, those that consist of given quantities with those of given quantities, and

[1 Appendix, Note 23.]

those of quantities not given with those of quantities not given, it will become

$$RGG - RFF + TFF : T^3 = -FFX : 3TTX + 3TXX - X^3$$
$$= -FF : 3TT + 3TX - XX.$$

Now since the orbit is supposed extremely near to a circle, let it coincide with a circle; and because in that case R and T become equal, and X is infinitely diminished, the last ratios will be

$$GG : T^2 = -FF : -3TT$$

and again, $GG : FF = TT : 3TT = 1 : 3$;

and therefore G is to F, that is, the angle VCp to the angle VCP, as 1 to $\sqrt{3}$. Therefore since the body, in a fixed ellipse, in descending from the upper to the lower apse, describes an angle, if I may so speak, of 180°, the other body in a movable ellipse, and therefore in the fixed plane we are treating of, will in its descent from the upper to the lower apse, describe an angle VCp of $\dfrac{180°}{\sqrt{3}}$. And this comes to pass by reason of the likeness of this orbit which a body acted upon by an uniform centripetal force describes, and of that orbit which a body performing its circuits in a revolving ellipse will describe in a fixed plane. By this collation of the terms, these orbits are made similar; not universally, indeed, but then only when they approach very near to a circular figure. A body, therefore, revolving with an uniform centripetal force in an orbit nearly circular, will always describe an angle of $\dfrac{180°}{\sqrt{3}}$, or 103° 55′ 23″ at the centre; moving from the upper apse to the lower apse when it has once described that angle, and thence returning to the upper apse when it has described that angle again; and so on *in infinitum*.

Exam. 2. Suppose the centripetal force to be as any power of the altitude A, as, for example, A^{n-3}, or $\dfrac{A^n}{A^3}$; where $n-3$ and n signify any indices of powers whatever, whether integers or fractions, rational or surd, affirmative or negative. That numerator A^n or $(T-X)^n$ being reduced to an indeterminate series by my method of converging series, will become

$$T^n - nXT^{n-1} + \frac{nn-n}{2} XXT^{n-2}, \&c.$$

And comparing these terms with the terms of the other numerator

$$RGG - RFF + TFF - FFX,$$

it becomes $\quad RGG - RFF + TFF : T^n = -FF : -nT^{n-1} + \dfrac{nn - n}{2} XT^{n-2}$, &c.

And taking the last ratios where the orbits approach to circles, it becomes

$$RGG : T^n = -FF : -nT^{n-1},$$

or, $\qquad GG : T^{n-1} = FF : nT^{n-1},$

and again, $\quad GG : FF = T^{n-1} : nT^{n-1} = 1 : n;$

and therefore G is to F, that is, the angle VCp to the angle VCP, as 1 to $\sqrt{n}$. Therefore since the angle VCP, described in the descent of the body from the upper apse to the lower apse in an ellipse, is of 180°, the angle VCp, described in the descent of the body from the upper apse to the lower apse in an orbit nearly circular which a body describes with a centripetal force proportional to the power A^{n-3}, will be equal to an angle of $\dfrac{180°}{\sqrt{n}}$, and this angle being repeated, the body will return from the lower to the upper apse, and so on *in infinitum*. As if the centripetal force be as the distance of the body from the centre, that is, as A, or $\dfrac{A^4}{A^3}$, n will be equal to 4, and $\sqrt{n}$ equal to 2; and therefore the angle between the upper and the lower apse will be equal to $\dfrac{180°}{2}$, or 90°. Therefore the body having performed a fourth part of one revolution, will arrive at the lower apse, and having performed another fourth part, will arrive at the upper apse, and so on *in infinitum*. This appears also from Prop. x. For a body acted on by this centripetal force will revolve in a fixed ellipse, whose centre is the centre of force. If the centripetal force is inversely as the distance, that is, directly as $\dfrac{1}{A}$ or $\dfrac{A^2}{A^3}$, n will be equal to 2; and therefore the angle between the upper and the lower apse will be $\dfrac{180°}{\sqrt{2}}$, or 127° 16′ 45″; and hence a body revolving with such a force will, by a continual repetition of this angle, move alternately from the upper to the lower and from the lower to the upper apse forever. So, also, if the centripetal force be inversely as the fourth root of the eleventh power of the altitude, that is, inversely as $A^{11/4}$, and therefore directly as $\dfrac{1}{A^{1\frac{1}{4}}}$, or as $\dfrac{A^{\frac{1}{4}}}{A^3}$, n will be equal to $\frac{1}{4}$, and $\dfrac{180°}{\sqrt{n}}$ will be equal to 360°; and therefore the body parting from the upper apse, and from thence continually descending, will arrive at the lower apse when it has completed one entire revolution; and thence ascending continually, when it has completed another entire revolution, it will arrive again at the upper apse; and so alternately forever.

Exam. 3. Taking m and n for any indices of the powers of the altitude, and b and c for any given numbers, suppose the centripetal force to be as $(bA^m + cA^n) \div A^3$, that is, as $[b(T-X)^m + c(T-X)^n] \div A^3$, or (by the method of converging series above mentioned) as

$$[bT^m + cT^n - mbXT^{m-1} - ncXT^{n-1} + \frac{mm - m}{2} bXXT^{m-2} + \frac{nn - n}{2}$$
$$- cXXT^{n-2}, \&c.] \div A^3;$$

and comparing the terms of the numerators, there will arise,

$$RGG - RFF + TFF : bT^m + cT^n = -FF : -mbT^{m-1} - ncT^{n-1} +$$
$$\frac{mm - m}{2} bXT^{m-2} + \frac{nn - n}{2} cXT^{n-2}, \&c.$$

And taking the last ratios that arise when the orbits come to a circular form, there will come forth

$$GG : bT^{m-1} + cT^{n-1} = FF : mbT^{m-1} + ncT^{n-1};$$

and again, $\quad GG : FF = bT^{m-1} + cT^{n-1} : mbT^{n-1} + ncT^{n-1}.$

This proportion, by expressing the greatest altitude CV or T arithmetically by unity, becomes, $GG : FF = b + c : mb + nc = 1 : \dfrac{mb + nc}{b + c}$. Whence G becomes to F, that is, the angle VCp to the angle VCP, as 1 to $\sqrt{\dfrac{mb + nc}{b + c}}$.

And therefore, since the angle VCP between the upper and the lower apse, in a fixed ellipse, is of 180°, the angle VCp between the same apsides in an orbit which a body describes with a centripetal force, that is, as $\dfrac{bA^m + cA^n}{A^3}$, will be equal to an angle of 180° $\sqrt{\dfrac{b + c}{mb + nc}}$. And by the same reasoning, if the centripetal force be as $\dfrac{bA^m - cA^n}{A^3}$, the angle between the apsides will be found equal to $180° \sqrt{\dfrac{b-c}{mb-nc}}$. After the same manner the Problem is solved in more difficult cases. The quantity to which the centripetal force is proportional must always be resolved into a converging series whose denominator is A^3. Then the given part of the numerator arising from that operation is to be supposed in the same ratio to that part of it which is not given, as the given part of this numerator RGG − RFF + TFF − FFX is to that part of the same numerator which is not given. And taking away the superfluous quantities, and writing unity for T, the proportion of G to F is obtained.

COR. I. Hence if the centripetal force be as any power of the altitude, that power may be found from the motion of the apsides; and conversely. That is, if the whole angular motion, with which the body returns to the same apse, be to the angular motion of one revolution, or 360°, as any number as m to another as n, and the altitude be called A; the force will be as the power $A^{\frac{nn}{mm}-3}$ of the altitude A; the index of which power is $\dfrac{nn}{mm} - 3$. This appears by the second Example. Hence it is plain that the force in its recess from the centre cannot decrease in a greater than a cubed ratio of the altitude. A body revolving with such a force, and parting from the apse, if it once begins to descend, can never arrive at the lower apse or least altitude, but will descend to the centre, describing the curved line treated of in Cor. III, Prop. XLI. But if it should, at its parting from the lower apse, begin to ascend ever so little, it will ascend *in infinitum,* and never come to the upper apse; but will describe the curved line spoken of in the same Cor., and Cor. VI, Prop. XLIV. So that where the force in its recess from the centre decreases in a greater than a cubed ratio of the altitude, the body at its parting from the apse, will either descend to the centre, or ascend *in infinitum,* according as it descends or ascends at the beginning of its motion. But if the force in its recess from the centre either decreases in a less than a cubed ratio of the altitude, or increases in any ratio of the altitude whatsoever, the body will never descend to the centre, but will at some time arrive at the lower apse; and, on the contrary, if the body alternately ascending and descending from one apse to another never comes to the centre, then either the force increases in the recess from the centre, or it decreases in a less than a cubed ratio of the altitude; and the sooner the body returns from one apse to another, the farther is the ratio of the forces from the cubed ratio. As if the body should return to and from the upper apse by an alternate descent and ascent in 8 revolutions, or in 4, or 2, or $1\frac{1}{2}$; that is, if m should be to n as 8, or 4, or 2, or $1\frac{1}{2}$ to 1, and therefore $\frac{nn}{mm}-3$, be $\frac{1}{64}-3$, or $\frac{1}{16}-3$, or $\frac{1}{4}-3$, or $\frac{4}{9}-3$; then the force will be as $A^{\frac{1}{64}-3}$, or $A^{\frac{1}{16}-3}$, or $A^{\frac{1}{4}-3}$, or $A^{\frac{4}{9}-3}$; that is, it will be inversely as $A^{3-\frac{1}{64}}$, or $A^{3-\frac{1}{16}}$, or $A^{3-\frac{1}{4}}$, or $A^{3-\frac{4}{9}}$. If the body after each revolution returns to the same apse, and the apse remains unmoved, then m will be to n as 1 to 1, and therefore $A^{\frac{nn}{mm}-3}$ will be equal to A^{-2}, or $\frac{1}{AA}$; and therefore the decrease of the forces will be in a squared ratio of the altitude; as was demonstrated above. If the body in three fourth parts, or two thirds, or one third, or one fourth part of an entire revolution, return to the same apse; m will be to n as $\frac{3}{4}$ or $\frac{2}{3}$ or $\frac{1}{3}$ or $\frac{1}{4}$ to 1, and therefore $A^{\frac{nn}{mm}-3}$ is equal to $A^{\frac{16}{9}-3}$, or $A^{\frac{9}{4}-3}$, or A^{9-3}, or A^{16-3}; and therefore the force is either inversely as $A^{1\frac{1}{9}}$ or $A^{\frac{3}{4}}$, or directly as A^6 or A^{13}. Lastly if the body in its progress from the upper apse to the same upper apse again, goes over one entire revolution and three degrees more, and therefore that apse in each revolution of the body moves forward three degrees, then m will be to n as $363°$ to $360°$, or as 121 to 120, and therefore $A^{\frac{nn}{mm}-3}$ will be equal to $A^{-\frac{29523}{14641}}$, and therefore the centripetal force will be inversely as $A^{\frac{29523}{14641}}$, or inversely as $A^{2\frac{4}{243}}$ very nearly. Therefore the centripetal force decreases in a ratio something greater than the squared ratio; but approaching $59\frac{3}{4}$ times nearer to the squared than the cubed.

COR. II. Hence also if a body, urged by a centripetal force which is inversely as the square of the altitude, revolves in an ellipse whose focus is in the centre of the forces; and a new and foreign force should be added to or subtracted from this centripetal force, the motion of the apsides arising from that foreign force may (by the third Example) be known; and conversely: If the force with which the body revolves in the ellipse be as $\frac{1}{AA}$;

and the foreign force as cA, and therefore the remaining force as $\frac{A-cA^4}{A^3}$;

then (by the third Example) b will be equal to 1, m equal to 1, and n equal to 4; and therefore the angle of revolution between the apsides is equal to $180° \sqrt{\frac{1-c}{1-4c}}$. Suppose that foreign force to be 357.45 times less than the other force with which the body revolves in the ellipse; that is, c to be $\frac{100}{35745}$, A or T being equal to 1; and then $180° \sqrt{\frac{1-c}{1-4c}}$ will be $180° \sqrt{\frac{35645}{35345}}$ or $180°.7623$, that is, $180°$ $45'$ $44''$. Therefore the body, parting from the upper apse, will arrive at the lower apse with an angular motion of $180°$ $45'$ $44''$, and this angular motion being repeated, will return to the upper apse; and therefore the upper apse in each revolution will go forward $1°$ $31'$ $28''$. The apse of the moon is about twice as swift.

So much for the motion of bodies in orbits whose planes pass through the centre of force. It now remains to determine those motions in eccentric planes. For those authors who treat of the motion of heavy bodies used to consider the ascent and descent of such bodies, not only in a perpendicular direction, but at all degrees of obliquity upon any given planes; and for the same reason we are to consider in this place the motions of bodies tending to centres by means of any forces whatsoever, when those bodies move in eccentric planes. These planes are supposed to be perfectly smooth and polished, so as not to retard the motion of the bodies in the least. Moreover, in these demonstrations, instead of the planes upon which those bodies roll or slide, and which are therefore tangent planes to the bodies, I shall use planes parallel to them, in which the centres of the bodies move, and by that motion describe orbits. And by the same method I afterwards determine the motions of bodies performed in curved surfaces.

SECTION X

*The motion of bodies in given surfaces; and the oscillating pendulous
motion of bodies.*

PROPOSITION XLVI. PROBLEM XXXII

*Any kind of centripetal force being supposed, and the centre of force, and
any plane whatsoever in which the body revolves, being given, and the
quadratures of curvilinear figures being allowed; it is required to determine the motion of a body going off from a given place with a given
velocity, in the direction of a given right line in that plane.*

Let S be the centre of force, SC the least distance of that centre from the
given plane, P a body issuing from the place P in the direction of the right
line PZ, Q the same body revolving in its curve, and PQR the curve itself

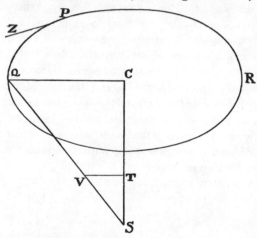

which is required to be
found, described in that
given plane. Join CQ,
QS, and if in QS we
take SV proportional to
the centripetal force
with which the body is
attracted towards the
centre S, and draw VT
parallel to CQ, and
meeting SC in T; then
will the force SV be re-
solved into two (by Cor.
II of the Laws of Mo-
tion), the force ST, and
the force TV; of which ST attracting the body in the direction of a line
perpendicular to that plane, does not at all change its motion in that
plane. But the action of the other force TV, coinciding with the position
of the plane itself, attracts the body directly towards the given point C in
that plane; and therefore causes the body to move in the plane in the same

manner as if the force ST were taken away, and the body were to revolve in free space about the centre C by means of the force TV alone. But there being given the centripetal force TV with which the body Q revolves in free space about the given centre C, there is given (by Prop. xlii) the curve PQR which the body describes; the place Q, in which the body will be found at any given time; and, lastly, the velocity of the body in that place Q. And conversely. Q.E.I.

PROPOSITION XLVII. THEOREM XV

Supposing the centripetal force to be proportional to the distance of the body from the centre; all bodies revolving in any planes whatsoever will describe ellipses, and complete their revolutions in equal times; and those which move in right lines, running backwards and forwards alternately, will complete their several periods of going and returning in the same times.

For letting all things stand as in the foregoing Proposition, the force SV, with which the body Q revolving in any plane PQR is attracted towards the centre S, is as the distance SQ; and therefore because SV and SQ, TV and CQ are proportional, the force TV with which the body is attracted towards the given point C in the plane of the orbit, is as the distance CQ. Therefore the forces with which bodies found in the plane PQR are attracted towards the point C, are in proportion to the distances equal to the forces with which the same bodies are attracted every way towards the centre S; and therefore the bodies will move in the same times, and in the same figures, in any plane PQR about the point C, as they would do in free spaces about the centre S; and therefore (by Cor. ii, Prop. x, and Cor. ii, Prop. xxxviii) they will in equal times either describe ellipses in that plane about the centre C, or move to and fro in right lines passing through the centre C in that plane; completing the same periods of time in all cases. Q.E.D.

SCHOLIUM

The ascent and descent of bodies in curved surfaces has a near relation to these motions we have been speaking of. Imagine curved lines to be described on any plane, and to revolve about any given axes passing through the centre of force, and by that revolution to describe curved surfaces; and that the bodies move in such sort that their centres may be always found in those surfaces. If those bodies oscillate to and fro with an oblique ascent and descent, their motions will be performed in planes passing through the axis, and therefore in the curved lines, by whose revolution those curved surfaces were generated. In those cases, therefore, it will be sufficient to consider the motion in those curved lines.

PROPOSITION XLVIII. THEOREM XVI

If a wheel stands upon the outside of a globe at right angles thereto, and revolving about its own axis goes forwards in a great circle, the length of the curvilinear path which any point, given in the perimeter of the wheel, hath described since the time that it touched the globe (which curvilinear path we may call the cycloid or epicycloid), will be to double the versed sine of half the arc which since that time hath touched the globe in passing over it, as the sum of the diameters of the globe and the wheel to the semidiameter of the globe.

PROPOSITION XLIX. THEOREM XVII

If a wheel stands upon the inside of a concave globe at right angles thereto, and revolving about its own axis goes forwards in one of the great circles of the globe, the length of the curvilinear path which any point, given in the perimeter of the wheel, hath described since it touched the globe, will be to the double of the versed sine of half the arc which in all that time hath touched the globe in passing over it, as the difference of the diameters of the globe and the wheel to the semidiameter of the globe.

Let ABL be the globe, C its centre, BPV the wheel resting on it, E the centre of the wheel, B the point of contact, and P the given point in the perimeter of the wheel. Imagine this wheel to proceed in the great circle ABL from A through B towards L, and in its progress to revolve in such a manner that the arcs AB, PB may be always equal one to the other, and the given point P in the perimeter of the wheel may describe in the meantime the curvilinear path AP. Let AP be the whole curvilinear path described since the wheel touched the globe in A, and the length of this path AP will be to twice the versed sine of the arc ½PB as 2CE to CB. For let the right line CE (produced if need be) meet the wheel in V, and join CP, BP, EP, VP; produce CP, and let fall thereon the perpendicular VF. Let PH, VH, meeting in H, touch the circle in P and V, and let PH cut VF in G, and to VP let fall the perpendiculars GI, HK. From the centre C with any radius

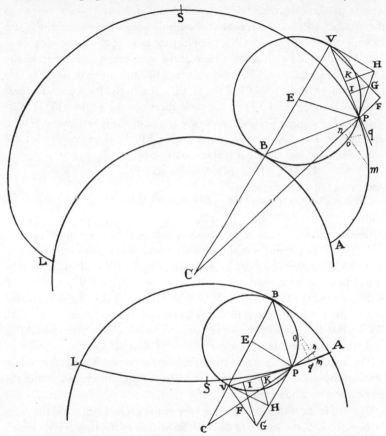

let there be described the circle *nom*, cutting the right line CP in *n*, the perimeter of the wheel BP in *o*, and the curvilinear path AP in *m*; and from the centre V with the radius V*o* let there be described a circle cutting VP produced in *q*.

Because the wheel in its progress always revolves about the point of contact B, it is manifest that the right line BP is perpendicular to that curved line AP which the point P of the wheel describes, and therefore that the right line VP will touch this curve in the point P. Let the radius of the circle *nom* be gradually increased or diminished so that at last it becomes equal to the distance CP; and by reason of the similitude of the evanescent figure P*nomq*, and the figure PFGVI, the ultimate ratio of the evanescent short lines P*m*, P*n*, P*o*, P*q*, that is, the ratio of the momentary increments of the curve AP, the right line CP, the circular arc BP, and the right line VP, will be the same as of the lines PV, PF, PG, PI, respectively. But since VF is perpendicular to CF, and VH to CV, and therefore the angles HVG, VCF equal; and the angle VHG (because the angles of the quadrilateral HVEP are right in V and P) is equal to the angle CEP, the triangles VHG, CEP will be similar; and thence it will come to pass that

$$EP : CE = HG : HV \text{ or } HP = KI : PK,$$

and by addition or subtraction,

$$CB : CE = PI : PK,$$

and $\qquad CB : 2CE = PI : PV = Pq : Pm.$

Therefore the decrement of the line VP, that is, the increment of the line BV − VP to the increment of the curved line AP is in a given ratio of CB to 2CE, and therefore (by Cor., Lem. iv) the lengths BV − VP and AP, generated by those increments, are in the same ratio. But if BV be radius, VP is the cosine of the angle BVP or ½BEP, and therefore BV − VP is the versed sine of the same angle, and therefore in this wheel, whose radius is ½BV, BV − VP will be double the versed sine of the arc ½BP. Therefore AP is to double the versed sine of the arc ½BP as 2CE to CB. Q.E.D.

The line AP in the former of these Propositions we shall name the cycloid without the globe, the other in the latter Proposition the cycloid within the globe, for distinction's sake.

Cor. i. Hence if there be described the entire cycloid ASL, and the same be bisected in S, the length of the part PS will be to the length PV (which is the double of the sine of the angle VBP, when EB is radius) as 2CE to CB, and therefore in a given ratio.

Cor. ii. And the length of the semidiameter of the cycloid AS will be equal to a right line which is to the diameter of the wheel BV as 2CE to CB.

PROPOSITION L. PROBLEM XXXIII

To cause a pendulous body to oscillate in a given cycloid.

Let there be given within the globe QVS described with the centre C, the cycloid QRS, bisected in R, and meeting the surface of the globe with its extreme points Q and S on either hand. Let there be drawn CR bisecting

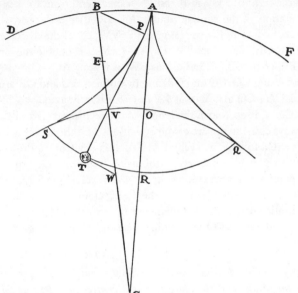

the arc QS in O, and let it be produced to A in such sort that CA may be to CO as CO to CR. About the centre C, with the radius CA, let there be described an exterior globe DAF; and within this globe, by a wheel whose diameter is AO, let there be described two semicycloids AQ, AS, touching the interior globe in Q and S, and meeting the exterior globe in A. From that point A, with a thread APT in length equal to the line AR, let the body T be suspended and oscillated in such manner between the two semicycloids AQ, AS, that, as often as the pendulum parts from the perpendicular AR, the upper part of the thread AP may be applied to that semicycloid APS towards which the motion tends, and fold itself round that curved line, as if it were some solid obstacle, the remaining part of the same thread PT which has not yet touched the semicycloid continuing straight. Then will the weight T oscillate in the given cycloid QRS. Q.E.F.

For let the thread PT meet the cycloid QRS in T, and the circle QOS in V, and let CV be drawn; and to the rectilinear part of the thread PT from the extreme points P and T let there be erected the perpendiculars BP, TW, meeting the right line CV in B and W. It is evident, from the construction and generation of the similar figures AS, SR, that those perpendiculars PB, TW, cut off from CV the lengths VB, VW, equal the diameters of the wheels OA, OR. Therefore TP is to VP (which is double the sine of the angle VBP when ½BV is radius) as BW to BV, or AO + OR to AO, that is (since CA and CO, CO and CR, and by division AO and OR are proportional), as CA + CO to CA, or, if BV be bisected in E, as 2CE to CB. Therefore (by Cor. I, Prop. XLIX), the length of the rectilinear part of the thread PT is always equal to the arc of the cycloid PS, and the whole thread APT is always equal to half the cycloid APS, that is (by Cor. II, Prop. XLIX), to the length AR. And conversely, if the string is always equal to the length AR, the point T will always move in the given cycloid QRS. Q.E.D.

Cor. The string AR is equal to the semicycloid AS, and therefore has the same ratio to AC, the semidiameter of the exterior globe, as the like semicycloid SR has to CO, the semidiameter of the interior globe.

PROPOSITION LI. THEOREM XVIII

If a centripetal force tending on all sides to the centre C of a globe, be in all places as the distance of the place from the centre; and, by this force alone acting upon it, the body T oscillate (in the manner above described) in the perimeter of the cycloid QRS: I say, that all the oscillations, howsoever unequal in themselves, will be performed in equal times.

For upon the tangent TW indefinitely produced let fall the perpendicular CX, and join CT. Because the centripetal force with which the body T is impelled towards C is as the distance CT, let this (by Cor. II of the Laws) be resolved into the parts CX, TX, of which CX impelling the body directly from P stretches the thread PT, and by the resistance the thread makes to it is totally employed, producing no other effect; but the other part TX, impelling the body transversely or towards X, directly accelerates the motion in the cycloid. Then it is plain that the acceleration of the body, propor-

tional to this accelerating force, will be every moment as the length TX,
that is (because CV, WV, and TX, TW proportional to them are given), as the length TW, that is (by Cor. I, Prop. XLIX), as the length of the arc of the cycloid TR. If therefore two pendulums APT, A*pt,* be unequally drawn aside from the perpendicular AR, and let fall together, their accelerations will be always as the arcs to be described TR, *t*R. But the parts described at the beginning of the motion are as the accelerations, that is, as the whole spaces that are to be described at the beginning, and therefore the parts which remain to be described, and the subsequent accelerations proportional to those parts, are also as the whole, and so on. Therefore the accelerations, and consequently the velocities generated, and the parts described with those velocities, and the parts to be described, are always as the whole; and therefore the parts to be described preserving a given ratio to each other will vanish together, that is, the two bodies oscillating will arrive together at the perpendicular AR. And since on the other hand the ascent of the pendulums from the lowest place R through the same cycloidal arcs with a retrograde motion, is retarded in the several places they pass through by the same forces by which their descent was accelerated, it is plain that the velocities of their ascent and descent through the same arcs are equal, and consequently performed in equal times; and, therefore, since the two parts of the cycloid RS and RQ lying on either side of the perpendicular are similar and equal, the two pendulums will perform as well the whole as the half of their oscillations in the same times. Q.E.D.

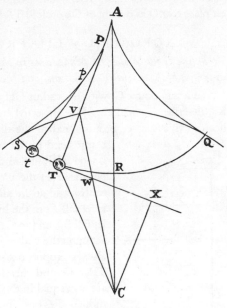

Cor. The force with which the body T is accelerated or retarded in any place T of the cycloid, is to the whole weight of the same body in the highest place S or Q as the arc of the cycloid TR is to the arc SR or QR.

PROPOSITION LII. PROBLEM XXXIV

To define the velocities of pendulums in the several places, and the times in which both the entire oscillations and their several parts are performed.

About any centre G, with the radius GH equal to the arc of the cycloid RS, describe a semicircle HKM bisected by the semidiameter GK. And if a centripetal force proportional to the distance of the places from the

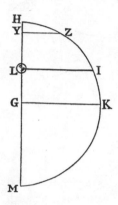

centre tend to the centre G, and it be in the perimeter HIK equal to the centripetal force in the perimeter of the globe QOS tending towards its centre, and at the same time that the pendulum T is let fall from the highest place S, a body, as L, is let fall from H to G; then because the forces which act upon the bodies are equal at the beginning, and always proportional to the spaces to be described TR, LG, and therefore if TR and LG are equal, are also equal in the places T and L, it is plain that those bodies describe at the beginning equal spaces ST, HL, and therefore are still acted upon equally, and continue to describe equal spaces. Therefore by Prop. xxxviii, the time in which the body describes the arc ST is to the time of one oscillation, as the arc HI the time in which the body H arrives at L, to the semiperiphery HKM, the time in which the body H will come to M. And the velocity of the pendulous body in the place T is to its velocity in the lowest place R, that is, the velocity of the body H in the place L to its velocity in the place G, or the momentary increment of the line HL to the momentary increment of the line HG (the arcs HI, HK increasing with an uniform velocity) as the ordinate LI to the radius GK, or as $\sqrt{(SR^2 - TR^2)}$ to SR. Hence, since in unequal oscillations there are described in equal times arcs proportional to the entire arcs of the oscillations, there are obtained, from the times given, both the velocities and the arcs described in all the oscillations universally. Which was first required.

Let now any pendulous bodies oscillate in different cycloids described within different globes, whose absolute forces are also different; and if the absolute force of any globe QOS be called V, the accelerative force with which the pendulum is acted on in the circumference of this globe, when it begins to move directly towards its centre, will be as the distance of the pendulous body from that centre and the absolute force of the globe conjointly, that is, as CO · V. Therefore the short line HY, which is as this accelerated force CO · V, will be described in a given time; and if there be erected the perpendicular YZ meeting the circumference in Z, the nascent arc HZ will denote that given time. But that nascent arc HZ varies as the square root of the rectangle GH · HY,

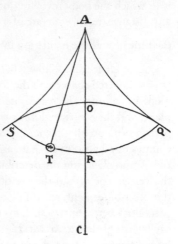

and therefore as $\sqrt{(GH \cdot CO \cdot V)}$. Whence the time of an entire oscillation in the cycloid QRS (it being as the semiperiphery HKM, which denotes that entire oscillation, directly; and as the arc HZ, which in like manner denotes a given time, inversely) will be as GH directly and $\sqrt{(GH \cdot CO \cdot V)}$ inversely; that is, because GH and SR are equal, as $\sqrt{\dfrac{SR}{CO \cdot V}}$, or (by Cor.,

Prop. L), as $\sqrt{\dfrac{AR}{AC \cdot V}}$. Therefore the oscillations in all globes and cycloids, performed with any absolute forces whatever, vary directly as the square root of the length of the string, and inversely as the square root of the distance between the point of suspension and the centre of the globe, and also inversely as the square root of the absolute force of the globe. Q.E.I.

Cor. I. Hence also the times of oscillating, falling, and revolving bodies may be compared among themselves. For if the diameter of the wheel with which the cycloid is described within the globe is supposed equal to the semidiameter of the globe, the cycloid will become a right line passing through the centre of the globe, and the oscillation will be changed into a descent and subsequent ascent in that right line. Hence there is given both the time of the descent from any place to the centre, and the time equal to it in which the body revolving uniformly about the centre of the globe at any distance describes an arc of a quadrant. For this time (by Case 2) is to the time of half the oscillation in any cycloid QRS as 1 to $\sqrt{\dfrac{AR}{AC}}$.

Cor. II. Hence also follow what Sir *Christopher Wren* and Mr. *Huygens* have discovered concerning the common cycloid. For if the diameter of the globe be infinitely increased, its spherical surface will be changed into a plane, and the centripetal force will act uniformly in the direction of lines perpendicular to that plane, and our cycloid will become the same with the common cycloid. But in that case the length of the arc of the cycloid between that plane and the describing point will become equal to four times the versed sine of half the arc of the wheel between the same plane and the describing point, as was discovered by Sir *Christopher Wren*. And a pendulum between two such cycloids will oscillate in a similar and equal cycloid in equal times, as Mr. *Huygens* demonstrated. The descent of heavy bodies also in the time of one oscillation will be the same as Mr. *Huygens* exhibited.

The Propositions here demonstrated are adapted to the true constitution of the earth, so far as wheels moving in any of its great circles will describe, by the motions of nails fixed in their perimeters, cycloids without the globe; and pendulums, in mines and deep caverns of the earth, must oscillate in cycloids within the globe, that those oscillations may be performed in equal times. For gravity (as will be shown in the third Book) decreases in its progress from the surface of the earth; upwards as the square root of the distances from the centre of the earth; downwards as these distances.

PROPOSITION LIII. PROBLEM XXXV

Granting the quadratures of curvilinear figures, it is required to find the forces with which bodies moving in given curved lines may always perform their oscillations in equal times.

Let the body T oscillate in any curved line STRQ, whose axis is AR passing through the centre of force C. Draw TX touching that curve in any place of the body T, and in that tangent TX take TY equal to the arc TR.

The length of that arc is known from the common methods used for the quadratures of figures. From the point Y draw the right line YZ perpendicular to the tangent. Draw CT meeting YZ in Z, and the centripetal force will be proportional to the right line TZ. Q.E.I.

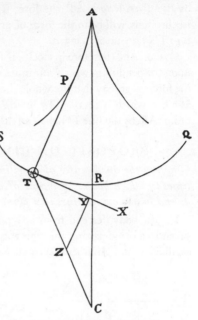

For if the force with which the body is attracted from T towards C be expressed by the right line TZ taken proportional to it, that force will be resolved into two forces TY, YZ, of which YZ, drawing the body in the direction of the length of the thread PT, does not at all change its motion; whereas the other force TY directly accelerates or retards its motion in the curve STRQ. Therefore since that force is as the space to be described TR, the accelerations or retardations of the body in describing two proportional parts (a greater and a less) of two oscillations, will be always as those parts, and therefore will cause those parts to be described together. But bodies which continually describe in the same time parts proportional to the whole, will describe the whole in the same time. Q.E.D.

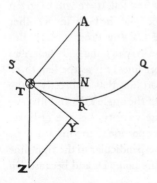

Cor. 1. Hence if the body T, hanging by a rectilinear thread AT from the centre A, describe the circular arc STRQ, and in the meantime be acted on by any force tending downwards with parallel directions, which is to the uniform force of gravity as the arc TR to its sine TN, the times of the several oscillations will be equal. For because TZ,

AR are parallel, the triangles ATN, ZTY are similar; and therefore TZ will be to AT as TY to TN; that is, if the uniform force of gravity be expressed by the given length AT, the force TZ, by which the oscillations become isochronous, will be to the force of gravity AT, as the arc TR equal to TY is to TN the sine of that arc.

Cor. ii. And therefore in clocks, if forces are impressed by some machine upon the pendulum which continues the motion, and so compounded with the force of gravity that the whole force tending downwards will be always as a line which is obtained by dividing the product of the arc TR and the radius AR, by the sine TN, then all the oscillations will become isochronous.

PROPOSITION LIV.   PROBLEM XXXVI

Granting the quadratures of curvilinear figures, it is required to find the times in which bodies by means of any centripetal force will descend or ascend in any curved lines in a plane passing through the centre of force.

Let the body descend from any place S, and move in any curve ST*t*R given in a plane passing through the centre of force C. Join CS, and let it be divided into innumerable equal parts, and let D*d* be one of those parts.

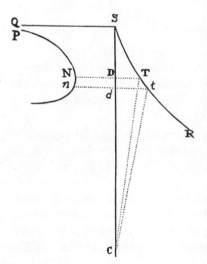

From the centre C, with the radii CD, C*d,* let the circles DT, *dt* be described, meeting the curved line ST*t*R in T and *t*. And because the law of centripetal force is given, and also the altitude CS from which the body at first fell, there will be given the velocity of the body in any other altitude CT (by Prop. xxxix). But the time in which the body describes the short line T*t* is as the length of that short line, that is, directly as the secant of the angle *t*TC and inversely as the velocity. Let the ordinate DN, proportional to this time, be made perpendicular to the right line CS at the point D, and because D*d*

is given, the rectangle Dd · DN, that is, the area DNnd, will be proportional to the same time. Therefore if PNn be a curved line which the point N continually touches, and its asymptote be the right line SQ standing upon the line CS at right angles, the area SQPND will be proportional to the time in which the body in its descent hath described the line ST; and therefore that area being found, the time is also given. Q.E.I.

PROPOSITION LV. THEOREM XIX

If a body move in any curved surface, whose axis passes through the centre of force, and from the body a perpendicular be let fall upon the axis; and a line parallel and equal thereto be drawn from any given point of the axis: I say, that this parallel line will describe an area proportional to the time.

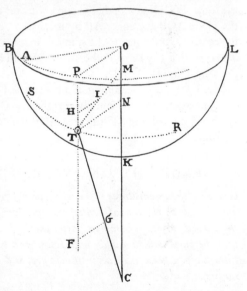

Let BKL be a curved surface, T a body revolving in it, STR a curve which the body describes in the same, S the beginning of the curve, OMK the axis of the curved surface, TN a right line let fall perpendicularly from the body to the axis; OP a line parallel and equal thereto drawn from the given point O in the axis; AP the path described by the point P in the plane AOP in which the revolving line OP is found; A the beginning of that path answering to the point S; TC a right line drawn from the body to the centre; TG a part thereof proportional to the centripetal force with which the body tends towards the centre C; TM a right line perpendicular to the curved surface; TI a part thereof proportional to the force of pressure with which the body urges the surface, and therefore with which it is again repelled by the surface towards M; PTF a right line parallel to the axis and passing through the body, and GF, IH right lines let fall perpendicularly from the points G and I upon that parallel PHTF. I say, now, that the area AOP, described by the radius OP from the beginning of the motion, is proportional to the time. For the force TG (by Cor. II of the Laws of Motion) is resolved into the forces TF, FG; and the force TI into the forces TH, HI; but the forces TF, TH, acting in the direction of the line PF perpendicular to the plane AOP, introduce no change in the motion of the body but in a direction perpendicular to that plane. Therefore its motion, so far as it hath the same direction with the position of the plane, that is, the motion of the point P, by which the projection AP of the curve is described in that plane, is the same as if the forces TF, TH were taken away, and the body were acted on by the forces FG, HI alone; that is, the same as if the body were to describe in the plane AOP the curve AP by means of a centripetal force tending to the centre O, and equal to the sum of the forces FG and HI. But with such a force as that (by Prop. 1) the area AOP will be described proportional to the time. Q.E.D.

Cor. By the same reasoning, if a body, acted on by forces tending to two or more centres in the same given right line CO, should describe in a free space any curved line ST, the area AOP would be always proportional to the time.

PROPOSITION LVI. PROBLEM XXXVII

Granting the quadratures of curvilinear figures, and supposing that there are given both the law of centripetal force tending to a given centre, and the curved surface whose axis passes through that centre; it is required to find the curve which a body will describe in that surface, when going off from a given place with a given velocity, and in a given direction in that surface.

The last construction remaining, let the body T go from the given place S, in the direction of a line given by position, and turn into the curve sought STR, whose orthographic projection in the plane BDO is AP. And from the given velocity of the body in the altitude SC, its velocity in any other altitude TC will be also given. With that velocity, in a given moment of time, let the body describe the segment T*t* of its curve and let P*p* be the projection of that segment described in the plane AOP. Join O*p*, and a little circle being described upon the curved surface about the centre T with the radius T*t*, let the projection of that little circle in the plane AOP be the ellipse *p*Q. And because the magnitude of that little circle T*t*, and TN or PO its distance from the axis CO is also given, the ellipse *p*Q will be given both in kind and magnitude, as also its position to the right line PO. And since the area PO*p* is proportional to the time, and

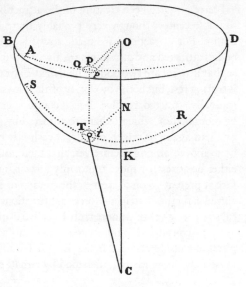

therefore given because the time is given, the angle PO*p* will be given. And thence will be given *p* the common intersection of the ellipse and the right line O*p*, together with the angle OP*p*, in which the projection AP*p* of the curve cuts the line OP. But from thence (by comparing Prop. xLi, with its Cor. II) the manner of determining the curve AP*p* easily appears. Then from the several points P of that projection erecting to the plane AOP, the perpendiculars PT meeting the curved surface in T, there will be given the several points T of the curve. Q.E.I.

SECTION XI

The motions of bodies tending to each other with centripetal forces.

I have hitherto been treating of the attractions of bodies towards an immovable centre; though very probably there is no such thing existent in nature. For attractions are made towards bodies, and the actions of the bodies attracted and attracting are always reciprocal and equal, by Law III; so that if there are two bodies, neither the attracted nor the attracting body is truly at rest, but both (by Cor. IV of the Laws of Motion), being as it were mutually attracted, revolve about a common centre of gravity. And if there be more bodies, which either are attracted by one body, which is attracted by them again, or which all attract each other mutually, these bodies will be so moved among themselves, that their common centre of gravity will either be at rest, or move uniformly forwards in a right line. I shall therefore at present go on to treat of the motion of bodies attracting each other; considering the centripetal forces as attractions; though perhaps in a physical strictness they may more truly be called impulses. But these Propositions are to be considered as purely mathematical; and therefore, laying aside all physical considerations, I make use of a familiar way of speaking, to make myself the more easily understood by a mathematical reader.

PROPOSITION LVII. THEOREM XX

Two bodies attracting each other mutually describe similar figures about their common centre of gravity, and about each other mutually.

For the distances of the bodies from their common centre of gravity are inversely as the bodies; and therefore in a given ratio to each other; and thence, by composition of ratios, in a given ratio to the whole distance between the bodies. Now these distances are carried round their common extremity with an uniform angular motion, because lying in the same right line they never change their inclination to each other. But right lines that are in a given ratio to each other, and are carried round their extremities with an uniform angular motion, describe upon planes, which either rest together with them, or are moved with any motion not angular, figures entirely similar round those extremities. Therefore the figures described by the revolution of these distances are similar. Q.E.D.

PROPOSITION LVIII. THEOREM XXI

If two bodies attract each other with forces of any kind, and revolve about the common centre of gravity: I say, that, by the same forces, there may be described round either body unmoved a figure similar and equal to the figures which the bodies so moving describe round each other.

Let the bodies S and P revolve about their common centre of gravity C, proceeding from S to T, and from P to Q. From the given point *s* let there be continually drawn *sp, sq,* equal and parallel to SP, TQ; and the curve *pqv,* which the point *p* describes in its revolution round the fixed point *s,*

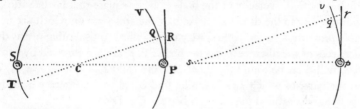

will be similar and equal to the curves which the bodies S and P describe about each other; and therefore, by Theor. xx, similar to the curves ST and PQV which the same bodies describe about their common centre of gravity C; and that because the proportions of the lines SC, CP, and SP or *sp,* to each other, are given.

CASE 1. The common centre of gravity C (by Cor. IV of the Laws of Motion) is either at rest, or moves uniformly in a right line. Let us first suppose it at rest, and in *s* and *p* let there be placed two bodies, one immovable in *s,* the other movable in *p,* similar and equal to the bodies S and P. Then let the right lines PR and *pr* touch the curves PQ and *pq* in P and *p,* and produc CQ and *sq* to R and *r.* And because the figures CPRQ, *sprq* are similar, RQ will be to *rq* as CP to *sp,* and therefore in a given ratio. Hence if the force with which the body P is attracted towards the body S, and by consequence towards the intermediate centre C, were to the force with which the body *p* is attracted towards the centre *s,* in the same given ratio, these forces would in equal times attract the bodies from the tangents PR, *pr* to the arcs PQ, *pq,* through the intervals proportional to them RQ, *rq;* and therefore this last force (tending to *s*) would make the body *p* revolve in the curve *pqv,* which would become similar to the curve PQV, in which the first force obliges the body P to revolve; and their revolutions would be completed in the same times. But because those forces are not to each other in the ratio of CP to *sp,* but (by reason of the similarity and equality of the bodies S and *s,* P and *p,* and the equality of the distances SP, *sp*) mutually equal, the bodies in equal times will be equally drawn from the tangents; and therefore that the body *p* may be attracted through the greater interval *rq,* there is required a greater time, which will vary as the square root of the intervals; because, by Lem. x, the spaces described at the beginning of the motion are as the square of the times. Suppose, then, the velocity of the body *p* to be to the velocity of the body P as the square root of the ratio of the distance *sp* to the distance CP, so that the arcs *pq,* PQ, which are in a simple proportion to each other, may be described in times that are as the square root of the distances; and the bodies P, *p,* always attracted by equal forces, will describe round the fixed centres C and *s* similar figures PQV, *pqv,* the latter of which *pqv* is similar and equal to the figure which the body P describes round the movable body S. Q.E.D.

CASE 2. Suppose now that the common centre of gravity, together with the space in which the bodies are moved among themselves, proceeds uniformly in a right line; and (by Cor. vi of the Laws of Motion) all the motions in this space will be performed in the same manner as before; and therefore the bodies will describe about each other the same figures as before, which will be therefore similar and equal to the figure *pqv.* Q.E.D.

COR. I. Hence two bodies attracting each other with forces proportional to their distance, describe (by Prop. x), both round their common centre of gravity, and round each other, concentric ellipses; and, conversely, if such figures are described, the forces are proportional to the distances.

COR. II. And two bodies, whose forces are inversely proportional to the square of their distance, describe (by Prop. xi, xii, xiii), both round their common centre of gravity, and round each other, conic sections having their focus in the centre about which the figures are described. And, conversely, if such figures are described, the centripetal forces are inversely proportional to the square of the distance.

COR. III. Any two bodies revolving round their common centre of gravity describe areas proportional to the times, by radii drawn both to that centre and to each other.

PROPOSITION LIX. THEOREM XXII

The periodic time of two bodies S *and* P *revolving round their common centre of gravity* C, *is to the periodic time of one of the bodies* P *revolving round the other* S *remaining fixed, and describing a figure similar and equal to those which the bodies describe about each other, as* $\sqrt{S}$ *is to* $\sqrt{(S+P)}$.

For, by the demonstration of the last Proposition, the times in which any similar arcs PQ and *pq* are described are as $\sqrt{CP}$ is to $\sqrt{SP}$, or $\sqrt{sp}$, that is, as $\sqrt{S}$ is to $\sqrt{(S+P)}$. And by composition of ratios, the sums of the times in which all the similar arcs PQ and *pq* are described, that is, the whole times in which the whole similar figures are described, are in the same ratio, $\sqrt{S}$ to $\sqrt{(S+P)}$. Q.E.D.

PROPOSITION LX. THEOREM XXIII

If two bodies S *and* P, *attracting each other with forces inversely proportional to the square of their distance, revolve about their common centre of gravity: I say, that the principal axis of the ellipse which either of the bodies, as* P, *describes by this motion about the other* S, *will be to the principal axis of the ellipse, which the same body* P *may describe in the same periodic time about the other body* S *fixed, as the sum of the two bodies* S+P *to the first of two mean proportionals[1] between that sum and the other body* S.

For if the ellipses described were equal to each other, their periodic times by the last Theorem would be as the square root of the ratio of the body S to the sum of the bodies S+P. Let the periodic time in the latter ellipse be diminished in that ratio, and the periodic times will become equal; but, by Prop. xv, the principal axis of the ellipse will be diminished in a ratio which is the 3/2th power of the former ratio; that is, in a ratio to which the ratio of S to S+P is the cube, and therefore that axis will be to the principal axis of the other ellipse as the first of two mean proportionals between S+P and S to S+P. And inversely the principal axis of the ellipse described about the movable body will be to the principal axis of that described round the immovable as S+P to the first of two mean proportionals between S+P and S. Q.E.D.

[1 Appendix, Note 24.]

PROPOSITION LXI. THEOREM XXIV

If two bodies attracting each other with any kind of forces, and not other-wise agitated or obstructed, are moved in any manner whatsoever, those motions will be the same as if they did not at all attract each other, but were both attracted with the same forces by a third body placed in their common centre of gravity; and the law of the attracting forces will be the same in respect of the distance of the bodies from the common centre, as in respect of the distance between the two bodies.

For those forces with which the bodies attract each other, by tending to the bodies, tend also to the common centre of gravity lying directly between them; and therefore are the same as if they proceeded from an intermediate body. Q.E.D.

And because there is given the ratio of the distance of either body from that common centre to the distance between the two bodies, there is given, of course, the ratio of any power of one distance to the same power of the other distance; and also the ratio of any quantity derived in any manner from one of the distances compounded in any manner with given quantities, to another quantity derived in like manner from the other distance, and as many given quantities having that given ratio of the distances to the first. Therefore if the force with which one body is attracted by another be directly or inversely as the distance of the bodies from each other, or as any power of that distance; or, lastly, as any quantity derived after any manner from that distance compounded with given quantities; then will the same force with which the same body is attracted to the common centre of grav-ity be in like manner directly or inversely as the distance of the attracted body from the common centre, or as any power of that distance; or, lastly, as a quantity derived in like sort from that distance compounded with anal-ogous given quantities. That is, the law of attracting force will be the same with respect to both distances. Q.E.D.

PROPOSITION LXII. PROBLEM XXXVIII

To determine the motions of two bodies which attract each other with forces inversely proportional to the squares of the distance between them, and are let fall from given places.

The bodies, by the last Theorem, will be moved in the same manner as if they were attracted by a third placed in the common centre of their gravity; and by the hypothesis that centre will be fixed at the beginning of their motion, and therefore (by Cor. IV of the Laws of Motion) will be always fixed. The motions of the bodies are therefore to be determined (by Prob. xxv) in the same manner as if they were impelled by forces tending to that centre; and then we shall have the motions of the bodies attracting each other. Q.E.I.

PROPOSITION LXIII. PROBLEM XXXIX

To determine the motions of two bodies attracting each other with forces inversely proportional to the squares of their distance, and going off from given places in given directions with given velocities.

The motions of the bodies at the beginning being given, there is given also the uniform motion of the common centre of gravity, and the motion of the space which moves along with this centre uniformly in a right line, and also the very first, or beginning motions of the bodies in respect of this space. Then (by Cor. V of the Laws, and the last Theorem) the subsequent motions will be performed in the same manner in that space, as if that space together with the common centre of gravity were at rest, and as if the bodies did not attract each other, but were attracted by a third body placed in that centre. The motion therefore in this movable space of each body going off from a given place, in a given direction, with a given velocity, and acted upon by a centripetal force tending to that centre, is to be determined by Prob. IX and XXVI, and at the same time will be obtained the motion of the other round the same centre. With this motion compound the uniform progressive motion of the entire system of the space and the bodies revolving in it, and there will be obtained the absolute motion of the bodies in immovable space. Q.E.I.

PROPOSITION LXIV. PROBLEM XL

Supposing forces with which bodies attract each other to increase in a simple ratio of their distances from the centres; it is required to find the motions of several bodies among themselves.

Suppose the first two bodies T and L to have their common centre of gravity in D. These, by Cor. I, Theor. XXI, will describe ellipses having their centres in D, the magnitudes of which ellipses are known by Prob. V.

Let now a third body S attract the two former T and L with the accelerative forces ST, SL, and let it be attracted again by them. The force ST (by Cor. II of the Laws of Motion) is resolved into the forces SD, DT; and the force SL into the forces SD and DL. Now the forces DT, DL, which are as

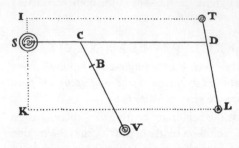

their sum TL, and therefore as the accelerative forces with which the bodies T and L attract each other, added to the forces of the bodies T and L, the first to the first, and the last to the last, compose forces proportional to the distances DT and DL as before, but only greater than those former forces; and therefore (by Cor. I, Prop. x, and Cor. I and VIII, Prop. IV) they will cause those bodies to describe ellipses as before, but with a swifter motion. The remaining accelerative forces SD and DL, by the motive forces SD · T and SD · L, which are as the bodies attracting those bodies equally and in the direction of the lines TI, LK parallel to DS, do not at all change their situations with respect to one another, but cause them equally to approach to the line IK; which must be imagined drawn through the middle of the body S, and perpendicular to the line DS. But that approach to the line IK will be hindered by causing the system of the bodies T and L on one side, and the body S on the other, with proper velocities, to revolve round the common centre of gravity C. With such a motion the body S, because the sum of the motive forces SD · T and SD · L is proportional to the distance CS, tends to the centre C, and will describe an ellipse round that centre; and the point D, because the lines CS and CD are proportional, will describe a like ellipse over against it. But the bodies T and L, attracted by the motive forces SD · T and SD · L, the first by the first, and the last by the last, equally and in the direction of the parallel lines TI and LK, as was said before, will (by Cor. v and vi of the Laws of Motion) continue to describe their ellipses round the movable centre D, as before. Q.E.I.

Let there be added a fourth body V, and, by the like reasoning, it will be demonstrated that this body and the point C will describe ellipses about the common centre of gravity B; the motions of the bodies T, L, and S round the centres D and C remaining the same as before, but accelerated. And by the same method one may add yet more bodies at pleasure. Q.E.I.

This would be the case, though the bodies T and L should attract each other with accelerative forces greater or less than those with which they attract the other bodies in proportion to their distance. Let all the accelerative attractions be to each other as the distances multiplied into the attracting bodies; and from what has gone before it will easily be concluded that all the bodies will describe different ellipses with equal periodic times about their common centre of gravity B, in an immovable plane. Q.E.I.

PROPOSITION LXV. THEOREM XXV

Bodies, whose forces decrease as the square of their distances from their centres, may move among themselves in ellipses; and by radii drawn to the foci may describe areas very nearly proportional to the times.

In the last Proposition we demonstrated that case in which the motions will be performed exactly in ellipses. The more distant the law of the forces is from the law in that case, the more will the bodies disturb each other's motions; neither is it possible that bodies attracting each other according to the law supposed in this Proposition should move exactly in ellipses, unless by keeping a certain proportion of distances from each other. However, in the following cases the orbits will not much differ from ellipses.

CASE 1. Imagine several lesser bodies to revolve about some very great one at different distances from it, and suppose absolute forces tending to every one of the bodies proportional to each. And because (by Cor. IV of the Laws) the common centre of gravity of them all is either at rest, or moves uniformly forwards in a right line, suppose the lesser bodies so small that the great body may be never at a sensible distance from that centre; and then the great body will, without any sensible error, be either at rest, or move uniformly forwards in a right line; and the lesser will revolve about that great one in ellipses, and by radii drawn thereto will describe areas proportional to the times; if we except the errors that may be introduced by the receding of the great body from the common centre of gravity, or by the actions of the lesser bodies upon each other. But the lesser bodies may be so far diminished, as that this recess and the actions of the bodies on each other may become less than any assignable; and therefore so as that the orbits may become ellipses, and the areas answer to the times, without any error that is not less than any assignable. Q.E.O.

CASE 2. Let us imagine a system of lesser bodies revolving about a very great one in the manner just described, or any other system of two bodies revolving about each other, to be moving uniformly forwards in a right line, and in the meantime to be impelled sideways by the force of another vastly greater body situate at a great distance. And because the equal accelerative forces with which the bodies are impelled in parallel directions do not change the situation of the bodies with respect to each other, but only oblige the whole system to change its place while the parts still retain their motions among themselves, it is manifest that no change in those motions of the attracted bodies can arise from their attractions towards the greater, unless by the inequality of the accelerative attractions, or by the inclinations of the lines towards each other, in whose directions the attractions are made. Suppose, therefore, all the accelerative attractions made towards the great body to be among themselves inversely as the squares of the distances; and then, by increasing the distance of the great body till the differences of the right lines drawn from that to the others in respect of their length, and the inclinations of those lines to each other, be less than any given, the motions of the parts of the system will continue without errors that are not less than any given. And because, by the small distance of those parts from each other, the whole system is attracted as if it were but one body, it will therefore be moved by this attraction as if it were one body; that is, its centre of gravity will describe about the great body one of the conic sections (that is, a parabola or hyperbola when the attraction is but languid and an ellipse when it is more vigorous); and by radii drawn thereto, it will describe areas proportional to the times, without any errors but those which arise from the distances of the parts, and these are by the supposition exceedingly small, and may be diminished at pleasure. Q.E.O.

By a like reasoning one may proceed to more complicated cases *in infinitum*.

COR. I. In the second Case, the nearer the very great body approaches to the system of two or more revolving bodies, the greater will the perturbation be of the motions of the parts of the system among themselves; because the inclinations of the lines drawn from that great body to those parts become greater; and the inequality of the proportion is also greater.

COR. II. But the perturbation will be greatest of all, if we suppose the accelerative attractions of the parts of the system towards the greatest body of all are not to each other inversely as the squares of the distances from that great body; especially if the inequality of this proportion be greater than the inequality of the proportion of the distances from the great body. For if the accelerative force, acting in parallel directions and equally, causes no perturbation in the motions of the parts of the system, it must of course, when it acts unequally, cause a perturbation somewhere, which will be greater or less as the inequality is greater or less. The excess of the greater impulses acting upon some bodies, and not acting upon others, must necessarily change their situation among themselves. And this perturbation, added to the perturbation arising from the inequality and inclination of the lines, makes the whole perturbation greater.

COR. III. Hence if the parts of this system move in ellipses or circles without any remarkable perturbation, it is manifest that, if they are at all impelled by accelerative forces tending to any other bodies, the impulse is very weak, or else is impressed very near equally and in parallel directions upon all of them.

PROPOSITION LXVI. THEOREM XXVI

If three bodies, whose forces decrease as the square of the distances, attract each other; and the accelerative attractions of any two towards the third be between themselves inversely as the squares of the distances; and the two least revolve about the greatest: I say, that the interior of the two revolving bodies will, by radii drawn to the innermost and greatest, describe round that body areas more proportional to the times, and a figure more approaching to that of an ellipse having its focus in the point of intersection of the radii, if that great body be agitated by those attractions, than it would do if that great body were not attracted at all by the lesser, but remained at rest; or than it would do if that great body were very much more or very much less attracted, or very much more or very much less agitated, by the attractions.

This appears plainly enough from the demonstration of the second Corollary of the foregoing Proposition; but it may be made out after this manner by a way of reasoning more distinct and more universally convincing.

CASE I. Let the lesser bodies P and S revolve in the same plane about the greatest body T, the body P describing the interior orbit PAB, and S the exterior orbit ESE. Let SK be the mean distance of the bodies P and S; and let the accelerative attraction of the body P towards S, at that mean distance, be expressed by that line SK. Make SL to SK as the square of SK to the square of SP, and SL will be the accelerative attraction of the body P towards S at any distance SP. Join PT, and draw LM parallel to it meet-

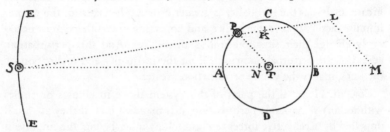

ing ST in M; and the attraction SL will be resolved (by Cor. II of the Laws of Motion) into the attractions SM, LM. And so the body P will be urged with a threefold accelerative force. One of these forces tends towards T, and arises from the mutual attraction of the bodies T and P. By this force alone the body P would describe round the body T, by the radius PT, areas proportional to the times, and an ellipse whose focus is in the centre of the body T; and this it would do whether the body T remained unmoved, or whether it were agitated by that attraction. This appears from Prop. XI, and Cor. II and III of Theor. XXI. The other force is that of the attraction LM, which, because it tends from P to T, will be superadded to and coincide with the former force; and cause the areas to be still proportional to the times, by Cor. III, Theor. XXI. But because it is not inversely proportional to the square of the distance PT, it will compose, when added to the former, a force varying from that proportion; this variation will be the greater by as much as the proportion of this force to the former is greater, other things remaining the same. Therefore, since by Prop. XI, and by Cor. II, Theor. XXI, the force with which the ellipse is described about the focus T ought to be directed to that focus, and to be inversely proportional to the square of the distance PT, that compounded force varying from that proportion will make the orbit PAB vary from the figure of an ellipse that has its focus in the point T; and so much the more by as much as the variation from that proportion is greater; and in consequence by as much as the proportion of the second force LM to the first force is greater, other things remaining the same. But now the third force SM, attracting the body P in a direction parallel to ST, composes with the other forces a new force which is no longer directed from P to T; and this varies so much more from this direction by as much as the proportion of the third force to the other forces is greater, other things remaining the same; and therefore causes the body P to describe, by the radius TP, areas no longer proportional to the times; and therefore makes the variation from that proportionality so much greater by as much as the proportion of this force to the others is greater. But this third force will increase the variation of the orbit PAB from the elliptical figure before mentioned upon two accounts: first, because that force is not directed from P to T; and, secondly, because it is not inversely proportional to the square of the distance PT. These things being premised, it is manifest that the areas are then most nearly proportional to the times, when that third force is the least possible, the rest preserving their former quantity; and that the orbit PAB does then approach nearest to the elliptical figure above mentioned, when both the second and third, but especially the third force, is the least possible; the first force remaining in its former quantity.

Let the accelerative attraction of the body T towards S be expressed by the line SN; then if the accelerative attractions SM and SN were equal, these, attracting the bodies T and P equally and in parallel directions, would not at all change their situation with respect to each other. The motions of the bodies between themselves would be the same in that case as if those attractions did not act at all, by Cor. vi of the Laws of Motion. And, by a like reasoning, if the attraction SN is less than the attraction SM, it will take away out of the attraction SM the part SN, so that there will remain only the part (of the attraction) MN to disturb the proportionality of the areas and times, and the elliptical figure of the orbit. And in like manner if the attraction SN be greater than the attraction SM, the perturbation of the orbit and proportion will be produced by the difference MN alone.

After this manner the attraction SN reduces always the attraction SM to the attraction MN, the first and second attractions remaining perfectly unchanged; and therefore the areas and times come then nearest to proportionality, and the orbit PAB to the above-mentioned elliptical figure, when the attraction MN is either none, or the least that is possible; that is, when the accelerative attractions of the bodies P and T approach as near as possible to equality; that is, when the attraction SN is neither none at all, nor less than the least of all the attractions SM, but is, as it were, a mean between the greatest and least of all those attractions SM, that is, not much greater nor much less than the attraction SK. Q.E.D.

CASE 2. Let now the lesser bodies P, S revolve about a greater T in different planes; and the force LM, acting in the direction of the line PT situated in the plane of the orbit PAB, will have the same effect as before; neither will it draw the body P from the plane of its orbit. But the other force NM, acting in the direction of a line parallel to ST (and therefore, when the body S is without the line of the nodes, inclined to the plane of the orbit PAB), besides the perturbation of the motion just now spoken of as to longitude, introduces another perturbation also as to latitude, attracting the body P out of the plane of its orbit. And this perturbation, in any given situation of the bodies P and T to each other, will be as the generating force MN; and therefore becomes least when the force MN is least, that is (as was just now shown), where the attraction SN is not much greater nor much less than the attraction SK. Q.E.D.

COR. I. Hence it may be easily inferred, that if several less bodies P, S, R, &c., revolve about a very great body T, the motion of the innermost revolving body P will be least disturbed by the attractions of the others, when the great body is as well attracted and agitated by the rest (according to the ratio of the accelerative forces) as the rest are by each other.

COR. II. In a system of three bodies T, P, S, if the accelerative attractions of any two of them towards a third be to each other inversely as the squares of the distances, the body P, by the radius PT, will describe its area about the body T swifter near the conjunction A and the opposition B than it will near the quadratures C and D. For every force with which the body P is acted on and the body T is not, and which does not act in the direction of the line PT, does either accelerate or retard the description of the area, according as its direction is the same as, or contrary to that of the motion of the body. Such is the force NM. This force in the passage of the body P from C to A tends in the direction in which the body is moving, and therefore accelerates it; then as far as D, it tends in the opposite direction, and retards the motion; then in the direction of the body, as far as B; and lastly in a contrary direction, as it moves from B to C.

Cor. III. And from the same reasoning it appears that the body P, other things remaining the same, moves more swiftly in the conjunction and opposition than in the quadratures.

Cor. IV. The orbit of the body P, other things remaining the same, is more curved at the quadratures than at the conjunction and opposition. For the swifter bodies move, the less they deflect from a rectilinear path. And besides, the force KL, or NM, at the conjunction and opposition, is contrary to the force with which the body T attracts the body P, and therefore diminishes that force; but the body P will deflect the less from a rectilinear path the less it is impelled towards the body T.

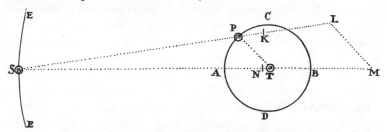

Cor. V. Hence the body P, other things remaining the same, goes farther from the body T at the quadratures than at the conjunction and opposition. This is said, however, when no account is taken of the variable eccentricity. For if the orbit of the body P be eccentric, its eccentricity (as will be shown presently by Cor. IX) will be greatest when the apsides are in the syzygies; and thence it may sometimes come to pass that the body P, in its near approach to the farther apse, may go farther from the body T at the syzygies than at the quadratures.

Cor. VI. Because the centripetal force of the central body T, by which the body P is retained in its orbit, is increased at the quadratures by the addition caused by the force LM, and diminished at the syzygies by the subtraction of the force KL, and, because the force KL is greater than LM, it is more diminished than increased; and, moreover, since that centripetal force (by Cor. II, Prop. IV) varies directly as the radius TP, and inversely as the square of the periodical time, it is plain that the resulting ratio is diminished by the action of the force KL; and therefore that the periodical time, supposing the radius of the orbit PT to remain the same, will be increased, and that as the square root of that ratio in which the centripetal force is diminished; and, therefore, supposing this radius increased or diminished, the periodical time will be increased more or diminished less than in the 3/2th power of this radius, by Cor. VI, Prop. IV. If that force of the central body should gradually decay, the body P being less and less attracted would go farther and farther from the centre T; and, on the contrary, if it were increased, it would draw nearer to it. Therefore if the action of the distant body S, by which that force is diminished, were to increase and decrease by turns, the radius TP would be also increased and diminished by turns; and the periodical time would be increased and diminished in a ratio compounded of the 3/2th power of the ratio of the radius, and of the square root of that ratio in which the centripetal force of the central body T was diminished or increased, by the increase or decrease of the action of the distant body S.

COR. VII. It also follows, from what was before laid down, that the axis of the ellipse described by the body P, or the line of the apsides, does as to its angular motion go forwards and backwards by turns, but more forwards than backwards, and by the excess of its direct motion is on the whole

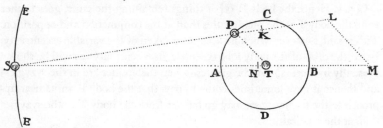

carried forwards. For the force with which the body P is urged to the body T at the quadratures, where the force MN vanishes, is compounded of the force LM and the centripetal force with which the body T attracts the body

P. The first force LM, if the distance PT be increased, is increased in nearly the same proportion with that distance, and the other force decreases as the square of the ratio of the distance; and therefore the sum of these two forces decreases in less than the square of the ratio of the distance PT; and therefore, by Cor. I, Prop. XLV, will make the line of the apsides, or, which is the same thing, the upper apse, to go backwards. But at the conjunction and opposition the force with which the body P is urged towards the body T is the difference of the force KL, and of the force with which the body T attracts the body P; and that difference, because the force KL is very nearly increased in the ratio of the distance PT, decreases in more than the square of the ratio of the distance PT; and therefore, by Cor. I, Prop. XLV, causes the line of the apsides to go forwards. In the places between the syzygies and the quadratures, the motion of the line of the apsides depends upon both of these causes conjointly, so that it either goes forwards or backwards in proportion to the excess of one of these causes above the other. Therefore since the force KL in the syzygies is almost twice as great as the force LM in the quadratures, the excess will be on the side of the force KL, and by consequence the line of the apsides will be carried forwards. The truth of this and the foregoing Corollary will be more easily understood by conceiving the system of the two bodies T and P to be surrounded on every side by several bodies S, S, S, &c., disposed about the orbit ESE. For by the actions of these bodies the action of the body T will be diminished on every side, and decrease in more than the square of the ratio of the distance.

Cor. VIII. But since the direct or retrograde motion of the apsides depends upon the decrease of the centripetal force, that is, upon its being in a greater or less ratio than the square of the ratio of the distance TP, in the passage of the body from the lower apse to the upper; and upon a like increase in its return to the lower apse again; and therefore becomes greatest where the proportion of the force at the upper apse to the force at the lower apse recedes farthest from the inverse square of the ratio of the distances; it is plain that, when the apsides are in the syzygies, they will, by reason of the subtracted force KL or NM − LM, go forwards more swiftly; and in the quadratures by the additional force LM go backwards more slowly. Because the velocity of the progression or the slowness of the retrogression is continued for a long time, this inequality becomes exceedingly great.

Cor. ix. If a body is obliged, by a force inversely proportional to the square of its distance from any centre, to revolve in an ellipse round that centre; and afterwards in its descent from the upper apse to the lower apse, that force by a continual accession of new force is increased in more than the square of the ratio of the diminished distance; it is manifest that the body, being impelled always towards the centre by the continual accession of this new force, will incline more towards that centre than if it were urged by that force alone which decreases as the square of the diminished distance, and therefore will describe an orbit interior to that elliptical orbit, and at the lower apse approaching nearer to the centre than before. Therefore the orbit by the accession of this new force will become more eccentric. If now, while the body is returning from the lower to the upper apse, it should decrease by the same degrees by which it increased before, the body would return to its first distance; and therefore if the force decreases in a yet greater ratio, the body, being now less attracted than before, will ascend to a still greater distance, and so, the eccentricity of the orbit will be increased still more. Therefore if the ratio of the increase and decrease of the centripetal force be augmented with each revolution, the eccentricity will be augmented also; and, on the contrary, if that ratio decrease, it will be diminished.

Now, therefore, in the system of the bodies T, P, S, when the apsides of the orbit PAB are in the quadratures, the ratio of that increase and decrease is least of all, and becomes greatest when the apsides are in the syzygies. If the apsides are placed in the quadratures, the ratio near the apsides is less, and near the syzygies greater, than the square of the ratio of the distances; and from that greater ratio arises a direct motion of the line of the apsides, as was just now said. But if we consider the ratio of the whole increase or decrease in the progress between the apsides, this is less than the square of the ratio of the distances. The force in the lower is to that in the upper apse in less than the square of the ratio of the distance of the upper apse from the focus of the ellipse to the distance of the lower apse from the same focus; and conversely, when the apsides are placed in the syzygies, the force in the lower apse is to the force in the upper apse in a greater than the square of the ratio of the distances. For the forces LM in the quadratures added to the forces of the body T, compose forces in a less ratio; and the forces KL in the syzygies subtracted from the forces of the body T, leave the forces in a greater ratio. Therefore the ratio of the whole increase and decrease in the passage between the apsides is least at the quadratures and greatest at the syzygies; and therefore in the passage of the apsides from the quadratures to the syzygies it is continually augmented, and increases the eccentricity of the ellipse; and in the passage from the syzygies to the quadratures it is continually decreasing, and diminishes the eccentricity.

Cor. x. That we may give an account of the errors of latitude, let us suppose the plane of the orbit EST to remain immovable; and from the cause of the errors above explained, it is manifest that, of the two forces NM, ML, which are the only and entire cause of them, the force ML acting always in the plane of the orbit PAB never disturbs the motions as to latitude; and that the force NM, when the nodes are in the syzygies, acting also in the same plane of the orbit, does not at that time affect those motions. But when the nodes are in the quadratures, it disturbs them very much,

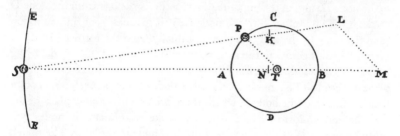

and, attracting the body P continually out of the plane of its orbit, it diminishes the inclination of the plane in the passage of the body from the quadratures to the syzygies, and again increases the same in the passage from the syzygies to the quadratures. Hence it comes to pass that when the body is in the syzygies, the inclination is then least of all, and returns to the first magnitude nearly, when the body arrives at the next node. But if the nodes are situated at the octants after the quadratures, that is, between C and A, D and B, it will appear, from what was just now shown, that in the passage of the body P from either node to the ninetieth degree from thence, the inclination of the plane is continually diminished; then, in the passage through the next 45 degrees to the next quadrature, the inclination is increased; and afterwards, again, in its passage through another 45 degrees to the next node, it is diminished. Therefore the inclination is more diminished than increased, and is therefore always less in the subsequent node than in the preceding one. And, by a like reasoning, the inclination is more increased than diminished when the nodes are in the other octants between A and D, B and C. The inclination, therefore, is the greatest of all when the nodes are in the syzygies. In their passage from the syzygies to the quadratures the inclination is diminished at each appulse of the body to the nodes; and becomes least of all when the nodes are in the quadratures, and the body in the syzygies; then it increases by the same degrees by which it decreased before; and, when the nodes come to the next syzygies, returns to its former magnitude.

Cor. xi. Because when the nodes are in the quadratures the body P is continually attracted from the plane of its orbit; and because this attraction is made towards S in its passage from the node C through the conjunction A to the node D; and in the opposite direction in its passage from the node D through the opposition B to the node C; it is manifest that, in its motion from the node C, the body recedes continually from the former plane CD of its orbit till it comes to the next node; and therefore at that node, being now at its greatest distance from the first plane CD, it will pass through the plane of the orbit EST not in D, the other node of that plane, but in a point that lies nearer to the body S, which therefore becomes a new place of the node behind its former place. And, by a like reasoning, the nodes will continue to recede in their passage from this node to the next. The nodes, therefore, when situated in the quadratures, recede continually; and at the syzygies, where no perturbation can be produced in the motion as to latitude, are quiescent; in the intermediate places they partake of both conditions, and recede more slowly; and, therefore, being always either retrograde or stationary, they will be carried backwards, or made to recede in each revolution.

Cor. xii. All the errors described in these Corollaries are a little greater at the conjunction of the bodies P, S than at their opposition; because the generating forces NM and ML are greater.

Cor. xiii. And since the causes and proportions of the errors and variations mentioned in these Corollaries do not depend upon the magnitude of the body S, it follows that all things before demonstrated will happen, if the magnitude of the body S be imagined so great that the system of the two bodies P and T may revolve about it. And from this increase of the body S, and the consequent increase of its centripetal force, from which the errors of the body P arise, it will follow that all these errors, at equal distances, will be greater in this case, than in the other where the body S revolves about the system of the bodies P and T.

Cor. xiv. But since the forces NM, ML, when the body S is exceedingly distant, are very nearly as the force SK and the ratio PT to ST conjointly; that is, if both the distance PT and the absolute force of the body S be given, inversely as ST^3; and since those forces NM, ML are the causes of all the errors and effects treated of in the foregoing Corollaries; it is manifest that all those effects, if the system of bodies T and P continue as before, and only the distance ST and the absolute force of the body S be changed, will be very nearly in a ratio compounded of the direct ratio of the absolute force of the body S, and the cubed inverse ratio of the distance ST. Hence if the system of bodies T and P revolve about a distant body S, those forces NM, ML, and their effects, will be (by Cor. ii and vi, Prop. iv) inversely as the square of the periodical time. And thence, also, if the magnitude of the body S be proportional to its absolute force, those forces NM, ML, and their effects, will be directly as the cube of the apparent diameter of the distant body S viewed from T; and conversely. For these ratios are the same as the compounded ratio above mentioned.

Cor. xv. If the orbits ESE and PAB, retaining their figure, proportions, and inclination to each other, should alter their magnitude, and if the forces of the bodies S and T should either remain unaltered or be changed in any given ratio, then these forces (that is, the force of the body T, which obliges the body P to deflect from a rectilinear course into the orbit PAB, and the force of the body S, which causes the body P to deviate from that orbit) will act always in the same manner, and in the same proportion. Consequently it follows, that all the effects will be similar and proportional, and the times of those effects will be proportional also; that is, that all the linear errors will be as the diameters of the orbits, the angular errors the same as before; and the times of similar linear errors, or equal angular errors, are as the periodical times of the orbits.

Cor. xvi. Therefore if the figures of the orbits and their inclination to each other be given, and the magnitudes, forces, and distances of the bodies be changed in any manner, we may, from the errors and times of those errors in one case, obtain very nearly the errors and times of the errors in any other case. But this may be done more expeditiously by the following method. The forces NM, ML, other things remaining unaltered, are as the radius TP; and their periodical effects (by Cor. ii, Lem. x) are as the forces and the square of the periodical time of the body P jointly. These are the linear errors of the body P; and hence the angular errors as they appear from the centre T (that is, the motion of the apsides and of the nodes, and all the apparent errors of longitude and latitude) are in each revolution of the body P as the square of the time of the revolution, very nearly. Let these ratios be compounded with the ratios in Cor. xiv, and in any system of bodies T, P, S, where P revolves about T very near to it, and T revolves about S at a great distance, the angular errors of the body P, observed from

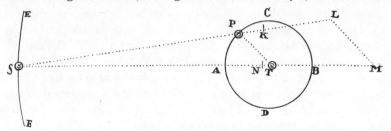

the centre T, will be in each revolution of the body P directly as the square of the periodical time of the body P, and inversely as the square of the periodical time of the body T. And therefore the mean motion of the line of the apsides will be in a given ratio to the mean motion of the nodes; and both those motions will be directly as the periodical time of the body P, and inversely as the square of the periodical time of the body T. The increase or diminution of the eccentricity and inclination of the orbit PAB makes no sensible variation in the motions of the apsides and nodes, unless that increase or diminution be very great indeed.

Cor. xvii. Since the line LM becomes sometimes greater and sometimes less than the radius PT, let the mean quantity of the force LM be expressed by that radius PT; and then that mean force will be to the mean force SK or SN (which may be also expressed by ST) as the length PT to the length ST. But the mean force SN or ST, by which the body T is retained in the orbit it describes about S, is to the force with which the body P is retained in its orbit about T in a ratio compounded of the ratio of the radius ST to the radius PT, and the squared ratio of the periodical time of the body P about T to the periodical time of the body T about S. And, consequently, the mean force LM is to the force by which the body P is retained in its orbit about T (or by which the same body P might revolve at the distance PT in the same periodical time about any immovable point T) in the same squared ratio of the periodical times. The periodical times therefore being given, together with the distance PT, the mean force LM is also given; and that force being given, there is given also the force MN, very nearly, by the analogy of the lines PT and MN.

COR. XVIII. By the same laws by which the body P revolves about the body T, let us suppose many fluid bodies to move round T at equal distances from it; and to be so numerous, that they may all become contiguous to each other, so as to form a fluid annulus, or ring, of a round figure, and concentric to the body T; and the several parts of this ring, performing their motions by the same law as the body P, will draw nearer to the body T, and move swifter in the conjunction and opposition of themselves and the body S, than in the quadratures. And the nodes of this ring or its intersections with the plane of the orbit of the body S or T, will rest at the syzygies; but out of the syzygies they will be carried backwards, or in a retrograde direction, with the greatest swiftness in the quadratures, and more slowly in other places. The inclination of this ring also will vary, and its axis will oscillate in each revolution, and when the revolution is completed will return to its former situation, except only that it will be carried round a little by the precession of the nodes.

COR. XIX. Suppose now the spherical body T, consisting of some matter not fluid, to be enlarged, and to extend itself on every side as far as that ring, and that a channel were cut all round its circumference containing water; and that this sphere revolves uniformly about its own axis in the same periodical time. This water being accelerated and retarded by turns (as in the last Corollary), will be swifter at the syzygies, and slower at the quadratures, than the surface of the globe, and so will ebb and flow in its channel after the manner of the sea. If the attraction of the body S were taken away, the water would acquire no motion of flux and reflux by revolving round the quiescent centre of the globe. The case is the same of a globe moving uniformly forwards in a right line, and in the meantime revolving about its centre (by Cor. v of the Laws of Motion), and of a globe uniformly attracted from its rectilinear course (by Cor. vi of the same Laws). But let the body S come to act upon it, and by its varying attraction the water will receive this new motion; for there will be a stronger attraction upon that part of the water that is nearest to the body, and a weaker upon that part which is more remote. And the force LM will attract the water downwards at the quadratures, and depress it as far as the syzygies; and the force KL will attract it upwards in the syzygies, and withhold its descent, and make it rise as far as the quadratures; except only so far as the motion of flux and reflux may be directed by the channel, and be a little retarded by friction.

Cor. xx. If, now, the ring becomes hard, and the globe is diminished, the motion of flux and reflux will cease; but the oscillating motion of the inclination and the precession of the nodes will remain. Let the globe have the same axis with the ring, and perform its revolutions in the same times, and at its surface touch the ring within, and adhere to it; then the globe partaking of the motion of the ring, this whole body will oscillate, and the nodes will go backwards, for the globe, as we shall show presently, is perfectly

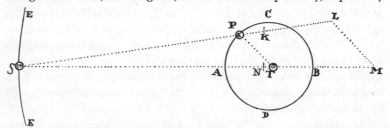

indifferent to the receiving of all impressions. The greatest angle of the inclination of the ring alone is when the nodes are in the syzygies. Thence in the progress of the nodes to the quadratures, it endeavors to diminish its inclination, and by that endeavor impresses a motion upon the whole globe. The globe retains this motion impressed, till the ring by a contrary endeavor destroys that motion, and impresses a new motion in a contrary direction.

And by this means the greatest motion of the decreasing inclination happens when the nodes are in the quadratures, and the least angle of inclination in the octants after the quadratures; and, again, the greatest motion of the reclination happens when the nodes are in the syzygies; and the greatest angle of inclination in the octants following. And the case is the same of a globe without this ring, if it be a little higher or a little denser in the equatorial than in the polar regions; for the excess of that matter in the regions near the equator supplies the place of the ring. And although we should suppose the centripetal force of this globe to be increased in any manner, so that all its parts tend downwards, as the parts of our earth gravitate to the centre, yet the phenomena of this and the preceding Corollary would scarce be altered; except that the places of the greatest and least height of the water will be different; for the water is now no longer sustained and kept in its orbit by its centrifugal force, but by the channel in which it flows. And, besides, the force LM attracts the water downwards most in the quadratures, and the force KL or NM – LM attracts it upwards most in the syzygies. And these forces conjoined cease to attract the water downwards, and begin to attract it upwards in the octants before the syzygies; and cease to attract the water upwards, and begin to attract the water downwards in the octants after the syzygies. And thence the greatest height of the water may happen about the octants after the syzygies; and the least height about the octants after the quadratures; excepting only so far as the motion of ascent or descent impressed by these forces may by the inertia of the water continue a little longer, or be stopped a little sooner by impediments in its channel.

Cor. xxi. For the same reason that redundant matter in the equatorial regions of a globe causes the nodes to go backwards, and therefore by the increase of that matter that retrograde motion is increased, by the diminution is diminished, and by the removal quite ceases; it follows, that, if more than that redundant matter be taken away, that is, if the globe be either more depressed, or of a rarer consistence near the equator than near the poles, there will arise a direct motion of the nodes.

Cor. xxii. And thence from the motion of the nodes is known the constitution of the globe. That is, if the globe retains unalterably the same poles, and the motion (of the nodes) is retrograde, there is a redundance of the matter near the equator; but if that motion is direct, a deficiency. Suppose a uniform and exactly spherical globe to be first at rest in a free space; then by some impulse made obliquely upon its surface to be driven from its place, and to receive a motion partly circular and partly straight forward. Since this globe is perfectly indifferent to all the axes that pass through its centre, nor has a greater propensity to one axis or to one situation of the axis than to any other, it is manifest that by its own force it will never change its axis, or the inclination of its axis. Let now this globe be impelled obliquely by a new impulse in the same part of its surface as before; and since the effect of an impulse is not at all changed by its coming sooner or later, it is manifest that these two impulses, successively impressed, will produce the same motion, as if they had been impressed at the same time; that is, the same motion, as if the globe had been impelled by a simple force compounded of them both (by Cor. II of the Laws), that is, a simple motion about an axis of a given inclination. And the case is the same if the second impulse were made upon any other place of the equator of the first motion; and also if the first impulse were made upon any place in the equator of the motion which would be generated by the second impulse alone; and therefore, also, when both impulses are made in any places whatsoever; for these impulses will generate the same circular motion as if they were impressed together, and at once, in the place of the intersections of the equators of those motions, which would be generated by each of them separately. Therefore, a homogeneous and perfect globe will not retain several motions distinct, but will unite all those that are impressed on it, and reduce them into one; revolving, as far as in it lies, always with a simple and uniform motion about one single given axis, with an inclination always invariable. And the inclination of the axis, or the velocity of the rotation, will not be changed by centripetal force. For if the globe be supposed to be divided into two hemispheres, by any plane whatsoever passing through its own centre, and the centre to which the force is directed, that force will always urge each hemisphere equally; and therefore will not incline the globe to any side with respect to its motion round its own axis. But let there be added anywhere between the pole and the equator a heap of new matter like a mountain, and this, by its continual endeavor to recede from the centre of its motion, will disturb the motion of the globe, and

cause its poles to wander about its surface describing circles about themselves and the points opposite to them. Neither can this enormous deviation
of the poles be corrected otherwise than by placing that mountain either
in one of the poles, in which case, by Cor. xxi, the nodes of the equator
will go forwards; or in the equatorial regions, in which case, by Cor. xx,
the nodes will go backwards; or, lastly, by adding on the other side of the
axis a new quantity of matter, by which the mountain may be balanced in
its motion; and then the nodes will either go forwards or backwards, as the
mountain and this newly added matter happen to be nearer to the pole or
to the equator.

PROPOSITION LXVII. THEOREM XXVII

*The same laws of attraction being supposed, I say, that the exterior body S
does, by radii drawn to the point O, the common centre of gravity of the
interior bodies P and T, describe round that centre areas more proportional
to the times, and an orbit more approaching to the form of an ellipse having
its focus in that centre, than it can describe round the innermost and greatest
body T by radii drawn to that body.*

For the attractions of the body S towards T and P compose its absolute
attraction, which is more directed towards O, the common centre of gravity
of the bodies T and P, than it is to the
greatest body T; and which approaches
nearer to the inverse proportion of
the square of the distance SO, than of
the square of the distance ST; as will
easily appear by a little consideration.

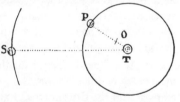

PROPOSITION LXVIII. THEOREM XXVIII

*The same laws of attraction supposed, I say, that the exterior body S will,
by radii drawn to O, the common centre of gravity of the interior bodies
P and T, describe round that centre areas more proportional to the times,
and an orbit more approaching to the form of an ellipse having its focus
in that centre, if the innermost and greatest body be agitated by these attractions as well as the rest, than it would do if that body either were at rest and
not attracted at all, or were much more or much less attracted, or were
much more or much less agitated.*

This may be demonstrated after the same manner as Prop. LXVI, but by a more prolix reasoning, which I therefore pass over. It will be sufficient to consider it after this manner. From the demonstration of the last Proposition it is plain, that the centre, towards which the body S is urged by the two forces conjointly, is very near to the common centre of gravity of those two other bodies. If this centre were to coincide with that common centre, and moreover the common centre of gravity of all the three bodies were at rest, the body S on one side, and the common centre of gravity of the other two bodies on the other side, would describe true ellipses about that quiescent common centre. This appears from Cor. II, Prop. LVIII, compared

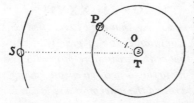

with what was demonstrated in Prop. LXIV, and LXV. Now this accurate elliptical motion will be disturbed a little by the distance of the centre of the two bodies from the centre towards which the third body S is attracted. Let there be added, moreover, a motion to the common centre of the three, and the perturbation will be increased yet more. Therefore the perturbation is least when the common centre of the three bodies is at rest; that is, when the innermost and greatest body T is attracted according to the same law as the rest are; and is always greatest when the common centre of the three, by the diminution of the motion of the body T, begins to be moved, and is more and more agitated.

Cor. And hence if several smaller bodies revolve about the great one, it may easily be inferred that the orbits described will approach nearer to ellipses; and the descriptions of areas will be more nearly uniform, if all the bodies attract and agitate each other with accelerative forces that are directly as their absolute forces, and inversely as the squares of the distances, and if the focus of each orbit be placed in the common centre of gravity of all the interior bodies (that is, if the focus of the first and innermost orbit be placed in the centre of gravity of the greatest and innermost body; the focus of the second orbit in the common centre of gravity of the two innermost bodies; the focus of the third orbit in the common centre of gravity of the three innermost; and so on), than if the innermost body were at rest, and was made the common focus of all the orbits.

PROPOSITION LXIX. THEOREM XXIX

In a system of several bodies A, B, C, D, &c., *if any one of those bodies, as* A, *attract all the rest,* B, C, D, &c., *with accelerative forces that are inversely as the squares of the distances from the attracting body; and another body, as* B, *attracts also the rest,* A, C, D, &c., *with forces that are inversely as the squares of the distances from the attracting body; the absolute forces of the attracting bodies* A *and* B *will be to each other as those very bodies* A *and* B *to which those forces belong.*

For the accelerative attractions of all the bodies B, C, D, towards A, are by the supposition equal to each other at equal distances; and in like manner the accelerative attractions of all the bodies towards B are also equal to each other at equal distances. But the absolute attractive force of the body A is to the absolute attractive force of the body B as the accelerative attraction of all the bodies towards A is to the accelerative attraction of all the bodies towards B at equal distances; and so is also the accelerative attraction of the body B towards A to the accelerative attraction of the body A towards B. But the accelerative attraction of the body B towards A is to the accelerative attraction of the body A towards B as the mass of the body A is to the mass of the body B; because the motive forces which (by the second, seventh and eighth Definitions) are as the accelerative forces and the bodies attracted conjointly are here equal to one another by the third Law. Therefore the absolute attractive force of the body A is to the absolute attractive force of the body B as the mass of the body A is to the mass of the body B. Q.E.D.

Cor. i. Therefore if each of the bodies of the system A, B, C, D, &c., does singly attract all the rest with accelerative forces that are inversely as the squares of the distances from the attracting body, the absolute forces of all those bodies will be to each other as the bodies themselves.

Cor. ii. By a like reasoning, if each of the bodies of the system A, B, C, D, &c., does singly attract all the rest with accelerative forces, which are either inversely or directly in the ratio of any power whatever of the distances from the attracting body; or which are defined by the distances from each of the attracting bodies according to any common law; it is plain that the absolute forces of those bodies are as the bodies themselves.

Cor. iii. In a system of bodies whose forces decrease as the square of the distances, if the lesser revolve about one very great one in ellipses, having their common focus in the centre of that great body, and of a figure exceedingly accurate; and moreover by radii drawn to that great body describe areas proportional to the times exactly; the absolute forces of those bodies to each other will be either accurately or very nearly in the ratio of the bodies. And so conversely. This appears from Cor. of Prop. xlviii, compared with the first Corollary of this Proposition.

SCHOLIUM

These Propositions naturally lead us to the analogy there is between centripetal forces and the central bodies to which those forces are usually directed; for it is reasonable to suppose that forces which are directed to bodies should depend upon the nature and quantity of those bodies, as we see they do in magnetical experiments. And when such cases occur, we are to compute the attractions of the bodies by assigning to each of their particles its proper force, and then finding the sum of them all. I here use the word *attraction* in general for any endeavor whatever, made by bodies to approach to each other, whether that endeavor arise from the action of the bodies themselves, as tending to each other or agitating each other by spirits emitted; or whether it arises from the action of the ether or of the air, or of any medium whatever, whether corporeal or incorporeal, in any manner impelling bodies placed therein towards each other. In the same general sense I use the word *impulse,* not defining in this treatise the species or physical qualities of forces, but investigating the quantities and mathematical proportions of them; as I observed before in the Definitions. In mathematics we are to investigate the quantities of forces with their proportions consequent upon any conditions supposed; then, when we enter upon physics, we compare those proportions with the phenomena of Nature, that we may know what conditions of those forces answer to the several kinds of attractive bodies. And this preparation being made, we argue more safely concerning the physical species, causes, and proportions of the forces. Let us see, then, with what forces spherical bodies consisting of particles endued with attractive powers in the manner above spoken of must act upon one another; and what kind of motions will follow from them.

SECTION XII

The attractive forces of spherical bodies.

PROPOSITION LXX. THEOREM XXX

If to every point of a spherical surface there tend equal centripetal forces decreasing as the square of the distances from those points, I say, that a corpuscle placed within that surface will not be attracted by those forces any way.

Let HIKL be that spherical surface, and P a corpuscle placed within. Through P let there be drawn to this surface two lines HK, IL, intercepting very small arcs HI, KL; and because (by Cor. III, Lem. VII) the triangles HPI, LPK are alike, those arcs will be proportional to the distances HP, LP; and any particles at HI and KL of the spherical surface, terminated by right lines passing through P, will be as the square of those distances. Therefore the forces of these particles exerted upon the body P are equal between themselves. For the forces are directly as the particles, and inversely as the square of the distances. And these two ratios compose the ratio of equality, 1 : 1. The attractions therefore, being equal, but exerted in opposite directions, destroy each other. And by a like reasoning all the attractions through the whole spherical surface are destroyed by contrary attractions. Therefore the body P will not be any way impelled by those attractions. Q.E.D.

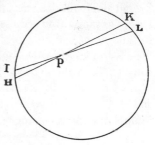

PROPOSITION LXXI. THEOREM XXXI

The same things supposed as above, I say, that a corpuscle placed without the spherical surface is attracted towards the centre of the sphere with a force inversely proportional to the square of its distance from that centre.

Let AHKB, *ahkb* be two equal spherical surfaces described about the centres S, *s;* their diameters AB, *ab;* and let P and *p* be two corpuscles situate

without the spheres in those diameters produced. Let there be drawn from
the corpuscles the lines PHK, PIL, *phk, pil*, cutting off from the great circles
AHB, *ahb*, the equal arcs HK, *hk*, IL, *il*; and to those lines let fall the perpendiculars SD, *sd*, SE, *se*, IR, *ir*; of which let SD, *sd*, cut PL, *pl*, in F and *f*.

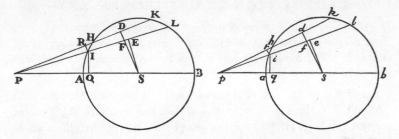

Let fall also to the diameters the perpendiculars IQ, *iq*. Let now the angles
DPE, *dpe* vanish; and because DS and *ds*, ES and *es* are equal, the lines PE,
PF, and *pe, pf*, and the short lines DF, *df* may be taken for equal; because
their last ratio, when the angles DPE, *dpe* vanish together, is the ratio of
equality. These things being thus determined, it follows that

$$\text{PI} : \text{PF} = \text{RI} : \text{DF}$$

and $pf : pi = df$ or DF : *ri*.

Multiplying corresponding terms,

$$\text{PI} \cdot pf : \text{PF} \cdot pi = \text{RI} : ri = \text{arc IH} : \text{arc } ih \text{ (by Cor. III, Lem. VII)}.$$

Again, $\text{PI} : \text{PS} = \text{IQ} : \text{SE}$

and $ps : pi = se$ or SE : *iq*.

Hence, $\text{PI} \cdot ps : \text{PS} \cdot pi = \text{IQ} : iq$.

Multiplying together corresponding terms of this and the similarly derived
preceding proportion,

$$\text{PI}^2 \cdot pf \cdot ps : pi^2 \cdot \text{PF} \cdot \text{PS} = \text{HI} \cdot \text{IQ} : ih \cdot iq,$$

that is, as the circular surface which is described by the arc IH, as the semicircle AKB revolves about the diameter AB, is to the circular surface described by the arc *ih* as the semicircle *akb* revolves about the diameter *ab*.
And the forces with which these surfaces attract the corpuscles P and *p* in
the direction of lines tending to those surfaces are directly, by the hypothesis,
as the surfaces themselves, and inversely as the squares of the distances of
the surfaces from those corpuscles; that is, as $pf \cdot ps$ to $\text{PF} \cdot \text{PS}$. And these

forces again are to the oblique parts of them which (by the resolution of forces as in Cor. II of the Laws) tend to the centres in the directions of the lines PS, ps, as PI to PQ, and pi to pq; that is (because of the like triangles PIQ and PSF, piq and psf), as PS to PF and ps to pf. Thence, the attraction of the corpuscle P towards S is to the attraction of the corpuscle p towards s as $\dfrac{\text{PF} \cdot pf \cdot ps}{\text{PS}}$ is to $\dfrac{pf \cdot \text{PF} \cdot \text{PS}}{ps}$, that is, as ps^2 to PS^2. And, by a like reasoning, the forces with which the surfaces described by the revolution of the arcs KL, kl attract those corpuscles, will be as ps^2 to PS^2. And in the same ratio will be the forces of all the circular surfaces into which each of the spherical surfaces may be divided by taking sd always equal to SD, and se equal to SE. And therefore, by composition, the forces of the entire spherical surfaces exerted upon those corpuscles will be in the same ratio. Q.E.D.

PROPOSITION LXXII. THEOREM XXXII

If to the several points of a sphere there tend equal centripetal forces decreasing as the square of the distances from those points; and there be given both the density of the sphere and the ratio of the diameter of the sphere to the distance of the corpuscle from its centre: I say, that the force with which the corpuscle is attracted is proportional to the semidiameter of the sphere.

For conceive two corpuscles to be severally attracted by two spheres, one by one, the other by the other, and their distances from the centres of the spheres to be proportional to the diameters of the spheres respectively; and the spheres to be resolved into like particles, disposed in a like situation to the corpuscles. Then the attractions of one corpuscle towards the several particles of one sphere will be to the attractions of the other towards as many analogous particles of the other sphere in a ratio compounded of the ratio of the particles directly, and the square of the distances inversely. But the particles are as the spheres, that is, as the cubes of the diameters, and the distances are as the diameters; and the first ratio directly with the last ratio taken twice inversely, becomes the ratio of diameter to diameter. Q.E.D.

Cor. I. Hence if corpuscles revolve in circles about spheres composed of matter equally attracting, and the distances from the centres of the spheres be proportional to their diameters, the periodic times will be equal.

Cor. II. And, *vice versa,* if the periodic times are equal, the distances will be proportional to the diameters. These two Corollaries appear from Cor. III, Prop. IV.

Cor. III. If to the several points of any two solids whatever, of like figure and equal density, there tend equal centripetal forces decreasing as the square of the distances from those points, the forces, with which corpuscles placed in a like situation to those two solids will be attracted by them, will be to each other as the diameters of the solids.

PROPOSITION LXXIII. THEOREM XXXIII

If to the several points of a given sphere there tend equal centripetal forces decreasing as the square of the distances from the points, I say, that a corpuscle placed within the sphere is attracted by a force proportional to its distance from the centre.

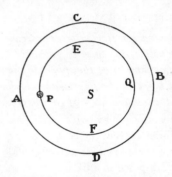

In the sphere ACBD, described about the centre S, let there be placed the corpuscle P; and about the same centre S, with the interval SP, conceive described an interior sphere PEQF. It is plain (by Prop. LXX) that the concentric spherical surfaces of which the difference AEBF of the spheres is composed, have no effect at all upon the body P, their attractions being destroyed by contrary attractions. There remains, therefore, only the attraction of the interior sphere PEQF. And (by Prop. LXXII) this is as the distance PS. Q.E.D.

SCHOLIUM

By the surfaces of which I here imagine the solids composed, I do not mean surfaces purely mathematical, but orbs so extremely thin, that their thickness is as nothing; that is, the evanescent orbs of which the sphere will at last consist, when the number of the orbs is increased, and their thickness diminished without end. In like manner, by the points of which lines, surfaces, and solids are said to be composed, are to be understood equal particles, whose magnitude is perfectly inconsiderable.

PROPOSITION LXXIV. THEOREM XXXIV

The same things supposed, I say, that a corpuscle situated without the sphere is attracted with a force inversely proportional to the square of its distance from the centre.

For suppose the sphere to be divided into innumerable concentric spherical surfaces, and the attractions of the corpuscle arising from the several surfaces will be inversely proportional to the square of the distance of the corpuscle from the centre of the sphere (by Prop. LXXI). And, by composition, the sum of those attractions, that is, the attraction of the corpuscle towards the entire sphere, will be in the same ratio. Q.E.D.

COR. I. Hence the attractions of homogeneous spheres at equal distances from the centres will be as the spheres themselves. For (by Prop. LXXII) if the distances be proportional to the diameters of the spheres, the forces will be as the diameters. Let the greater distance be diminished in that ratio; and the distances now being equal, the attraction will be increased as the square of that ratio; and therefore will be to the other attraction as the cube of that ratio; that is, in the ratio of the spheres.

COR. II. At any distances whatever the attractions are as the spheres applied to the squares of the distances.

COR. III. If a corpuscle placed without an homogeneous sphere is attracted by a force inversely proportional to the square of its distance from the centre, and the sphere consists of attractive particles, the force of every particle will decrease as the square of the distance from each particle.

PROPOSITION LXXV. THEOREM XXXV

If to the several points of a given sphere there tend equal centripetal forces decreasing as the square of the distances from the point, I say, that another similar sphere will be attracted by it with a force inversely proportional to the square of the distance of the centres.[1]

For the attraction of every particle is inversely as the square of its distance from the centre of the attracting sphere (by Prop. LXXIV), and is therefore the same as if that whole attracting force issued from one single corpuscle placed in the centre of this sphere. But this attraction is as great as

[¹ Appendix, Note 25.]

on the other hand the attraction of the same corpuscle would be, if that were itself attracted by the several particles of the attracted sphere with the same force with which they are attracted by it. But that attraction of the corpuscle would be (by Prop. LXXIV) inversely proportional to the square of its distance from the centre of the sphere; therefore the attraction of the sphere, equal thereto, is also in the same ratio. Q.E.D.

COR. I. The attractions of spheres towards other homogeneous spheres are as the attracting spheres applied to the squares of the distances of their centres from the centres of those which they attract.

COR. II. The case is the same when the attracted sphere does also attract. For the several points of the one attract the several points of the other with the same force with which they themselves are attracted by the others again; and therefore since in all attractions (by Law III) the attracted and attracting point are both equally acted on, the force will be doubled by their mutual attractions, the proportions remaining.

COR. III. Those several truths demonstrated above concerning the motion of bodies about the focus of the conic sections will take place when an attracting sphere is placed in the focus, and the bodies move without the sphere.

COR. IV. Those things which were demonstrated before of the motion of bodies about the centre of the conic sections take place when the motions are performed within the sphere.

PROPOSITION LXXVI. THEOREM XXXVI

If spheres be however dissimilar (as to density of matter and attractive force) in the same ratio onwards from the centre to the circumference; but everywhere similar, at every given distance from the centre, on all sides round about; and the attractive force of every point decreases as the square of the distance of the body attracted: I say, that the whole force with which one of these spheres attracts the other will be inversely proportional to the square of the distance of the centres.

Imagine several concentric similar spheres AB, CD, EF, &c., the innermost of which added to the outermost may compose a matter more dense towards the centre, or subtracted from them may leave the same more lax and rare. Then, by Prop. LXXV, these spheres will attract other similar concentric spheres GH, IK, LM, &c., each the other, with forces inversely proportional to the square of the distance SP. And, by addition or subtraction, the sum of all those forces, or the excess of any of them above the others; that is, the entire force with which the whole sphere AB (composed of any concentric spheres or of their differences) will attract the whole sphere GH (composed of any concentric spheres or their differences) in the same ratio. Let the number of the concentric spheres be increased *in infinitum,* so that the density of the matter together with the attractive force may, in the progress from the circumference to the centre, increase or decrease according to any given law; and by the addition of matter not attractive, let the deficient density be supplied, that so the spheres may acquire any form desired; and the force with which one of these attracts the other will be still, by the former reasoning, in the same inverse ratio of the square of the distance. Q.E.D.

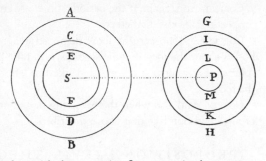

Cor. I. Hence if many spheres of this kind, similar in all respects, attract each other, the accelerative attractions of each to each, at any equal distances of the centres, will be as the attracting spheres.

Cor. II. And at any unequal distances, as the attracting spheres divided by the squares of the distances between the centres.

Cor. III. The motive attractions, or the weights of the spheres towards one another, will be at equal distances of the centres conjointly as the attracting and attracted spheres; that is, as the products arising from multiplying the spheres into each other.

Cor. IV. And at unequal distances directly as those products and inversely as the squares of the distances between the centres.

Cor. V. These proportions hold true also when the attraction arises from the attractive power of both spheres exerted upon each other. For the attraction is only doubled by the conjunction of the forces, the proportions remaining as before.

Cor. vi. If spheres of this kind revolve about others at rest, each about each, and the distances between the centres of the quiescent and revolving bodies are proportional to the diameters of the quiescent bodies, the periodic times will be equal.

Cor. vii. And, again, if the periodic times are equal, the distances will be proportional to the diameters.

Cor. viii. All those truths above demonstrated, relating to the motions of bodies about the foci of conic sections, will take place when an attracting sphere, of any form and condition like that above described, is placed in the focus.

Cor. ix. And also when the revolving bodies are also attracting spheres of any condition like that above described.

PROPOSITION LXXVII. THEOREM XXXVII

If to the several points of spheres there tend centripetal forces proportional to the distances of the points from the attracted bodies, I say, that the compounded force with which two spheres attract each other is as the distance between the centres of the spheres.

Case i. Let AEBF be a sphere; S its centre; P a corpuscle attracted; PASB the axis of the sphere passing through the centre of the corpuscle; EF, *ef* two planes cutting the sphere, and perpendicular to the axis, and equidis-

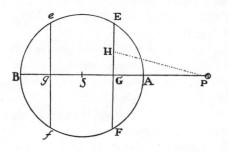

tant, one on one side, the other on the other, from the centre of the sphere; G and *g* the intersections of the planes and the axis; and H any point in the plane EF. The centripetal force of the point H upon the corpuscle P, exerted in the direction of the line PH, is as the distance PH; and (by Cor. ii of the Laws) the same exerted in the direction of the line PG, or towards the centre S, is as the length PG. Therefore the force of all the points in the plane EF (that is, of that whole plane) by which the corpuscle P is attracted towards the centre S is as the distance PG multiplied by the number of those points, that is, as the solid contained under that plane EF and the distance PG. And in like manner the force of the plane *ef,* by which the corpuscle P is attracted towards the centre S, is as that plane multiplied by its distance P*g,* or as the equal plane EF multiplied by that distance P*g*; and the sum of the forces of both planes as the plane EF multiplied by the sum of the distances PG + P*g,* that is, as that plane multiplied by twice the distance PS of the centre and the corpuscle; that is, as twice the plane EF multiplied by the distance PS, or as the sum of the equal planes EF + *ef* multiplied by the same distance. And, by a like reasoning, the forces of all the planes in the whole sphere, equidistant on each side from the centre of the sphere, are as the sum of those planes multiplied by the distance PS, that is, as the whole sphere and the distance PS conjointly. Q.E.D.

CASE 2. Let now the corpuscle P attract the sphere AEBF. And, by the same reasoning, it will appear that the force with which the sphere is attracted is as the distance PS. Q.E.D.

CASE 3. Imagine another sphere composed of innumerable corpuscles P; and because the force with which every corpuscle is attracted is as the distance of the corpuscle from the centre of the first sphere, and as the same sphere conjointly, and is therefore the same as if it all proceeded from a single corpuscle situated in the centre of the sphere, the entire force with which all the corpuscles in the second sphere are attracted, that is, with which that whole sphere is attracted, will be the same as if that sphere were attracted by a force issuing from a single corpuscle in the centre of the first sphere; and is therefore proportional to the distance between the centres of the spheres. Q.E.D.

CASE 4. Let the spheres attract each other, and the force will be doubled, but the proportion will remain. Q.E.D.

CASE 5. Let the corpuscle *p* be placed within the sphere AEBF; and because the force of the plane *ef* upon the corpuscle is as the solid contained under that plane and the distance *pg*; and the contrary force of the plane

EF as the solid contained under that plane and the distance pG; the force compounded of both will be as the difference of the solids, that is, as the sum of the equal planes multiplied by half the difference of the distances; that is, as that sum multiplied by pS, the distance of the corpuscle from the centre of the sphere. And, by a like reasoning, the attraction of all the planes EF, ef, throughout the whole sphere, that is, the attraction of the whole sphere, is conjointly as the sum of all the planes, or as the whole sphere, and as pS, the distance of the corpuscle from the centre of the sphere. Q.E.D.

CASE 6. And if there be composed a new sphere out of innumerable corpuscles such as p, situated within the first sphere AEBF, it may be proved, as before, that the attraction, whether single of one sphere towards the other, or mutual of both towards each other, will be as the distance pS of the centres. Q.E.D.

PROPOSITION LXXVIII. THEOREM XXXVIII

If spheres in the progress from the centre to the circumference be however dissimilar and unequable, but similar on every side round about at all given distances from the centre; and the attractive force of every point be as the distance of the attracted body: I say, that the entire force with which two spheres of this kind attract each other mutually is proportional to the distance between the centres of the spheres.

This is demonstrated from the foregoing Proposition, in the same manner as Prop. LXXVI was demonstrated from Prop. LXXV.

COR. Those things that were above demonstrated in Prop. X and LXIV, of the motion of bodies round the centres of conic sections, take place when all the attractions are made by the force of spherical bodies of the condition above described, and the attracted bodies are spheres of the same kind.

SCHOLIUM

I have now explained the two principal cases of attractions; to wit, when the centripetal forces decrease as the square of the ratio of the distances, or increase in a simple ratio of the distances, causing the bodies in both cases to revolve in conic sections, and composing spherical bodies whose centripetal forces observe the same law of increase or decrease in the recess from the

centre as the forces of the particles themselves do; which is very remarkable. It would be tedious to run over the other cases, whose conclusions are
less elegant and important, so particularly as I have done these. I choose
rather to comprehend and determine them all by one general method as
follows.

LEMMA XXIX

If about the centre S there be described any circle as AEB, *and about the
centre* P *there be also described two circles* EF, ef, *cutting the first in* E *and*
e, *and the line* PS *in* F *and* f; *and there be let fall to* PS *the perpendiculars*
ED, ed: *I say, that if the distance of the arcs* EF, ef *be supposed to be infinitely diminished, the last ratio of the evanescent line* Dd *to the evanescent
line* Ff *is the same as that of the line* PE *to the line* PS.

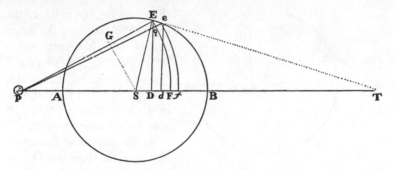

For if the line Pe cut the arc EF in q; and the right line Ee, which coincides with the evanescent arc Ee, be produced, and meet the right line PS
in T; and there be let fall from S to PE the perpendicular SG; then, because
of the like triangles DTE, dTe, DES,

$$Dd : Ee = DT : TE = DE : ES;$$

and because the triangles, Eeq, ESG (by Lem. VIII, and Cor. III, Lem. VII)
are similar,

$$Ee : eq \text{ or } Ff = ES : SG.$$

Multiplying together corresponding terms of the two proportions,

$$Dd : Ff = DE : SG = PE : PS$$

(because of the similar triangles PDE, PGS). Q.E.D.

PROPOSITION LXXIX. THEOREM XXXIX

Suppose a surface as EFfe *to have its breadth infinitely diminished, and to be just vanishing; and that the same surface by its revolution round the axis* PS *describes a spherical concavoconvex solid, to the several equal particles of which there tend equal centripetal forces: I say, that the force with which that solid attracts a corpuscle situated in* P *is in a ratio compounded of the ratio of the solid* DE² · Ff *and the ratio of the force with which the given particle in the place* Ff *would attract the same corpuscle.*

For if we consider, first, the force of the spherical surface FE which is generated by the revolution of the arc FE, and is cut anywhere, as in *r*, by the line *de,* the annular part of the surface generated by the revolu-

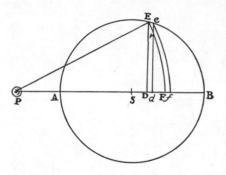

tion of the arc *r*E will be as the short line D*d,* the radius of the sphere PE remaining the same; as *Archimedes* has demonstrated in his *Book of the Sphere and Cylinder.* And the force of this surface exerted in the direction of the lines PE or P*r* situated all round in the conical surface, will be as this annular surface itself; that is, as the short line D*d,* or, which is the same, as the rectangle under the given radius PE of the sphere and the short line D*d*; but that force, exerted in the direction of the line PS tending to the centre S, will be less in the ratio PD to PE, and therefore will be as PD · D*d*. Suppose now the line DF to be divided into innumerable little equal particles, each of which call D*d,* and then the surface FE will be divided into so many equal annuli, whose forces will be as the sum of all the rectangles PD · D*d*, that is, as ½PF² − ½PD², and therefore as DE². Let now the surface FE be multiplied by the altitude F*f*; and the force of the solid EF*fe* exerted upon the corpuscle P will be as DE² · F*f*; that is, if the force be given which any given particle as F*f* exerts upon the corpuscle P at the distance PF. But if that force be not given, the force of the solid EF*fe* will be conjointly as the solid DE² · F*f* and that force not given. Q.E.D.

PROPOSITION LXXX. THEOREM XL

If to the several equal parts of a sphere ABE *described about the centre* S *there tend equal centripetal forces; and from the several points* D *in the axis of the sphere* AB *in which a corpuscle, as* P, *is placed, there be erected the perpendiculars* DE *meeting the sphere in* E, *and if in those perpendiculars the lengths* DN *be taken as the quantity* $\dfrac{DE^2 \cdot PS}{PE}$, *and as the force which a particle of the sphere situated in the axis exerts at the distance* PE *upon the corpuscle* P *conjointly: I say, that the whole force with which the corpuscle* P *is attracted towards the sphere is as the area* ANB, *comprehended under the axis of the sphere* AB, *and the curved line* ANB, *the locus of the point* N.

For supposing the construction in the last Lemma and Theorem to stand, conceive the axis of the sphere AB to be divided into innumerable equal particles D*d*, and the whole sphere to be divided into so many spherical

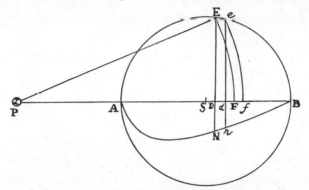

concavoconvex laminæ EF*fe*; and erect the perpendicular *dn*. By the last Theorem, the force with which the laminæ EF*fe* attract the corpuscle P is as DE² · F*f* and the force of one particle exerted at the distance PE or PF, conjointly. But (by the last Lemma) D*d* is to F*f* as PE to PS, and therefore F*f* is equal to $\dfrac{PS \cdot Dd}{PE}$; and DE² · F*f* is equal to $Dd \cdot \dfrac{DE^2 \cdot PS}{PE}$; and therefore the force of the lamina EF*fe* is as $Dd \cdot \dfrac{DE^2 \cdot PS}{PE}$ and the force of a particle exerted at the distance PF conjointly; that is, by the supposition, as DN · Dd, or as the evanescent area DNnd. Therefore the forces of all the laminæ exerted upon the corpuscle P are as all the areas DNnd, that is, the whole force of the sphere will be as the whole area ANB. Q.E.D.

Cor. I. Hence if the centripetal force tending to the several particles remain always the same at all distances, and DN be made as $\dfrac{DE^2 \cdot PS}{PE}$, the whole force with which the corpuscle is attracted by the sphere is as the area ANB.

Cor. II. If the centripetal force of the particles be inversely as the distance of the corpuscle attracted by it, and DN be made as $\dfrac{DE^2 \cdot PS}{PE^2}$, the force with which the corpuscle P is attracted by the whole sphere will be as the area ANB.

Cor. III. If the centripetal force of the particles be inversely as the cube of the distance of the corpuscle attracted by it, and DN be made as $\dfrac{DE^2 \cdot PS}{PE^4}$, the force with which the corpuscle is attracted by the whole sphere will be as the area ANB.

Cor. IV. And universally if the centripetal force tending to the several particles of the sphere be supposed to be inversely as the quantity V; and DN be made as $\dfrac{DE^2 \cdot PS}{PE \cdot V}$; the force with which a corpuscle is attracted by the whole sphere will be as the area ANB.

PROPOSITION LXXXI. PROBLEM XLI

The things remaining as above, it is required to measure the area ANB.

From the point P let there be drawn the right line PH touching the sphere in H; and to the axis PAB, letting fall the perpendicular HI, bisect PI in L; and (by Prop. XII, Book II, *Elem.* of *Euclid*) PE^2 is equal to PS^2 + SE^2 + 2PS · SD. But because the triangles SPH, SHI are alike, SE^2 or SH^2 is equal to the rectangle PS · IS. Therefore PE^2 is equal to the rectangle contained under PS and PS + SI + 2SD; that is, under PS and 2LS + 2SD; that is, under PS and 2LD. Moreover DE^2 is equal to SE^2 − SD^2, or

$$SE^2 - LS^2 + 2LS \cdot LD - LD^2,$$

that is, $2LS \cdot LD - LD^2 - LA \cdot LB.$

For $LS^2 - SE^2$ or $LS^2 - SA^2$ (by Prop. vi, Book ii, *Elem.* of *Euclid*) is equal to the rectangle $LA \cdot LB$. Therefore if instead of DE^2 we write

$$2LS \cdot LD - LD^2 - LA \cdot LB,$$

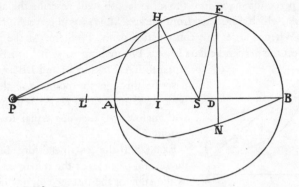

the quantity $\dfrac{DE^2 \cdot PS}{PE \cdot V}$, which (by Cor. iv of the foregoing Prop.) is as the length of the ordinate DN, will now resolve itself into three parts

$$\frac{2SLD \cdot PS}{PE \cdot V} - \frac{LD^2 \cdot PS}{PE \cdot V} - \frac{ALB \cdot PS}{PE \cdot V};$$

where if instead of V we write the inverse ratio of the centripetal force, and instead of PE the mean proportional between PS and $2LD$, those three parts will become ordinates to so many curved lines, whose areas are discovered by the common methods. Q.E.D.

EXAM. 1. If the centripetal force tending to the several particles of the sphere be inversely as the distance; instead of V write PE the distance, then $2PS \cdot LD$ for PE^2; and DN will become as $SL - \frac{1}{2}LD - \dfrac{LA \cdot LB}{2LD}$. Suppose DN equal to its double $2SL - LD - \dfrac{LA \cdot LB}{LD}$; and $2SL$ the given part of the ordinate drawn into the length AB will describe the rectangular area $2SL \cdot AB$; and the indefinite part LD, drawn perpendicularly into the same length with a continued motion, in such sort as in its motion one way or another it may either by increasing or decreasing remain always equal to the length LD, will describe the area $\dfrac{LB^2 - LA^2}{2}$, that is, the area $SL \cdot AB$;

which taken from the former area $2SL \cdot AB$, leaves the area $SL \cdot AB$. But the third part $\dfrac{LA \cdot LB}{LD}$, drawn after the same manner with a continued motion perpendicularly into the same length, will describe the area of an hyperbola, which subtracted from the area $SL \cdot AB$ will leave ANB the area sought. Whence arises this construction of the Problem. At the points L, A, B, erect the perpendiculars Ll, Aa, Bb; making Aa equal to LB, and Bb

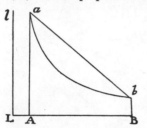

equal to LA. Making Ll and LB asymptotes, describe through the points a, b the hyperbolic curve ab. And the chord ba being drawn, will inclose the area aba equal to the area sought ANB.

EXAM. 2. If the centripetal force tending to the several particles of the sphere be inversely as the cube of the distance, or (which is the same thing) as that cube applied to any given plane; write $\dfrac{PE^3}{2AS^2}$ for V, and $2PS \cdot LD$ for PE^2; and DN will become as

$$\frac{SL \cdot AS^2}{PS \cdot LD} - \frac{AS^2}{2PS} - \frac{LA \cdot LB \cdot AS^2}{2PS \cdot LD^2},$$

that is (because PS, AS, SI are continually proportional), as

$$\frac{LSI}{LD} - \tfrac{1}{2}SI - \frac{LA \cdot LB \cdot SI}{2LD^2}.$$

If we draw then these three parts into the length AB, the first $\dfrac{SL \cdot SI}{LD}$ will generate the area of an hyperbola; the second $\tfrac{1}{2}SI$ the area $\tfrac{1}{2}AB \cdot SI$; the third $\dfrac{LA \cdot LB \cdot SI}{2LD^2}$ the area $\dfrac{LA \cdot LB \cdot SI}{2LA} - \dfrac{LA \cdot LB \cdot SI}{2LB}$, that is, $\tfrac{1}{2}AB \cdot SI$. From the first subtract the sum of the second and third, and there will remain ANB the area sought. Whence arises this construction of the Problem. At the points L, A, S, B, erect the perpendiculars Ll, Aa, Ss, Bb, of which suppose Ss equal to SI; and through the point s, to the asymptotes Ll, LB, describe the hyperbola asb meeting the perpendiculars Aa, Bb in a and b; and the rectangle 2SA · SI subtracted from the hyperbolic area AasbB, will leave ANB the area sought.

EXAM. 3. If the centripetal force tending to the several particles of the spheres decrease as the fourth power of the distance from the particles; write $\dfrac{PE^4}{2AS^3}$ for V, then $\sqrt{(2PS+LD)}$ for PE, and DN will become as

$$\frac{SI^2 \cdot SL}{\sqrt{2SI}} \cdot \frac{1}{\sqrt{LD^3}} - \frac{SI^2}{2\sqrt{2SI}} \cdot \frac{1}{\sqrt{LD}} - \frac{SI^2 \cdot LA \cdot LB}{2\sqrt{2SI}} \cdot \frac{1}{\sqrt{LD^5}}.$$

These three parts drawn into the length AB, produce so many areas, viz.,

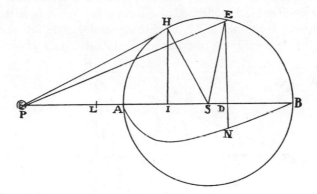

$\dfrac{2SI^2 \cdot SL}{\sqrt{2SI}}$ into $\left(\dfrac{1}{\sqrt{LA}} - \dfrac{1}{\sqrt{LB}}\right)$; $\dfrac{SI^2}{\sqrt{2SI}}$ into $\sqrt{(LB - \sqrt{LA})}$; and

$\dfrac{SI^2 \cdot LA \cdot LB}{3\sqrt{2SI}}$ into $\left(\dfrac{1}{\sqrt{LA^3}} - \dfrac{1}{\sqrt{LB^3}}\right)$. And these after due reduction come

forth $\dfrac{2SI^2 \cdot SL}{LI}$, SI^2, and $SI^2 + \dfrac{2SI^3}{3LI}$. And these by subtracting the last from

the first, become $\dfrac{4SI^3}{3LI}$. Therefore the entire force with which the corpuscle

P is attracted towards the centre of the sphere is as $\dfrac{SI^3}{PI}$, that is, inversely

as $PS^3 \cdot PI$. Q.E.I.

By the same method one may determine the attraction of a corpuscle situated within the sphere, but more expeditiously by the following Theorem.

PROPOSITION LXXXII. THEOREM XLI

In a sphere described about the centre S with the radius SA, if there be taken SI, SA, SP continually proportional: I say, that the attraction of a corpuscle within the sphere in any place I is to its attraction without the sphere in the place P in a ratio compounded of the square root of the ratio of IS, PS, the distances from the centre, and the square root of the ratio of the centripetal forces tending to the centre in those places P and I.

As, if the centripetal forces of the particles of the sphere be inversely as the distances of the corpuscle attracted by them, the force with which the corpuscle situated in I is attracted by the entire sphere will be to the force with which it is attracted in P in a ratio compounded of the square root of the ratio of the distance SI to the distance SP, and the square root of the ratio

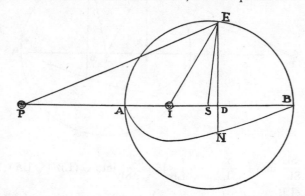

of the centripetal force in the place I arising from any particle in the centre to the centripetal force in the place P arising from the same particle in the centre; that is, inversely as the square root of the ratio of the distances SI, SP to each other. These two square roots of ratios compose the ratio of equality, and therefore the attractions in I and P produced by the whole sphere are equal. By the like calculation, if the forces of the particles of the sphere are inversely as the square of the ratio of the distances, it will be found that the attraction in I is to the attraction in P as the distance SP to the semidiameter SA of the sphere. If those forces are inversely as the cube of the ratio of the distances, the attractions in I and P will be to each other as SP^2 to SA^2; if as the fourth power of the ratio, as SP^3 to SA^3. Therefore since the attraction in P was found in this last case to be inversely as $PS^3 \cdot PI$, the attraction in I will be inversely as $SA^3 \cdot PI$, that is, because SA^3 is given, inversely as PI. And the progression is the same *in infinitum*. The demonstration of this Theorem is as follows:

The things remaining as above constructed, and a corpuscle being in any place P, the ordinate DN was found to be as $\dfrac{DE^2 \cdot PS}{PE \cdot V}$. Therefore if IE be drawn, that ordinate for any other place of the corpuscle, as I, will become (other things being equal) as $\dfrac{DE^2 \cdot IS}{IE \cdot V}$. Suppose the centripetal forces flowing from any point of the sphere, as E, to be to each other at the distances IE and PE as PE^n to IE^n (where the number n denotes the index of the powers of PE and IE), and those ordinates will become as $\dfrac{DE^2 \cdot PS}{PE \cdot PE^n}$ and $\dfrac{DE^2 \cdot IS}{IE \cdot IE^n}$, whose ratio to each other is as $PS \cdot IE \cdot IE^n$ to $IS \cdot PE \cdot PE^n$. Because SI, SE, SP are in continued proportion, the triangles SPE, SEI are alike; and thence IE is to PE as IS to SE or SA. For the ratio of IE to PE write the ratio of IS to SA; and the ratio of the ordinates becomes that of $PS \cdot IE^n$ to $SA \cdot PE^n$. But the ratio of PS to SA is the square root of that of the distances PS, SI; and the ratio of IE^n to PE^n (because IE is to PE as IS to SA) is the square root of that of the forces at the distances PS, IS. Therefore the ordinates, and consequently the areas which the ordinates describe, and the attractions proportional to them, are in a ratio compounded of the square root of those ratios. Q.E.D.

PROPOSITION LXXXIII. PROBLEM XLII

To find the force with which a corpuscle placed in the centre of a sphere is attracted towards any segment of that sphere whatsoever.

Let P be a body in the centre of that sphere, and RBSD a segment thereof contained under the plane RDS and the spherical surface RBS. Let DB be cut in F by a spherical surface EFG described from the centre P, and let the segment be divided into the parts BREFGS, FEDG. Let us suppose that segment to be not a purely mathematical but a physical surface, having

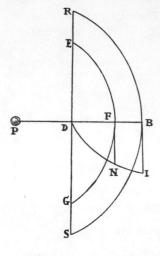

some, but a perfectly inconsiderable thickness. Let that thickness be called O, and (by what *Archimedes* hath demonstrated) that surface will be as $PF \cdot DF \cdot O$. Let us suppose, besides, the attractive forces of the particles of the sphere to be inversely as that power of the distances, of which n is index; and the force with which the surface EFG attracts the body P will be (by Prop. LXXIX) as $\dfrac{DE^2 \cdot O}{PF^n}$, that is, as $\dfrac{2DF \cdot O}{PF^{n-1}} - \dfrac{DF^2 \cdot O}{PF^n}$. Let the perpendicular FN multiplied by O be proportional to this quantity; and the curvilinear area BDI, which the ordinate FN, drawn through the length DB with a continued motion will describe, will be as the whole force with which the whole segment RBSD attracts the body P. Q.E.I.

PROPOSITION LXXXIV. PROBLEM XLIII

To find the force with which a corpuscle, placed without the centre of a sphere in the axis of any segment, is attracted by that segment.

Let the body P placed in the axis ADB of the segment EBK be attracted by that segment. About the centre P, with the radius PE, let the spherical surface EFK be described; and let it divide the segment into two parts EBKFE and EFKDE. Find the force of the first of those parts by Prop. LXXXI, and the force of the latter part by

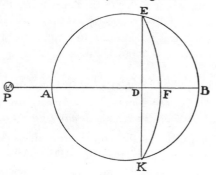

Prop. LXXXIII, and the sum of the forces will be the force of the whole segment EBKDE. Q.E.I.

SCHOLIUM

The attractions of spherical bodies being now explained, it comes next in order to treat of the laws of attraction in other bodies consisting in like manner of attractive particles; but to treat of them particularly is not necessary to my design. It will be sufficient to add some general Propositions relating to the forces of such bodies, and the motions thence arising, because the knowledge of these will be of some little use in philosophical inquiries.

SECTION XIII

The attractive forces of bodies which are not spherical.

PROPOSITION LXXXV. THEOREM XLII

If a body be attracted by another, and its attraction be vastly stronger when it is contiguous to the attracting body than when they are separated from each other by a very small interval; the forces of the particles of the attracting body decrease, in the recess of the body attracted, in more than the squared ratio of the distance of the particles.

For if the forces decrease as the square of the distances from the particles, the attraction towards a spherical body being (by Prop. LXXIV) inversely as the square of the distance of the attracted body from the centre of the sphere, will not be sensibly increased by the contact, and it will be still less increased by it, if the attraction, in the recess of the body attracted, decreases in a still less proportion. The Proposition, therefore, is evident concerning attractive spheres. And the case is the same of concave spherical orbs attracting external bodies. And much more does it appear in orbs that attract bodies placed within them, because there the attractions diffused through the cavities of those orbs are (by Prop. LXX) destroyed by contrary attractions, and therefore have no effect even in the place of contact. Now if from these spheres and spherical orbs we take away any parts remote from the place of contact, and add new parts anywhere at pleasure, we may change the figures of the attractive bodies at pleasure; but the parts added or taken away, being remote from the place of contact, will cause no remarkable excess of the attraction arising from the contact of the two bodies. Therefore the Proposition holds good in bodies of all figures. Q.E.D.

PROPOSITION LXXXVI. THEOREM XLIII

If the forces of the particles of which an attractive body is composed decrease, in the recession of the attractive body, as the third or more than the third power of the distance from the particles, the attraction will be vastly stronger in the point of contact than when the attracting and attracted bodies are separated from each other, though by ever so small an interval.

For that the attraction is infinitely increased when the attracted corpuscle comes to touch an attracting sphere of this kind, appears, by the solution of Problem xli, exhibited in the second and third Examples. The same will also appear (by comparing those Examples and Theor. xli together) of attractions of bodies made towards concavoconvex orbs, whether the attracted bodies be placed without the orbs, or in the cavities within them. And by adding to or taking from those spheres and orbs any attractive matter anywhere without the place of contact, so that the attractive bodies may receive any assigned figure, the Proposition will hold good of all bodies universally. Q.E.D.

PROPOSITION LXXXVII. THEOREM XLIV

If two bodies similar to each other, and consisting of matter equally attractive, attract separately two corpuscles proportional to those bodies, and in a like situation to them, the accelerative attractions of the corpuscles towards the entire bodies will be as the accelerative attractions of the corpuscles towards particles of the bodies proportional to the wholes, and similarly situated in them.

For if the bodies are divided into particles proportional to the wholes, and alike situated in them, it will be, as the attraction towards any particle of one of the bodies to the attraction towards the correspondent particle in the other body, so are the attractions towards the several particles of the first body, to the attractions towards the several correspondent particles of the other body; and, by composition, so is the attraction towards the first whole body to the attraction towards the second whole body. Q.E.D.

Cor. i. Therefore if, as the distances of the corpuscles attracted increase, the attractive forces of the particles decrease in the ratio of any power of the distances, the accelerative attractions towards the whole bodies will be directly as the bodies, and inversely as those powers of the distances. As if the forces of the particles decrease as the square of the distances from the corpuscles attracted, and the bodies are as A^3 and B^3, and therefore both the cubic sides of the bodies, and the distance of the attracted corpuscles from the bodies, are as A and B; the accelerative attractions towards the bodies will be as $\dfrac{A^3}{A^2}$ and $\dfrac{B^3}{B^2}$, that is, as A and B the cubic sides of those bodies. If the forces of the particles decrease as the cube of the distances from the attracted corpuscles, the accelerative attractions towards the whole bodies will be as $\dfrac{A^3}{A^3}$ and $\dfrac{B^3}{B^3}$, that is, equal. If the forces decrease as the fourth

power, the attractions towards the bodies will be as $\dfrac{A^3}{A^4}$ and $\dfrac{B^3}{B^4}$, that is,

inversely as the cubic sides A and B. And so in other cases.

Cor. II. Hence, on the other hand, from the forces with which like bodies attract corpuscles similarly situated, may be obtained the ratio of the decrease of the attractive forces of the particles as the attracted corpuscle recedes from them; if only that decrease is directly or inversely in any ratio of the distances.

PROPOSITION LXXXVIII. THEOREM XLV

If the attractive forces of the equal particles of any body be as the distance of the places from the particles, the force of the whole body will tend to its centre of gravity; and will be the same with the force of a globe, consisting of similar and equal matter, and having its centre in the centre of gravity.

Let the particles A, B of the body RSTV attract any corpuscle Z with forces which, supposing the particles to be equal between themselves, are as the distances AZ, BZ; but, if they are supposed unequal, are as those

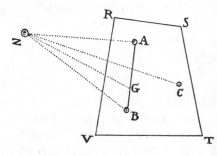

particles and their distances AZ, BZ conjointly, or (if I may so speak) as those particles multiplied by their distances AZ, BZ respectively. And let those forces be expressed by the contents under A · AZ, and B · BZ. Join AB, and let it be cut in G, so that AG may be to BG as the particle B to the particle A; and G will be the common centre of gravity of the particles A and B. The force A · AZ will (by Cor. II of the Laws) be resolved into the forces A · GZ and A · AG; and the force B · BZ into the forces B · GZ and B · BG. Now the forces A · AG and B · BG, because A is proportional to B, and BG to AG, are equal, and therefore having contrary directions destroy one another. There remain then the forces $A \cdot GZ$ and $B \cdot GZ$. These tend from Z towards the centre G, and compose the force $(A + B) \cdot GZ$; that is, the same force as if the attractive particles A and B were placed in their common centre of gravity G, composing there a little globe.

By the same reasoning, if there be added a third particle C, and the force of it be compounded with the force $(A + B) \cdot GZ$ tending to the centre G, the force thence arising will tend to the common centre of gravity of that globe in G and of the particle C; that is, to the common centre of gravity of the three particles A, B, C; and will be the same as if that globe and the particle C were placed in that common centre composing a greater globe there; and so we may go on *in infinitum*. Therefore the whole force of all the particles of any body whatever RSTV is the same as if that body, without removing its centre of gravity, were to put on the form of a globe. Q.E.D.

Cor. Hence the motion of the attracted body Z will be the same as if the attracting body RSTV were spherical; and therefore if that attracting body be either at rest, or proceed uniformly in a right line, the body attracted will move in an ellipse having its centre in the centre of gravity of the attracting body.

PROPOSITION LXXXIX. THEOREM XLVI

If there be several bodies consisting of equal particles whose forces are as the distances of the places from each, the force compounded of all the forces by which any corpuscle is attracted will tend to the common centre of gravity of the attracting bodies; and will be the same as if those attracting bodies, preserving their common centre of gravity, should unite there, and be formed into a globe.

This is demonstrated after the same manner as the foregoing Proposition.

Cor. Therefore the motion of the attracted body will be the same as if the attracting bodies, preserving their common centre of gravity, should unite there, and be formed into a globe. And, therefore, if the common centre of gravity of the attracting bodies be either at rest, or proceed uniformly in a right line, the attracted body will move in an ellipse having its centre in the common centre of gravity of the attracting bodies.

PROPOSITION XC. PROBLEM XLIV

If to the several points of any circle there tend equal centripetal forces, increasing or decreasing in any ratio of the distances; it is required to find the force with which a corpuscle is attracted, that is, situated anywhere in a right line which stands at right angles to the plane of the circle at its centre.

Suppose a circle to be described about the centre A with any radius AD in a plane to which the right line AP is perpendicular; and let it be required to find the force with which a corpuscle P is attracted towards the same. From any point E of the circle, to the attracted corpuscle P, let there be drawn the right line PE. In the right line PA take PF equal to PE, and

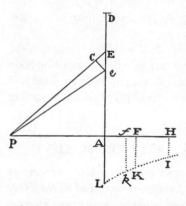

make a perpendicular FK, erected at F, to be as the force with which the point E attracts the corpuscle P. And let the curved line IKL be the locus of the point K. Let that curve meet the plane of the circle in L. In PA take PH equal to PD, and erect the perpendicular HI meeting that curve in I; and the attraction of the corpuscle P towards the circle will be as the area AHIL multiplied by the altitude AP. Q.E.I.

For let there be taken in AE a very small line E*e*. Join P*e*, and in PE, PA take PC, P*f*, both equal to P*e*. And because the force, with which any point E of the ring described about the centre A with the radius AE in the aforesaid plane attracts to itself the body P, is supposed to be as FK; and, therefore, the force with which that point attracts the body P towards A is as $\dfrac{AP \cdot FK}{PE}$; and the force with which the whole ring attracts the body P towards A is as the ring and $\dfrac{AP \cdot FK}{PE}$ conjointly; and that ring also is as the rectangle under the radius AE and the breadth E*e*, and this rectangle (because PE and AE, E*e* and CE are proportional) is equal to the rectangle PE · CE or PE · F*f*; the force with which that ring attracts the body P towards A will be as PE · Ff and $\dfrac{AP \cdot FK}{PE}$ conjointly; that is, as the content under Ff · FK · AP, or as the area FKkf multiplied by AP. And therefore the sum of the forces with which all the rings, in the circle described about the centre A with the radius AD, attract the body P towards A, is as the whole area AHIKL multiplied by AP. Q.E.D.

Cor. I. Hence if the forces of the points decrease as the square of the distances, that is, if FK be as $\dfrac{1}{PF^2}$, and therefore the area AHIKL as $\dfrac{1}{PA} - \dfrac{1}{PH}$; the attraction of the corpuscle P towards the circle will be as

$$1 - \frac{PA}{PH}; \text{ that is, as } \frac{AH}{PH}.$$

Cor. II. And universally if the forces of the points at the distances D be inversely as any power D^n of the distances; that is, if FK be as $\dfrac{1}{D^n}$, and therefore the area AHIKL as $\dfrac{1}{PA^{n-1}} - \dfrac{1}{PH^{n-1}}$; the attraction of the corpuscle P towards the circle will be as $\dfrac{1}{PA^{n-2}} - \dfrac{PA}{PH^{n-1}}$.

Cor. III. And if the diameter of the circle be increased *in infinitum,* and the number n be greater than unity; the attraction of the corpuscle P towards the whole infinite plane will be inversely as PA^{n-2}, because the other term $\dfrac{PA}{PH^{n-1}}$ vanishes.

PROPOSITION XCI. PROBLEM XLV

To find the attraction of a corpuscle situated in the axis of a round solid, to whose several points there tend equal centripetal forces decreasing in any ratio of the distances whatsoever.

Let the corpuscle P, situated in the axis AB of the solid DECG, be attracted towards that solid. Let the solid be cut by any circle as RFS, perpendicular to the axis; and in its semidiameter FS, in any plane PALKB

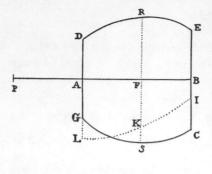

passing through the axis, let there be taken (by Prop. xc) the length FK proportional to the force with which the corpuscle P is attracted towards that circle. Let the locus of the point K be the curved line LKI, meeting the planes of the outermost circles AL and BI in L and I; and the attraction of the corpuscle P towards the solid will be as the area LABI. Q.E.I.

COR. I. Hence if the solid be a cylinder described by the parallelogram ADEB revolved about the axis AB, and the centripetal forces tending to the several points be inversely as the squares of the distances from the points; the attraction of the corpuscle P towards this cylinder will be as $AB - PE + PD$. For the ordinate FK (by Cor. I, Prop. xc) will be as $1 - \dfrac{PF}{PR}$. The part 1 of this quantity, multiplied by the length AB, describes the area $1 \cdot AB$; and the other part $\dfrac{PF}{PR}$, multiplied by the length PB, describes the area

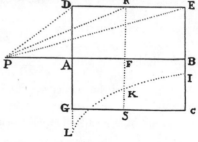

$1 \cdot (PE - AD)$ (as may be easily shown from the quadrature of the curve LKI); and, in like manner, the same part multiplied by the length PA describes the area $1 \cdot (PD - AD)$, and multiplied by AB, the difference of PB and PA, describes $1 \cdot (PE - PD)$, the difference of the areas. From the first content $1 \cdot AB$ take away the last content $1 \cdot (PE - PD)$, and there will remain the area LABI equal to $1 \cdot (AB - PE + PD)$. Therefore the force, being proportional to this area, is as $AB - PE + PD$.

COR. II. Hence also is known the force by which a spheroid AGBC attracts any body P situate externally in its axis AB. Let NKRM be a conic section whose ordinate ER perpendicular to PE may be always equal to the length of the line PD, continually drawn to the point D in which that ordinate cuts the spheroid. From the vertices A, B of the spheroid, let there be erected to its axis AB the perpendiculars AK, BM, respectively equal to AP, BP, and therefore meeting the conic section in K and M; and join KM

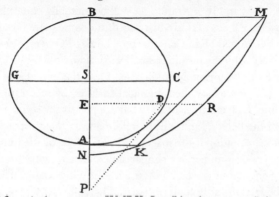

cutting off from it the segment KMRK. Let S be the centre of the spheroid, and SC its greatest semidiameter; and the force with which the spheroid attracts the body P will be to the force with which a sphere described with the diameter AB attracts the same body as $\dfrac{AS \cdot CS^2 - PS \cdot KMRK}{PS^2 + CS^2 - AS^2}$ is to $\dfrac{AS^3}{3PS^2}$. And by a calculation founded on the same principles may be found the forces of the segments of the spheroid.

Cor. III. If the corpuscle be placed within the spheroid and in its axis, the attraction will be as its distance from the centre. This may be easily inferred from the following reasoning, whether the particle be in the axis or in any

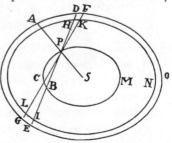

other given diameter. Let AGOF be an attracting spheroid, S its centre, and P the body attracted. Through the body P let there be drawn the semidiameter SPA, and two right lines DE, FG meeting the spheroid in D and E, F and G; and let PCM, HLN be the surfaces of two interior spheroids similar and concentric to the exterior, the first of which passes through the body P, and cuts the right lines DE, FG in B and C; the latter cuts the same right lines in H and I, K and L. Let the spheroids have all one common axis, and the parts of the right lines intercepted on both sides DP and BE, FP and CG, DH and IE, FK and LG, will be mutually equal; because the right lines DE, PB, and HI are bisected in the same point, as are also the right lines FG, PC, and KL. Conceive now DPF, EPG to represent opposite cones described with the infinitely small vertical angles DPF, EPG, and the lines DH, EI to be infinitely small also. Then the particles of the cones DHKF, GLIE, cut off by the spheroidal surfaces, by reason of the equality of the lines DH and EI, will be to one another as the squares of the distances from the body P, and will therefore attract that corpuscle equally. And by a like reasoning if the spaces DPF, EGCB be divided into particles by the surfaces of innumerable similar spheroids concentric to the former and having one common axis, all these particles will equally attract on both sides the body P towards contrary parts. Therefore the forces of the cone DPF, and of the conic segment EGCB, are equal, and by their opposed actions destroy each other. And the case is the same of the forces of all the matter that lies without the interior spheroid PCBM. Therefore the body P is attracted by the interior spheroid PCBM alone, and therefore (by Cor. iii, Prop. lxxii) its attraction is to the force with which the body A is attracted by the whole spheroid AGOD as the distance PS is to the distance AS. Q.E.D.

PROPOSITION XCII. PROBLEM XLVI

An attracting body being given, it is required to find the ratio of the decrease of the centripetal forces tending to its several points.

The body given must be formed into a sphere, a cylinder, or some regular figure, whose law of attraction answering to any ratio of decrease may be found by Prop. lxxx, lxxxi, and xci. Then, by experiments, the force of the attractions must be found at several distances, and the law of attraction towards the whole, made known by that means, will give the ratio of the decrease of the forces of the several parts; which was to be found.

PROPOSITION XCIII. THEOREM XLVII

If a solid be plane on one side, and infinitely extended on all other sides, and consist of equal particles equally attractive, whose forces decrease, in receding from the solid, in the ratio of any power greater than the square of the distances; and a corpuscle placed towards either part of the plane is attracted by the force of the whole solid: I say, that the attractive force of the whole solid, in receding from its plane surface will decrease in the ratio of a power whose side is the distance of the corpuscle from the plane, and its index less by 3 than the index of the power of the distances.

CASE I. Let LG*l* be the plane by which the solid is terminated. Let the solid lie on that side of the plane that is towards I, and let it be resolved into innumerable planes *m*HM, *n*IN, *o*KO, &c., parallel to GL. And first let the attracted body C be placed without the solid. Let there be drawn CGHI perpendicular to those innumerable planes, and let the attractive forces of the points of the solid decrease in the ratio of a power of the distances whose index is the number n not less than 3. Therefore (by Cor. III, Prop. xc) the force with which any plane *m*HM attracts the point C is inversely as CH^{n-2}. In the plane *m*HM take the length HM inversely proportional to CH^{n-2}, and that force will be as HM. In like manner in the several planes *l*GL, *n*IN, *o*KO, &c., take the lengths GL, IN, KO, &c., inversely proportional to CG^{n-2}, CI^{n-2}, CK^{n-2}, &c., and the forces of those planes will be as the lengths so taken, and therefore the sum of the forces as the sum of the lengths, that is, the force of the whole solid as

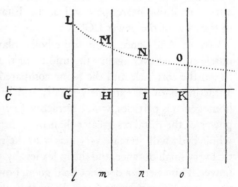

the area GLOK produced infinitely towards OK. But that area (by the known methods of quadratures) is inversely as CG^{n-3}, and therefore the force of the whole solid is inversely as CG^{n-3}. Q.E.D.

CASE 2. Let the corpuscle C be now placed on that side of the plane *l*GL that is within the solid, and take the distance CK equal to the distance CG.

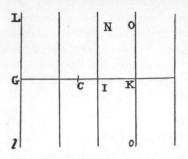

And the part of the solid LG*loKO* terminated by the parallel planes *l*GL, *o*KO, will attract the corpuscle C, situated in the middle, neither one way nor another, the contrary actions of the opposite points destroying one another by reason of their equality. Therefore the corpuscle C is attracted by the force only of the solid situated beyond the plane OK. But this force (by Case 1) is inversely as CK^{n-3}, that is (because CG, CK are equal), inversely as CG^{n-3}. Q.E.D.

COR. I. Hence if the solid LGIN be terminated on each side by two infinite parallel planes LG, IN, its attractive force is known, subtracting from the attractive force of the whole infinite solid LGKO the attractive force of the more distant part NIKO infinitely produced towards KO.

COR. II. If the more distant part of this solid be rejected, because its attraction compared with the attraction of the nearer part is inconsiderable, the attraction of that nearer part will, as the distance increases, decrease nearly in the ratio of the power CG^{n-3}.

COR. III. And hence if any finite body, plane on one side, attract a corpuscle situated over against the middle of that plane, and the distance between the corpuscle and the plane compared with the dimensions of the attracting body be extremely small; and the attracting body consist of homogeneous particles, whose attractive forces decrease in the ratio of any power of the distances greater than the fourth; the attractive force of the whole body will decrease very nearly in the ratio of a power whose side is that very small distance, and the index less by 3 than the index of the former power. This assertion does not hold good, however, of a body consisting of particles whose attractive forces decrease in the ratio of the third power of the distances; because, in that case, the attraction of the remoter part of the infinite body in the second Corollary is always infinitely greater than the attraction of the nearer part.

SCHOLIUM

If a body is attracted perpendicularly towards a given plane, and from the law of attraction given, the motion of the body be required; the Problem will be solved by seeking (by Prop. xxxix) the motion of the body descending in a right line towards that plane, and (by Cor. ii of the Laws) compounding that motion with an uniform motion performed in the direction of lines parallel to that plane. And, on the contrary, if there be required the law of the attraction tending towards the plane in perpendicular directions, by which the body may be caused to move in any given curved line, the Problem will be solved by working after the manner of the third Problem.

But the operations may be contracted by resolving the ordinates into converging series. As if to a base A the length B be ordinately applied in any given angle, and that length be as any power of the base $A^{\frac{m}{n}}$; and there be sought the force with which a body, either attracted towards the base or driven from it in the direction of that ordinate, may be caused to move in the curved line which that ordinate always describes with its superior extremity; I suppose the base to be increased by a very small part O, and I resolve the ordinate $(A+O)^{\frac{m}{n}}$ into an infinite series

$$A^{\frac{m}{n}} + \frac{m}{n} OA^{\frac{m-n}{n}} + \frac{mm-mn}{2nn} OOA^{\frac{m-2n}{n}} \text{ \&c.,}$$

and I suppose the force proportional to the term of this series in which O is of two dimensions, that is, to the term $\frac{mm-mn}{2nn} OOA^{\frac{m-2n}{n}}$. Therefore the force sought is as $\frac{mm-mn}{nn} A^{\frac{m-2n}{n}}$, or, which is the same thing, as $\frac{mm-mn}{nn} B^{\frac{m-2n}{m}}$. As if the ordinate describe a parabola, m being $= 2$, and $n = 1$, the force will be as the given quantity $2B^0$, and therefore is given. Therefore with a given force the body will move in a parabola, as *Galileo* hath demonstrated. If the ordinate describe an hyperbola, m being $= 0 - 1$, and $n = 1$, the force will be as $2A^{-3}$ or $2B^3$; and therefore a force which is as the cube of the ordinate will cause the body to move in an hyperbola. But leaving Propositions of this kind, I shall go on to some others relating to motion which I have not yet touched upon.

SECTION XIV

The motion of very small bodies when agitated by centripetal forces tending to the several parts of any very great body.

PROPOSITION XCIV. THEOREM XLVIII

If two similar mediums be separated from each other by a space terminated on both sides by parallel planes, and a body in its passage through that space be attracted or impelled perpendicularly towards either of those mediums, and not agitated or hindered by any other force; and the attraction be everywhere the same at equal distances from either plane, taken towards the same side of the plane: I say, that the sine of incidence upon either plane will be to the sine of emergence from the other plane in a given ratio.

CASE I. Let A*a* and B*b* be two parallel planes, and let the body light upon the first plane A*a* in the direction of the line GH, and in its whole passage through the intermediate space let it be attracted or impelled towards the medium of incidence, and by that action let it be made to describe a curved

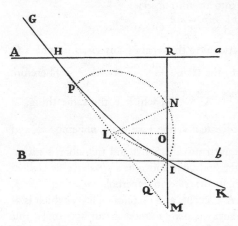

line HI, and let it emerge in the direction of the line IK. Let there be erected IM perpendicular to B*b* the plane of emergence, and meeting the line of incidence GH prolonged in M, and the plane of incidence A*a* in R; and let the line of emergence KI be produced and meet HM in L. About the centre L, with the radius LI, let a circle be described cutting both HM in P and Q, and MI produced in N; and, first, if the attraction or impulse be supposed uniform, the curve HI (by what *Galileo* hath demonstrated) will be a parabola, whose property is that of a rectangle under its given latus rectum, and the line IM equal to the square of HM; and moreover the line HM will be bisected in L. Hence if to MI there be let fall the perpendicular LO, then MO, OR will be equal; and adding the equal lines ON, OI, the wholes MN, IR will be equal also. Therefore since IR is given, MN is also given, and the rectangle MI · MN is to the rectangle under the latus rectum and IM, that is, to HM^2 in a given ratio. But the rectangle MI · MN is equal to the rectangle MP · MQ, that is, to the difference of the squares ML^2, and PL^2 or LI^2; and HM^2 hath a given ratio to its fourth part ML^2; therefore the ratio of $ML^2 - LI^2$ to ML^2 is given, and by conversion the ratio of LI^2 to ML^2, and its square root, the ratio of LI to ML. But in every triangle, as LMI, the sines of the angles are proportional to the opposite sides. Therefore the ratio of the sine of the angle of incidence LMR to the sine of the angle of emergence LIR is given. Q.E.D.

CASE 2. Let now the body pass successively through several spaces terminated with parallel planes A*ab*B, B*bc*C, &c., and let it be acted on by a force which is uniform in each of them separately, but different in the different spaces; and by what was just demonstrated, the sine of the angle of incidence on the first plane A*a* is to the sine of emergence from the second plane B*b* in a given ratio; and this sine of incidence upon the second plane B*b* will be to the sine of emergence from the third plane

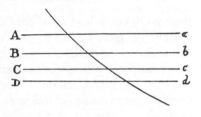

C*c* in a given ratio; and this sine to the sine of emergence from the fourth plane D*d* in a given ratio; and so on *in infinitum;* and, by multiplication of equals, the sine of incidence on the first plane is to the sine of emergence from the last plane in a given ratio. Let now the intervals of the planes be diminished, and their number be infinitely increased, so that the action of attraction or impulse, exerted according to any assigned law, may become continual, and the ratio of the sine of incidence on the first plane to the sine of emergence from the last plane being all along given, will be given then also. Q.E.D.

PROPOSITION XCV. THEOREM XLIX

*The same things being supposed, I say, that the velocity of the body before
its incidence is to its velocity after emergence as the sine of emergence to
the sine of incidence.*

Make AH and I*d* equal, and erect the perpendiculars AG, *d*K meeting
the lines of incidence and emergence GH, IK in G and K. In GH take
TH equal to IK, and to the plane A*a* let fall a perpendicular T*v*. And (by

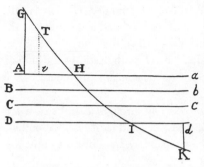

Cor. II of the Laws of Motion) let
the motion of the body be resolved
into two, one perpendicular to the
planes A*a*, B*b*, C*c*, &c., and another
parallel to them. The force of attraction or impulse, acting in directions perpendicular to those
planes, does not at all alter the
motion in parallel directions; and
therefore the body proceeding
with this motion will in equal
times go through those equal parallel intervals that lie between the line
AG and the point H, and between the point I and the line *d*K; that is,
they will describe the lines GH, IK in equal times. Therefore the velocity
before incidence is to the velocity after emergence as GH to IK or TH,
that is, as AH or I*d* to *v*H, that is (supposing TH or IK radius), as the sine
of emergence to the sine of incidence.[1] Q.E.D.

PROPOSITION XCVI. THEOREM L

*The same things being supposed, and that the motion before incidence is
swifter than afterwards: I say, that if the line of incidence be inclined continually, the body will be at last reflected, and the angle of reflection will
be equal to the angle of incidence.*

For conceive the body passing between the parallel planes A*a*, B*b*, C*c*,
&c., to describe parabolic arcs as above; and let those arcs be HP, PQ, QR,
&c. And let the obliquity of the line of incidence GH to the first plane A*a*

[1 Appendix, Note 26.]

be such that the sine of incidence may be to the radius of the circle whose
sine it is, in the same ratio which the same sine of incidence hath to the sine
of emergence from the plane D*d* into the space D*de*E; and because the sine
of emergence is now be-
come equal to the radius,
the angle of emergence
will be a right one, and
therefore the line of emergence will coincide with
the plane D*d*. Let the body

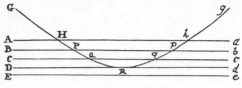

come to this plane in the point R; and because the line of emergence coincides with that plane, it is manifest that the body can proceed no farther
towards the plane E*e*. But neither can it proceed in the line of emergence
R*d*; because it is perpetually attracted or impelled towards the medium of
incidence. It will return, therefore, between the planes C*c*, D*d*, describing
an arc of a parabola QR*q*, whose principal vertex (by what *Galileo* hath
demonstrated) is in R, cutting the plane C*c* in the same angle at *q*, that it
did before at Q; then going on in the parabolic arcs *qp*, *ph*, &c., similar and
equal to the former arcs QP, PH, &c., it will cut the rest of the planes in
the same angles at *p*, *h*, &c., as it did before in P, H, &c., and will emerge at
last with the same obliquity at *h* with which it first impinged on that plane
at H. Conceive now the intervals of the planes A*a*, B*b*, C*c*, D*d*, E*e*, &c., to
be infinitely diminished, and the number infinitely increased, so that the
action of attraction or impulse, exerted according to any assigned law, may
become continual; and, the angle of emergence remaining all along equal
to the angle of incidence, will be equal to the same also at last. Q.E.D.

SCHOLIUM

These attractions bear a great resemblance to the reflections and refractions of light made in a given ratio of the secants, as was discovered by
Snell; and consequently in a given ratio of the sines, as was exhibited by
Descartes. For it is now certain from the phenomena of Jupiter's satellites, confirmed by the observations of different astronomers, that light is
propagated in succession, and requires about seven or eight minutes to
travel from the sun to the earth. Moreover, the rays of light that are in our

air (as lately was discovered by *Grimaldi,* by the admission of light into a dark room through a small hole, which I have also tried) in their passage near the angles of bodies, whether transparent or opaque (such as the circular and rectangular edges of gold, silver, and brass coins, or of knives,

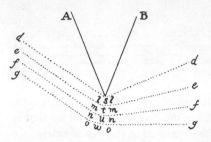

or broken pieces of stone or glass), are bent or inflected[1] round those bodies as if they were attracted to them; and those rays which in their passage come nearest to the bodies are the most inflected, as if they were most attracted; which thing I myself have also carefully observed. And those which pass at greater distances are less inflected; and those at still greater distances are a little inflected the contrary way, and form three fringes of colors. In the figure *s* represents the edge of a knife, or any kind of wedge A*s*B; and *gowog, fnunf, emtme, dlsld* are rays inflected towards the knife in the arcs *owo, nvn, mtm, lsl*; which inflection is greater or less according to their distance from the knife. Now since this inflection of the rays is performed in the air without the knife, it follows that the rays which fall upon the knife are first inflected in the air before they touch the knife. And the case is the same of the rays falling upon glass. The refraction, therefore, is made not in the point of incidence, but gradually, by a continual inflection of the rays; which is done partly in the air before they touch the glass, partly (if I mistake not) within the glass, after they have entered it; as is represented in the rays *ckzc, biyb, ahxa,* falling upon *r, q, p,* and inflected between *k* and *z, i* and *y, h* and *x.* Therefore because of the an-

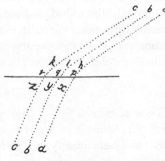

alogy there is between the propagation of the rays of light and the motion of bodies, I thought it not amiss to add the following Propositions for optical uses; not at all considering the nature of the rays of light, or inquiring

[1 Appendix, Note 27.]

whether they are bodies or not; but only determining the curves of bodies which are extremely like the curves of the rays.

PROPOSITION XCVII. PROBLEM XLVII

Supposing the sine of incidence upon any surface to be in a given ratio to the sine of emergence; and that the inflection of the paths of those bodies near that surface is performed in a very short space, which may be considered as a point; it is required to determine such a surface as may cause all the corpuscles issuing from any one given place to converge to another given place.

Let A be the place from whence the corpuscles diverge; B the place to which they should converge; CDE the curved line which by its revolution round the axis AB describes the surface sought; D, E any two points of that curve; and EF, EG perpendiculars let fall on the paths of the bodies AD, DB. Let the point D approach to and coalesce with the point E; and the

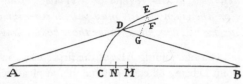

ultimate ratio of the line DF by which AD is increased, to the line DG by which DB is diminished, will be the same as that of the sine of incidence to the sine of emergence. Therefore the ratio of the increment of the line AD to the decrement of the line DB is given; and therefore if in the axis AB there be taken anywhere the point C through which the curve CDE must pass, and CM the increment of AC be taken in that given ratio to CN the decrement of BC, and from the centres A, B, with the radii AM, BN, there be described two circles cutting each other in D; that point D will touch the curve sought CDE, and, by touching it anywhere at pleasure, will determine that curve. Q.E.I.

Cor. 1. By causing the point A or B to go off sometimes *in infinitum,* and sometimes to move towards other parts of the point C, will be obtained all those figures which *Descartes* has exhibited in his *Optics* and *Geometry* relating to refractions. The invention of which *Descartes* having thought fit to conceal, is here laid open in this Proposition.

CoR. II. If a body lighting on any surface CD in the direction of a right line AD, drawn according to any law, should emerge in the direction of another right line DK; and from the point C there be drawn curved lines CP, CQ, always perpendicular to AD, DK; the increments of the lines PD,

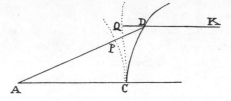

QD, and therefore the lines themselves PD, QD, generated by those increments, will be as the sines of incidence and emergence to each other, and conversely.

PROPOSITION XCVIII. PROBLEM XLVIII

The same things supposed; if round the axis AB *any attractive surface be described, as* CD, *regular or irregular, through which the bodies issuing from the given place* A *must pass; it is required to find a second attractive surface* EF, *which may make those bodies converge to a given place* B.

Let a line joining AB cut the first surface in C and the second in E, the point D being taken in any manner at pleasure. And supposing the sine of incidence on the first surface to the sine of emergence from the same, and the sine of emergence from the second surface to the sine of incidence on the same, to be as any given quantity M to another given quantity N; then

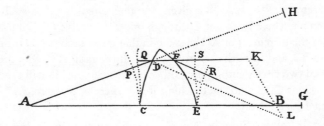

produce AB to G, so that BG may be to CE as M−N to N; and AD to H, so that AH may be equal to AG; and DF to K, so that DK may be to DH as N to M. Join KB, and about the centre D with the radius DH describe a circle meeting KB produced in L, and draw BF parallel to DL; and the point F will touch the line EF, which, being turned round the axis AB, will describe the surface sought. Q.E.F.

For conceive the lines CP, CQ to be everywhere perpendicular to AD, DF, and the lines ER, ES to FB, FD respectively, and therefore QS to be always equal to CE; and (by Cor. II, Prop. xcvii) PD will be to QD as M to N, and therefore as DL to DK, or FB to FK; and by subtraction, as DL – FB or PH – PD – FB to FD or FQ – QD; and by addition as PH – FB to FQ, that is (because PH and CG, QS and CE, are equal), as CE + BG – FR to CE – FS. But (because BG is to CE as M – N to N) it comes to pass also that CE + BG is to CE as M to N; and therefore, by subtraction, FR is to FS as M to N; and therefore (by Cor. II, Prop. xcvii) the surface EF compels a body, falling upon it in the direction DF, to go on in the line FR to the place B. Q.E.D.

SCHOLIUM

In the same manner one may go on to three or more surfaces. But of all figures the spherical is the most proper for optical uses. If the object glasses of telescopes were made of two glasses of a spherical figure, containing water between them, it is not unlikely that the errors of the refractions made in the extreme parts of the surfaces of the glasses may be accurately enough corrected by the refractions of the water. Such object glasses are to be preferred before elliptic and hyperbolic glasses, not only because they may be formed with more ease and accuracy, but because the pencils of rays situated without the axis of the glass would be more accurately refracted by them. But the different refrangibility of different rays is the real obstacle that hinders optics from being made perfect by spherical or any other figures. Unless the errors thence arising can be corrected, all the labor spent in correcting the others is quite thrown away.

Book Two

THE MOTION OF BODIES

(IN RESISTING MEDIUMS)

SECTION I

The motion of bodies that are resisted in the ratio of the velocity.

PROPOSITION I. THEOREM I

If a body is resisted in the ratio of its velocity, the motion lost by resistance is as the space gone over in its motion.

For since the motion lost in each equal interval of time is as the velocity, that is, as the small increment of space gone over, then, by composition, the motion lost in the whole time will be as the whole space gone over. Q.E.D.

Cor. Therefore if the body, destitute of all gravity, move by its innate force only in free spaces, and there be given both its whole motion at the beginning, and also the motion remaining after some part of the way is gone over, there will be given also the whole space which the body can describe in an infinite time. For that space will be to the space now described as the whole motion at the beginning is to the part lost of that motion.

LEMMA I

Quantities proportional to their differences are continually proportional.

Let $\qquad A : A - B = B : B - C = C : C - D = \&c.$;

then, by subtraction,

$\qquad A : B = B : C = C : D = \&c.$ $\qquad\qquad$ Q.E.D.

PROPOSITION II. THEOREM II

If a body is resisted in the ratio of its velocity, and moves, by its inertia only, through an homogeneous medium, and the times be taken equal, the velocities in the beginning of each of the times are in a geometrical progression, and the spaces described in each of the times are as the velocities.

CASE 1. Let the time be divided into equal intervals; and if at the very beginning of each interval we suppose the resistance to act with one single impulse which is as the velocity, the decrement of the velocity in each of the intervals of time will be as the same velocity. Therefore the velocities are proportional to their differences, and therefore (by Lem. 1, Book 11) continually proportional. Therefore if out of an equal number of intervals there be compounded any equal portions of time, the velocities at the beginning of those times will be as terms in a continued progression, which are taken by jumps, omitting everywhere an equal number of intermediate terms. But the ratios of these terms are compounded of the equal ratios of the intermediate terms equally repeated, and therefore are equal. Therefore the velocities, being proportional to those terms, are in geometrical progression. Let those equal intervals of time be diminished, and their number increased *in infinitum,* so that the impulse of resistance may become continual; and the velocities at the beginnings of equal times, always continually proportional, will be also in this case continually proportional. Q.E.D.[1]

CASE 2. And, by division, the differences of the velocities, that is, the parts of the velocities lost in each of the times, are as the wholes; but the spaces

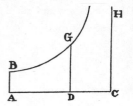

described in each of the times are as the lost parts of the velocities (by Prop. 1, Book 1), and therefore are also as the wholes. Q.E.D.

COR. Hence if to the rectangular asymptotes AC, CH, the hyperbola BG is described, and AB, DG be drawn perpendicular to the asymptote AC, and both the velocity of the body, and the resistance of the medium, at the very beginning of the motion, be expressed by any given line AC, and, after some time is elapsed, by the indefinite line DC; the time may be expressed by the area ABGD, and the space described

[¹ Appendix, Note 28.]

in that time by the line AD. For if that area, by the motion of the point D, be uniformly increased in the same manner as the time, the right line DC will decrease in a geometrical ratio in the same manner as the velocity; and the parts of the right line AC, described in equal times, will decrease in the same ratio.

PROPOSITION III. PROBLEM I

To define the motion of a body which, in an homogeneous medium, ascends or descends in a right line, and is resisted in the ratio of its velocity, and acted upon by an uniform force of gravity.

The body ascending, let the gravity be represented by any given rectangle BACH; and the resistance of the medium, at the beginning of the ascent, by the rectangle BADE, taken on the contrary side of the right line AB.

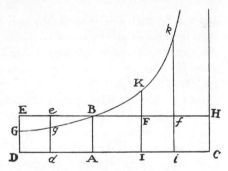

Through the point B, with the rectangular asymptotes AC, CH, describe an hyperbola, cutting the perpendiculars DE, *de* in G, *g;* and the body ascending will in the time DG*gd* describe the space EG*ge*; in the time DGBA, the space of the whole ascent EGB; in the time ABKI, the space of descent BFK; and in the time IK*ki* the space of descent KF*fk*; and the velocities of the bodies (proportional to the resistance of the medium) in these periods of time will be ABED, AB*ed*, o, ABFI, AB*fi* respectively; and the greatest velocity which the body can acquire by descending will be BACH.

For let the rectangle BACH be resolved into innumerable rectangles A*k*, K*l*, L*m*, M*n*, &c., which shall be as the increments of the velocities produced in so many equal times; then will o, A*k*, A*l*, A*m*, A*n*, &c., be as the whole velocities, and therefore (by supposition) as the resistances of the medium in the beginning of each of the equal times. Make AC to AK, or ABHC to AB*k*K, as the force of gravity to the resistance in the beginning of the second time; then from the force of gravity subtract the resistances, and ABHC, K*k*HC, L*l*HC, M*m*HC, &c., will be as the absolute forces with which the body is acted upon in the beginning of each of the times, and therefore (by Law 1) as the increments of the velocities, that is, as the rectangles A*k*, K*l*, L*m*, M*n*, &c., and therefore (by Lem. 1, Book 11) in a geometrical progression. Therefore, if the right lines K*k*, L*l*, M*m*, N*n*, &c., are produced so as to meet the hyperbola in *q*, *r*, *s*, *t*, &c., the areas AB*q*K, K*q*rL, L*r*sM, M*s*tN, &c., will be equal, and therefore analogous to the equal times and equal gravitating forces. But the area AB*q*K (by Cor. 111, Lem. 111 and 1111, Book 1) is to the area B*k*q as K*q* to ½*kq*, or AC to ½AK, that is, as the force of gravity to the resistance in the middle of the first time. And by the like reasoning, the areas *q*KL*r*, *r*LM*s*, *s*MN*t*, &c., are to the areas *qklr*, *rlms*, *smnt*, &c., as the gravitating forces to the resistances in the middle of the second, third, fourth time, and so on. Therefore since the equal areas BAK*q*, *q*KL*r*, *r*LM*s*, *s*MN*t*, &c., are analogous to the gravitating forces, the areas B*kq*, *qklr*, *rlms*, *smnt*, &c., will be analogous to the resistances in the middle of each of the times, that is (by supposition), to the velocities, and so to the spaces described. Take the sums of the analogous quantities, and the areas B*kq*, B*lr*, B*ms*, B*nt*, &c., will be analogous to the whole spaces described; and also the areas AB*q*K, AB*r*L, AB*s*M, AB*t*N, &c., to the times. Therefore the body, in descending, will in any time AB*r*L describe the space B*lr*, and in the time L*rt*N the space *rlnt*. Q.E.D. And the like demonstration holds in ascending motion.

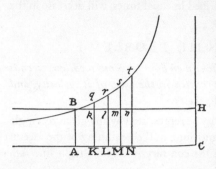

Cor. 1. Therefore the greatest velocity that the body can acquire by falling is to the velocity acquired in any given time as the given force of gravity which continually acts upon it to the resisting force which opposes it at the end of that time.

Cor. 11. But the time being augmented in an arithmetical progression, the sum of that greatest velocity and the velocity in the ascent, and also their difference in the descent, decreases in a geometrical progression.

COR. III. Also the differences of the spaces, which are described in equal differences of the times, decrease in the same geometrical progression.

COR. IV. The space described by the body is the difference of two spaces, whereof one is as the time taken from the beginning of the descent, and the other as the velocity; which [spaces] also at the beginning of the descent are equal among themselves.

PROPOSITION IV. PROBLEM II

Supposing the force of gravity in any homogeneous medium to be uniform, and to tend perpendicularly to the plane of the horizon: to define the motion of a projectile therein, which suffers resistance proportional to its velocity.

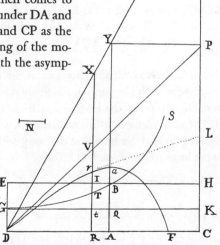

Let the projectile go from any place D in the direction of any right line DP, and let its velocity at the beginning of the motion be represented by the length DP. From the point P let fall the perpendicular PC on the horizontal line DC, and cut DC in A, so that DA may be to AC as the vertical component of the resistance of the medium arising from the motion upwards at the beginning, to the force of gravity; or (which comes to the same) so that the rectangle under DA and DP may be to that under AC and CP as the whole resistance at the beginning of the motion, to the force of gravity. With the asymptotes DC, CP describe any hyperbola GTBS cutting the perpendiculars DG, AB in G and B; complete the parallelogram DGKC, and let its side GK cut AB in Q. Take a line N in the same ratio to QB as DC is in to CP; and from any point R of the right line DC erect RT perpendicular to it, meeting the hyperbola in T, and the right lines EH, GK, DP in I, t, and V; in that perpendicular take Vr equal to $\dfrac{t\text{GT}}{\text{N}}$, or, which is the same thing, take Rr equal to $\dfrac{\text{GTIE}}{\text{N}}$; and the projectile in the time DRTG will arrive at the point r, describing the curved line DraF, the locus of the point r; thence it will come to its greatest height a in the perpendicular AB; and afterwards ever approach to the asymptote PC. And its velocity in any point r will be as the tangent rL to the curve. Q.E.I.

For   N : QB = DC : CP = DR : RV,

and therefore RV is equal to $\dfrac{\text{DR} \cdot \text{QB}}{\text{N}}$, and R$r$ $\Big($ that is, RV $-$ Vr, or $\dfrac{\text{DR} \cdot \text{QB} - t\text{GT}}{\text{N}} \Big)$ is equal to $\dfrac{\text{DR} \cdot \text{AB} - \text{RDGT}}{\text{N}}$. Now let the time be represented by the area RDGT, and (by Laws, Cor. II) distinguish the motion of the body into two others, one of ascent, the other lateral. And since the resistance is as the motion, let that also be distinguished into two parts proportional and contrary to the parts of the motion: and therefore the length described by the lateral motion will be (by Prop. II, Book II) as the line DR, and the height (by Prop. III, Book II) as the area DR $\cdot$ AB $-$ RDGT, that is, as the line Rr. But in the very beginning of the motion the area RDGT is equal to the rectangle DR $\cdot$ AQ, and therefore that line Rr $\Big($ or $\dfrac{\text{DR} \cdot \text{AB} - \text{DR} \cdot \text{AQ}}{\text{N}} \Big)$ will then be to DR as AB $-$ AQ or QB to N, that is, as CP to DC; and therefore as the motion upwards to the motion lengthwise at the beginning. Since, therefore, Rr is always as the height, and DR always as the length, and Rr is to DR at the beginning as the height to the length, it follows, that Rr is always to DR as the height to the length; and therefore that the body will move in the line DraF, which is the locus of the point r. Q.E.D.

Cor. I. Therefore Rr is equal to $\dfrac{\text{DR} \cdot \text{AB}}{\text{N}} - \dfrac{\text{RDGT}}{\text{N}}$; and therefore if RT be produced to X so that RX may be equal to $\dfrac{\text{DR} \cdot \text{AB}}{\text{N}}$, that is, if the parallelogram ACPY be completed, and DY cutting CP in Z be drawn, and

RT be produced till it meets DY in X; Xr will be equal to $\dfrac{RDGT}{N}$, and therefore proportional to the time.

Cor. ii. Whence if innumerable lines CR, or, which is the same, innumerable lines ZX, be taken in a geometrical progression, there will be as many lines Xr in an arithmetical progression. And hence the curve DraF is easily delineated by the table of logarithms.

Cor. iii. If a parabola be constructed to the vertex D, and the diameter DG produced downwards, and its latus rectum is to 2DP as the whole resistance at the beginning of the motion to the gravitating force, the velocity with which the body ought to go from the place D, in the direction of the right line DP, so as in an uniform resisting medium to describe the curve DraF, will be the same as that with which it ought to go from the same place D in the direction of the same right line DP, so as to describe a parabola in a nonresisting medium. For the latus rectum of this parabola, at the very beginning of the motion, is $\dfrac{DV^2}{Vr}$; and Vr is $\dfrac{tGT}{N}$ or $\dfrac{DR \cdot Tt}{2N}$. But a right

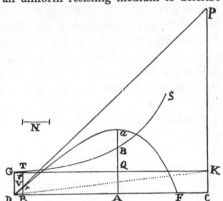

line which, if drawn, would touch the hyperbola GTS in G, is parallel to DK, and therefore Tt is $\dfrac{CK \cdot DR}{DC}$, and N is $\dfrac{QB \cdot DC}{CP}$. And therefore Vr is equal to $\dfrac{DR^2 \cdot CK \cdot CP}{2DC^2 \cdot QB}$, that is (because DR and DC, DV and DP are proportionals), to $\dfrac{DV^2 \cdot CK \cdot CP}{2DP^2 \cdot QB}$; and the latus rectum $\dfrac{DV^2}{Vr}$ comes out $\dfrac{2DP^2 \cdot QB}{CK \cdot CP}$, that is (because QB and CK, DA and AC are proportional), $\dfrac{2DP^2 \cdot DA}{AC \cdot CP}$, and therefore is to 2DP as DP · DA to CP · AC; that is, as the resistance to the gravity. Q.E.D.

Cor. iv. Hence if a body be projected from any place D with a given velocity, in the direction of a right line DP given by position, and the resistance of the medium, at the beginning of the motion, be given, the curve DraF, which that body will describe, may be found. For the velocity being given, the latus rectum of the parabola is given, as is well known. And taking 2DP to that latus rectum, as the force of gravity to the resisting force, DP is also given. Then cutting DC in A, so that CP · AC may be to DP · DA in the same ratio of the gravity to the resistance, the point A will be given. And hence the curve DraF is also given.

Cor. v. And conversely, if the curve DraF be given, there will be given both the velocity of the body and the resistance of the medium in each of the places r. For the ratio of CP · AC to DP · DA being given, there is given both the resistance of the medium at the beginning of the motion, and the latus rectum of the parabola; and thence the velocity at the beginning of the motion is given also. Then from the length of the tangent rL there is given both the velocity proportional to it, and the resistance proportional to the velocity in any place r.

Cor. vi. But since the length 2DP is to the latus rectum of the parabola as the gravity to the resistance in D, and, from the velocity augmented, the resistance is augmented in the same ratio, but the latus rectum of the parabola is augmented as the square of that ratio, it is plain that the length 2DP is augmented in that simple ratio only; and is therefore always proportional to the velocity; nor will it be augmented or diminished by the change of the angle CDP, unless the velocity be also changed.

Cor vii. Hence appears the method of determining the curve D*ra*F nearly from the phenomena, and thence finding the resistance and velocity with which the body is projected. Let two similar and equal bodies be pro-

jected with the same velocity, from the place D, in different angles CDP, CD*p*; and let the places F, *f,* where they fall upon the horizontal plane DC, be known. Then taking any length for DP or D*p* suppose the resistance in D to be to the gravity in any ratio whatsoever, and let that ratio be represented by any length SM. Then, by computation, from that assumed length DP, find the lengths DF, D*f*;

and from the ratio $\frac{Ff}{DF}$, found

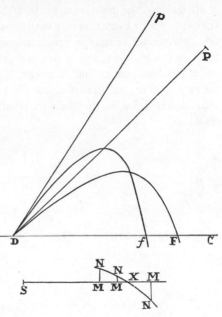

by calculation, subtract the same ratio as found by experiment; and let the difference be represented by the perpendicular MN. Repeat the same a second and a third time, by assuming always a new ratio SM of the resistance to the gravity, and collecting a new difference MN. Draw the positive differences on one side of the right line SM, and the negative on the other side; and through the points N, N, N, draw a regular curve NNN, cutting the right line SMMM in X, and SX will be the true ratio of the resistance to the gravity, which was to be found. From this ratio the length DF is to be found by calculation; and a length, which is to the assumed length DP as the length DF known by experiment to the length DF just now found, will be the true length DP. This being known, you will have both the curved line D*ra*F which the body describes, and also the velocity and resistance of the body in each place.

SCHOLIUM

However, that the resistance of bodies is in the ratio of the velocity, is more a mathematical hypothesis than a physical one. In mediums void of all tenacity, the resistances made to bodies are as the square of the velocities. For by the action of a swifter body, a greater motion in proportion to a greater velocity is communicated to the same quantity of the medium in a less time; and in an equal time, by reason of a greater quantity of the disturbed medium, a motion is communicated as the square of the ratio greater; and the resistance (by Law II and III) is as the motion communicated. Let us, therefore, see what motions arise from this law of resistance.

SECTION II

The motion of bodies that are resisted as the square of their velocities.

PROPOSITION V. THEOREM III

If a body is resisted as the square of its velocity, and moves by its innate force only through an homogeneous medium; and the times be taken in a geometrical progression, proceeding from less to greater terms: I say, that the velocities at the beginning of each of the times are in the same geometrical progression inversely; and that the spaces are equal, which are described in each of the times.

For since the resistance of the medium is proportional to the square of the velocity, and the decrement of the velocity is proportional to the resistance: if the time be divided into innumerable equal intervals, the squares of the velocities at the beginning of each of the times will be proportional to the differences of the same velocities. Let those intervals of time be AK,

KL, LM, &c., taken in the right line CD; and erect the perpendiculars AB, Kk, Ll, Mm, &c., meeting the hyperbola BklmG, described with the centre C, and the rectangular asymptotes CD, CH, in B, k, l, m, &c.; then AB will be to Kk as CK to CA, and, by division, AB−Kk to Kk as AK to CA, and alternately, AB−Kk to AK as Kk to CA; and therefore as AB·Kk to AB·CA. Therefore since AK and

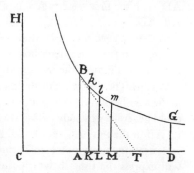

AB·CA are given, AB−Kk will be as AB·Kk; and, lastly, when AB and Kk coincide, as AB². And, by the like reasoning, Kk−Ll, Ll−Mm, &c., will be as Kk², Ll², &c. Therefore the squares of the lines AB, Kk, Ll, Mm, &c., are as their differences; and, therefore, since the squares of the velocities were shown above to be as their differences, the progression of both will be alike. This being demonstrated it follows also that the areas described by these lines are in a like progression with the spaces described by these velocities. Therefore if the velocity at the beginning of the first time AK be represented by the line AB, and the velocity at the beginning of the second time KL by the line Kk, and the length described in the first time by the area AKkB, all the following velocities will be represented by the following lines Ll, Mm, &c., and the lengths described by the areas Kl, Lm, &c. And, by composition, if the whole time be represented by AM, the sum of its parts, the whole length described will be represented by AMmB, the sum of its parts. Now conceive the time AM to be divided into the parts AK, KL, LM, &c., so that CA, CK, CL, CM, &c., may be in a geometrical progression; and those parts will be in the same progression, and the velocities AB, Kk, Ll, Mm, &c., will be in the same progression inversely, and the spaces described Ak, Kl, Lm, &c., will be equal. Q.E.D.

Cor. i. Hence it appears, that if the time be represented by any part AD of the asymptote, and the velocity in the beginning of the time by the ordinate AB, the velocity at the end of the time will be represented by the ordinate DG; and the whole space described by the adjacent hyperbolic area ABGD; and the space which any body can describe in the same time AD, with the first velocity AB, in a nonresisting medium, by the rectangle AB · AD.

Cor. ii. Hence the space described in a resisting medium is given, by taking it to the space described with the uniform velocity AB in a non-resisting medium, as the hyperbolic area ABGD to the rectangle AB · AD.

Cor. iii. The resistance of the medium is also given, by making it equal, in the very beginning of the motion, to an uniform centripetal force, which could generate, in a body falling through a nonresisting medium, the velocity AB in the time AC. For if BT be drawn touching the hyperbola in B, and meeting the asymptote in T, the right line AT will be equal to AC, and will express the time in which the first resistance, uniformly continued, may take away the whole velocity AB.

Cor. iv. And thence is also given the proportion of this resistance to the force of gravity, or any other given centripetal force.

Cor. v. And, conversely, if there is given the proportion of the resistance to any given centripetal force, the time AC is also given, in which a centripetal force equal to the resistance may generate any velocity as AB; and thence is given the point B, through which the hyperbola, having CH,

CD for its asymptotes, is to be described; as also the space ABGD, which a body, by beginning its motion with that velocity AB, can describe in any time AD, in an homogeneous resisting medium.

PROPOSITON VI. THEOREM IV

Homogeneous and equal spherical bodies, opposed by resistances that are as the square of the velocities, and moving on by their innate force only, will, in times which are inversely as the velocities at the beginning, describe equal spaces, and lose parts of their velocities proportional to the wholes.

To the rectangular asymptotes CD, CH describe any hyperbola B*b*E*e*, cutting the perpendiculars AB, *ab*, DE, *de* in B, *b*, E, *e*; let the initial veloci-

ties be represented by the perpendiculars AB, DE, and the times by the lines A*a*, D*d*. Therefore as A*a* is to D*d*, so (by the hypothesis) is DE to AB, and so (from the nature of the hyperbola) is CA to CD; and, by composition, so is C*a* to C*d*. Therefore the areas AB*ba*, DE*ed*, that is, the spaces described, are equal among themselves, and the first velocities AB, DE are proportional to the last *ab, de*; and there-

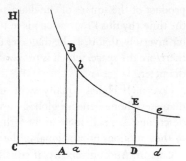

fore, by subtraction, proportional to the parts of the velocities lost, AB − *ab*, DE − *de*. Q.E.D.

PROPOSITION VII. THEOREM V

If spherical bodies are resisted as the squares of their velocities, in times which are directly as the first motions, and inversely as the first resistances, they will lose parts of their motions proportional to the wholes, and will describe spaces proportional to the product of those times and the first velocities.[1]

For the parts of the motions lost are as the product of the resistances and times. Therefore, that those parts may be proportional to the wholes, the product of the resistance and time ought to be as the motion. Therefore the

[1 Appendix, Note 29.]

time will be as the motion directly and the resistance inversely. Therefore the intervals of the times being taken in that ratio, the bodies will always lose parts of their motions proportional to the wholes, and therefore will retain velocities always proportional to their first velocities. And because of the given ratio of the velocities, they will always describe spaces which are as the product of the first velocities and the times. Q.E.D.

Cor. I. Therefore if bodies equally swift are resisted as the square of their diameters, homogeneous globes moving with any velocities whatsoever, by describing spaces proportional to their diameters, will lose parts of their motions proportional to the wholes. For the motion of each globe will be as the product of its velocity and mass, that is, as the product of the velocity and the cube of its diameter; the resistance (by supposition) will be as the product of the square of the diameter and the square of the velocity; and the time (by this Proposition) is in the former ratio directly, and in the latter inversely, that is, as the diameter directly and the velocity inversely; and therefore the space, which is proportional to the time and velocity, is as the diameter.

Cor. II. If bodies equally swift are resisted as the $\frac{3}{2}$th power of their diameters, homogeneous globes, moving with any velocities whatsoever, by describing spaces that are as the $\frac{3}{2}$th power of the diameters, will lose parts of their motions proportional to the wholes.

Cor. III. And universally, if equally swift bodies are resisted in the ratio of any power of the diameters, the spaces, in which homogeneous globes, moving with any velocity whatsoever, will lose parts of their motions proportional to the wholes, will be as the cubes of the diameters applied to that power. Let those diameters be D and E; and if the resistances, where the velocities are supposed equal, are as D^n and E^n; the spaces in which the globes, moving with any velocities whatsoever, will lose parts of their motions proportional to the wholes, will be as D^{3-n} and E^{3-n}. And therefore homogeneous globes, in describing spaces proportional to D^{3-n} and E^{3-n}, will retain their velocities in the same ratio to one another as at the beginning.

Cor. IV. Now if the globes are not homogeneous, the space described by the denser globe must be augmented in the ratio of the density. For the motion, with an equal velocity, is greater in the ratio of the density, and the time (by this Proposition) is augmented in the ratio of motion directly, and the space described in the ratio of the time.

Cor. v. And if the globes move in different mediums, the space, in a medium which, other things being equal, resists the most, must be diminished in the ratio of the greater resistance. For the time (by this Proposition) will be diminished in the ratio of the augmented resistance, and the space in the ratio of the time.

LEMMA II

The moment of any genitum is equal to the moments of each of the generating sides multiplied by the indices of the powers of those sides, and by their coefficients continually.

I call any quantity a *genitum* which is not made by addition or subtraction of divers parts, but is generated or produced in arithmetic by the multiplication, division, or extraction of the root of any terms whatsoever; in geometry by the finding of contents and sides, or of the extremes and means of proportionals. Quantities of this kind are products, quotients, roots, rectangles, squares, cubes, square and cubic sides, and the like. These quantities I here consider as variable and indetermined, and increasing or decreasing, as it were, by a continual motion or flux; and I understand their momentary increments or decrements by the name of moments; so that the increments may be esteemed as added or affirmative moments; and the decrements as subtracted or negative ones. But take care not to look upon finite particles as such.[1] Finite particles are not moments, but the very quantities generated by the moments. We are to conceive them as the just nascent principles of finite magnitudes. Nor do we in this Lemma regard the magnitude of the moments, but their first proportion, as nascent. It will be the same thing, if, instead of moments, we use either the velocities of the increments and decrements (which may also be called the motions, mutations, and fluxions of quantities), or any finite quantities proportional to those velocities. The coefficient of any generating side is the quantity which arises by applying the genitum to that side.

Wherefore the sense of the Lemma is, that if the moments of any quantities A, B, C, &c., increasing or decreasing by a continual flux, or the velocities of the mutations which are proportional to them, be called *a, b, c,* &c.,

[1 Appendix, Note 30.]

the moment or mutation of the generated rectangle AB will be aB + bA; the moment of the generated content ABC will be aBC + bAC + cAB; and the moments of the generated powers A^2, A^3, A^4, A$^{1/2}$, A$^{3/2}$, A$^{1/3}$, A$^{2/3}$, A^{-1}, A^{-2}, A$^{-1/2}$ will be $2a$A, $3a$A^2, $4a$A^3, $\frac{1}{2}a$A$^{-1/2}$, $\frac{3}{2}a$A$^{1/2}$, $\frac{1}{3}a$A$^{-2/3}$, $\frac{2}{3}a$A$^{-1/3}$, $-a$A^{-2}, $-2a$A^{-3}, $-\frac{1}{2}a$A$^{-3/2}$ respectively; and, in general, that the moment of any power A$^{\frac{n}{m}}$ will be $\frac{n}{m} a$A$^{\frac{n-m}{m}}$. Also, that the moment of the generated quantity A^{2}B will be $2a$AB + bA^2; the moment of the generated quantity A^{3}B^{4}C^2 will be $3a$A^{2}B^{4}C^2 + $4b$A^{3}B^{3}C^2 + $2c$A^{3}B^{4}C; and the moment of the generated quantity $\frac{A^3}{B^2}$ or A^{3}B^{-2} will be $3a$A^{2}B^{-2} − $2b$A^{3}B^{-3}; and so on. The Lemma is thus demonstrated.[1]

CASE 1. Any rectangle, as AB, augmented by a continual flux, when, as yet, there wanted of the sides A and B half their moments $\frac{1}{2}a$ and $\frac{1}{2}b$, was A − $\frac{1}{2}a$ into B − $\frac{1}{2}b$, or AB − $\frac{1}{2}a$ B − $\frac{1}{2}b$ A + $\frac{1}{4}ab$; but as soon as the sides A and B are augmented by the other half-moments, the rectangle becomes A + $\frac{1}{2}a$ into B + $\frac{1}{2}b$, or AB + $\frac{1}{2}a$ B + $\frac{1}{2}b$ A + $\frac{1}{4}ab$. From this rectangle subtract the former rectangle, and there will remain the excess aB + bA. Therefore with the whole increments a and b of the sides, the increment aB + bA of the rectangle is generated. Q.E.D.

CASE 2. Suppose AB always equal to G, and then the moment of the content ABC or GC (by Case 1) will be gC + cG, that is (putting AB and aB + bA for G and g), aBC + bAC + cAB. And the reasoning is the same for contents under ever so many sides. Q.E.D.

CASE 3. Suppose the sides A, B, and C, to be always equal among themselves; and the moment aB + bA, of A^2, that is, of the rectangle AB, will be $2a$A; and the moment aBC + bAC + cAB of A^3, that is, of the content ABC, will be $3a$A^2. And by the same reasoning the moment of any power A^n is naA^{n-1}. Q.E.D.

CASE 4. Therefore since $\frac{1}{A}$ into A is 1, the moment of $\frac{1}{A}$ multiplied by A, together with $\frac{1}{A}$ multiplied by a, will be the moment of 1, that is, nothing. Therefore the moment of $\frac{1}{A}$, or of A^{-1}, is $\frac{-a}{A^2}$. And generally since $\frac{1}{A^n}$ into A^n is 1, the moment of $\frac{1}{A^n}$ multiplied by A^n together with $\frac{1}{A^n}$ into

[1 Appendix, Note 31.]

$na\text{A}^{n-1}$ will be nothing. And, therefore, the moment of $\dfrac{1}{\text{A}^n}$ or A^{-n} will be $-\dfrac{na}{\text{A}^{n+1}}$. Q.E.D.

CASE 5. And since $\text{A}^{\frac{1}{2}}$ into $\text{A}^{\frac{1}{2}}$ is A, the moment of $\text{A}^{\frac{1}{2}}$ multiplied by $2\text{A}^{\frac{1}{2}}$ will be a (by Case 3); and, therefore, the moment of $\text{A}^{\frac{1}{2}}$ will be $\dfrac{a}{2\text{A}^{\frac{1}{2}}}$ or $\frac{1}{2}a\text{A}^{-\frac{1}{2}}$. And generally, putting $\text{A}^{\frac{m}{n}}$ equal to B, then A^m will be equal to B^n, and therefore $ma\text{A}^{m-1}$ equal to $nb\text{B}^{n-1}$, and $ma\text{A}^{-1}$ equal to $nb\text{B}^{-1}$, or $nb\text{A}^{-\frac{m}{n}}$; and therefore $\dfrac{m}{n}a\text{A}^{\frac{n-m}{n}}$ is equal to b, that is, equal to the moment of $\text{A}^{\frac{m}{n}}$. Q.E.D.

CASE 6. Therefore the moment of any generated quantity A^mB^n is the moment of A^m multiplied by B^n, together with the moment of B^n multiplied by A^m, that is, $ma\text{A}^{m-1}\text{B}^n + nb\text{B}^{n-1}\text{A}^m$; and that whether the indices m and n of the powers be whole numbers or fractions, affirmative or negative. And the reasoning is the same for higher powers. Q.E.D.

COR. I. Hence in quantities continually proportional,[1] if one term is given, the moments of the rest of the terms will be as the same terms multiplied by the number of intervals between them and the given term. Let A, B, C, D, E, F be continually proportional; then if the term C is given, the moments of the rest of the terms will be among themselves as -2A, $-$B, D, 2E, 3F.

COR. II. And if in four proportionals the two means are given, the moments of the extremes will be as those extremes. The same is to be understood of the sides of any given rectangle.

COR. III. And if the sum or difference of two squares is given, the moments of the sides will be inversely as the sides.

SCHOLIUM[2]

In a letter of mine to Mr. *J. Collins,* dated *December* 10, 1672, having described a method of tangents, which I suspected to be the same with *Sluse's* method, which at that time was not made public, I added these words: *This is one particular, or rather a Corollary, of a general method, which extends itself, without any troublesome calculation, not only to the drawing of tangents to any curved lines, whether geometrical or mechan-*

[1 Appendix, Note 32.]   [2 Appendix, Note 33.]

ical or in any manner respecting right lines or other curves, but also to the resolving other abstruser kinds of problems about the crookedness, areas, lengths, centres of gravity of curves, &c.; nor is it (as Hudden's *method* de maximis et minimis) *limited to equations which are free from surd quantities. This method I have interwoven with that other of working in equations, by reducing them to infinite series.* So far that letter. And these last words relate to a treatise I composed on that subject in the year 1671. The foundation of that general method is contained in the preceding Lemma.

PROPOSITION VIII. THEOREM VI

If a body in an uniform medium, being uniformly acted upon by the force of gravity, ascends or descends in a right line; and the whole space described be divided into equal parts, and in the beginning of each of the parts (by adding or subtracting the resisting force of the medium to or from the force of gravity, when the body ascends or descends) you derive the absolute forces: I say, that those absolute forces are in a geometrical progression.

Let the force of gravity be represented by the given line AC; the force of resistance by the indefinite line AK; the absolute force in the descent of the body by the difference KC; the velocity of the body by a line AP, which

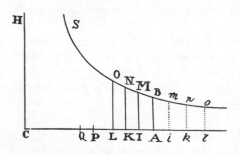

shall be a mean proportional between AK and AC, and therefore as the square root of the resistance; the increment of the resistance made in a given interval of time by the short line KL, and the contemporaneous increment of the velocity by the short line PQ; and with the centre C, and rectangular asymptotes CA, CH, describe any hyperbola BNS meeting the erected perpendiculars AB, KN, LO in B, N, and O. Because AK is as AP^2, the moment KL of the one will be as the moment $2AP \cdot PQ$ of the other, that is, as $AP \cdot KC$; for the increment PQ of the velocity is (by Law II) proportional to the generating force KC. Let the ratio of KL be

multiplied by the ratio KN, and the rectangle KL · KN will become as
AP · KC · KN; that is (because the rectangle KC · KN is given), as AP.
But the ultimate ratio of the hyperbolic area KNOL to the rectangle
KL · KN becomes, when the points K and L coincide, the ratio of equality.
Therefore that hyperbolic evanescent area is as AP. Therefore the whole
hyperbolic area ABOL is composed of intervals KNOL which are always
proportional to the velocity AP; and therefore is itself proportional to the
space described with that velocity. Let that area be now divided into equal
parts, as ABMI, IMNK, KNOL, &c., and the absolute forces AC, IC, KC,
LC, &c., will be in a geometrical progression. Q.E.D. And by a like reasoning, in the ascent of the body, taking, on the contrary side of the point A,
the equal areas AB*mi, imnk, knol,* &c., it will appear that the absolute forces
AC, *i*C, *k*C, *l*C, &c., are continually proportional. Therefore if all the spaces
in the ascent and descent are taken equal, all the absolute forces *l*C, *k*C, *i*C,
AC, IC, KC, LC, &c., will be continually proportional. Q.E.D.

Cor. i. Hence if the space described be represented by the hyperbolic area
ABNK, the force of gravity, the velocity of the body, and the resistance of
the medium, may be represented by the lines AC, AP, and AK respectively;
and conversely.

Cor. ii. And the greatest velocity which the body can ever acquire in an
infinite descent will be represented by the line AC.

Cor. iii. Therefore if the resistance of the medium answering to any
given velocity be known, the greatest velocity will be found, by taking it
to that given velocity, as the square root of the ratio which the force of
gravity bears to that known resistance of the medium.

PROPOSITION IX. THEOREM VII

*Supposing what is above demonstrated, I say, that if the tangents of the
angles of the sector of a circle, and of an hyperbola, be taken proportional
to the velocities, the radius being of a fit magnitude, all the time of the
ascent to the highest place will be as the sector of the circle, and all the time
of descending from the highest place as the sector of the hyperbola.*

To the right line AC, which expresses the force of gravity, let AD be
drawn perpendicular and equal. From the centre D, with the semidiameter
AD describe as well the quadrant A*t*E of a circle, as the rectangular hyper-

bola AVZ, whose axis is AK, principal vertex A, and asymptote DC. Let Dp, DP be drawn; and the circular sector AtD will be as all the time of the ascent to the highest place; and the hyperbolic sector ATD as all the time of descent from the highest place; if so be that the tangents Ap, AP of those sectors be as the velocities.

CASE I. Draw D$v$$q$ cutting off the moments or least intervals tDv and qDp, described in the same time, of the sector ADt and of the triangle

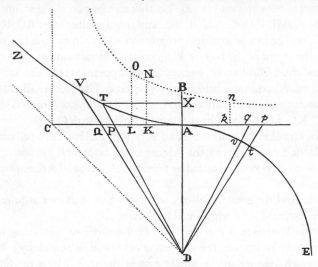

ADp. Since those intervals (because of the common angle D) are as the square of the sides, the interval tDv will be as $\dfrac{q\mathrm{D}p \cdot t\mathrm{D}^2}{p\mathrm{D}^2}$, that is (because tD is given), as $\dfrac{q\mathrm{D}p}{p\mathrm{D}^2}$. But $p\mathrm{D}^2$ is $\mathrm{AD}^2 + \mathrm{A}p^2$, that is, $\mathrm{AD}^2 + \mathrm{AD} \cdot \mathrm{A}k$, or $\mathrm{AD} \cdot \mathrm{C}k$; and $q\mathrm{D}p$ is $\frac{1}{2}\mathrm{AD} \cdot pq$. Therefore tDv, the interval of the sector, is as $\dfrac{pq}{\mathrm{C}k}$; that is, directly as the least decrement pq of the velocity, and inversely as the force Ck which diminishes the velocity; and therefore as the interval of time answering to the decrement of the velocity. And, by composition, the sum of all the intervals tDv in the sector ADt will be as the sum of the intervals of time answering to each of the lost intervals pq of the decreasing velocity Ap, till that velocity, being diminished into nothing, vanishes; that is, the whole sector ADt is as the whole time of ascent to the highest place. Q.E.D.

CASE 2. Draw DQV cutting off the least intervals TDV and PDQ of the sector DAV, and of the triangle DAQ; and these intervals will be to each other as DT^2 to DP^2, that is (if TX and AP are parallel), as DX^2 to DA^2 or TX^2 to AP^2; and, by subtraction, as DX^2 − TX^2 to DA^2 − AP^2. But, from the nature of the hyperbola, DX^2 − TX^2 is AD^2; and, by the supposition, AP^2 is AD · AK. Therefore the intervals are to each other as AD^2 to AD^2 − AD · AK; that is, as AD to AD − AK or AC to CK; and therefore the interval TDV of the sector is $\dfrac{\text{PDQ} \cdot \text{AC}}{\text{CK}}$; and therefore (because AC and AD are given) as $\dfrac{\text{PQ}}{\text{CK}}$; that is, directly as the increment of the velocity, and inversely as the force generating the increment; and therefore as the interval of the time answering to the increment. And, by composition, the

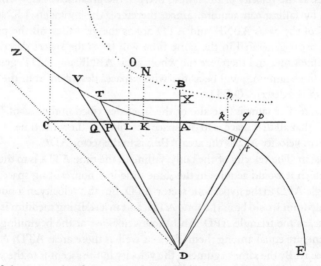

sum of the intervals of time, in which all the intervals PQ of the velocity AP are generated, will be as the sum of the intervals of the sector ATD; that is, the whole time will be as the whole sector. Q.E.D.

COR. I. Hence if AB be equal to a fourth part of AC, the space which a body will describe by falling in any time will be to the space which the body could describe, by moving uniformly on in the same time with its greatest velocity AC, as the area ABNK, which expresses the space described in falling to the area ATD, which expresses the time. For since

$$AC : AP = AP : AK,$$

and by Cor. I, Lem. II, of this Book,

$$LK : PQ = 2AK : AP = 2AP : AC,$$

therefore $LK : \frac{1}{2}PQ = AP : \frac{1}{4}AC$ or AB,

and since $KN : AC$ or $AD = AD : CK$,

multiplying together corresponding terms,

$$LKNO : DPQ = AP : CK.$$

As shown above,

$$DPQ : DTV = CK : AC.$$

Hence, $LKNO : DTV = AP : AC$;

that is, as the velocity of the falling body to the greatest velocity which the body by falling can acquire. Since, therefore, the moments LKNO and DTV of the areas ABNK and ATD are as the velocities, all the parts of those areas generated in the same time will be as the spaces described in the same time; and therefore the whole areas ABNK and ADT, generated from the beginning, will be as the whole spaces described from the beginning of the descent. Q.E.D.

COR. II. The same is true also of the space described in the ascent. That is to say, that all that space is to the space described in the same time, with the uniform velocity AC, as the area AB*nk* is to the sector AD*t*.

COR. III. The velocity of the body, falling in the time ATD, is to the velocity which it would acquire in the same time in a nonresisting space, as the triangle APD to the hyperbolic sector ATD. For the velocity in a nonresisting medium would be as the time ATD, and in a resisting medium is as AP, that is, as the triangle APD. And those velocities, at the beginning of the descent, are equal among themselves, as well as those areas ATD, APD.

COR. IV. By the same argument, the velocity in the ascent is to the velocity with which the body in the same time, in a nonresisting space, would lose all its motion of ascent, as the triangle A*p*D to the circular sector A*t*D; or as the right line A*p* to the arc A*t*.

Cor. v. Therefore the time in which a body, by falling in a resisting medium, would acquire the velocity AP, is to the time in which it would acquire its greatest velocity AC, by falling in a nonresisting space, as the sector ADT to the triangle ADC; and the time in which it would lose its velocity A*p*, by ascending in a resisting medium, is to the time in which it would lose the same velocity by ascending in a nonresisting space, as the arc A*t* to its tangent A*p*.

Cor. vi. Hence from the given time there is given the space described in the ascent or descent. For the greatest velocity of a body descending *in infinitum* is given (by Cor. ii and iii, Theor. vi, of this Book); and thence the time is given in which a body would acquire that velocity by falling in nonresisting space. Taking the sector ADT or AD*t* to the triangle ADC in the ratio of the given time to the time just found, there will be given both the velocity AP or A*p*, and the area ABNK or AB*nk*, which is to the sector ADT, or AD*t*, as the space sought to that which would, in the given time, be uniformly described with that greatest velocity found just before.

Cor. vii. And by going backwards, from the given space of ascent or descent AB*nk* or ABNK, there will be given the time AD*t* or ADT.

PROPOSITION X. PROBLEM III[1]

Suppose the uniform force of gravity to tend directly to the plane of the horizon, and the resistance to be as the product of the density of the medium and the square of the velocity: it is proposed to find the density of the medium in each place, which shall make the body move in any given curved line, the velocity of the body, and the resistance of the medium in each place.

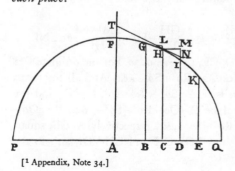

Let PQ be a plane perpendicular to the plane of the scheme itself; PFHQ a curved line meeting that plane in the points P and Q; G, H, I, K four places of the body going on in this curve from F to Q; and GB, HC, ID, KE four parallel ordinates let fall from these points to the horizon, and

[1 Appendix, Note 34.]

standing on the horizontal line PQ at the points B, C, D, E; and let the distances BC, CD, DE of the ordinates be equal among themselves. From the points G and H let the right lines GL, HN be drawn touching the curve in G and H, and meeting the ordinates CH, DI, produced upwards, in L and N; and complete the parallelogram HCDM. And the times in which the body describes the arcs GH, HI, will be as the square root of the altitudes LH, NI, which the bodies would describe in those times, by falling from the tangents; and the velocities will be directly as the lengths described GH, HI, and inversely as the times. Let the times be represented by T and t, and the velocities by $\dfrac{GH}{T}$ and $\dfrac{HI}{t}$; and the decrement of the velocity produced in the time t will be represented by $\dfrac{GH}{T} - \dfrac{HI}{t}$. This decrement arises from the resistance which retards the body, and from the gravity which accelerates it. Gravity, in a falling body, which in its fall describes the space NI, produces a velocity with which it would be able to describe twice that space in the same time, as *Galileo* hath demonstrated; that is, the velocity $\dfrac{2NI}{t}$: but if the body describes the arc HI, it augments that arc only by the length HI − HN or $\dfrac{MI \cdot NI}{HI}$; and therefore generates only the velocity $\dfrac{2MI \cdot NI}{t \cdot HI}$. Let this velocity be added to the before-mentioned decrement, and we shall have the decrement of the velocity arising from the resistance alone, that is, $\dfrac{GH}{T} - \dfrac{HI}{t} + \dfrac{2MI \cdot NI}{t \cdot HI}$. Therefore since, in the same time, the action of gravity generates, in a falling body, the velocity $\dfrac{2NI}{t}$, the resistance will be to the gravity as

$$\frac{GH}{T} - \frac{HI}{t} + \frac{2MI \cdot NI}{t \cdot HI} \text{ to } \frac{2NI}{t} \text{ or as } \frac{t \cdot GH}{T} - HI + \frac{2MI \cdot NI}{HI} \text{ to } 2NI.$$

Now, for the abscissas CB, CD, CE, put −o, o, $2o$. For the ordinate CH put P; and for MI put any series $Qo + Ro^2 + So^3 +$, &c. And all the terms of the series after the first, that is, $Ro^2 + So^3 +$, &c., will be NI; and the ordinates DI, EK, and BG will be $P - Qo - Ro^2 - So^3 -$, &c., $P - 2Qo - 4Ro^2 - 8So^3 -$, &c., and $P + Qo - Ro^2 + So^3 -$, &c., respectively. And by squaring the differences of the ordinates BG − CH and CH − DI, and to the squares thence produced adding the squares of BC and CD themselves, you will have $oo + QQoo - 2QRo^3 +$, &c., and $oo + QQoo + 2QRo^3 +$, &c., the squares of the arcs GH, HI; whose roots $o\sqrt{(1+QQ)} - \dfrac{QRoo}{\sqrt{(1+QQ)}}$ and $o\sqrt{(1+QQ)} + \dfrac{QRoo}{\sqrt{(1+QQ)}}$ are the arcs GH and HI. Moreover, if from the ordinate CH there be subtracted half the sum of the ordinates BG and DI, and from the ordinate DI there be subtracted half the sum of the ordinates CH and EK, there will remain Roo and $Roo + 3So^3$, the versed sines of the arcs GI and HK. And these are proportional to the short lines LH and NI, and therefore are as the squares of the infinitely small times T and t:

and thence the ratio $\dfrac{t}{T}$ varies as the square root of $\dfrac{R+3So}{R}$ or $\dfrac{R+\frac{3}{2}So}{R}$; and $\dfrac{t \times GH}{T} - HI + \dfrac{2MI \times NI}{HI}$ by substituting the values of $\dfrac{t}{T}$, GH, HI, MI, and NI just found, becomes $\dfrac{3Soo}{2R} \cdot \sqrt{(1+QQ)}$. And since 2NI is $2Roo$, the resistance will be now to the gravity as $\dfrac{3Soo}{2R}\sqrt{(1+QQ)}$ to $2Roo$, that is, as $3S\sqrt{(1+QQ)}$ to $4RR$.

And the velocity will be such, that a body going off therewith from any place H, in the direction of the tangent HN, would describe, in a vacuum, a parabola, whose diameter is HC, and its latus rectum $\dfrac{HN^2}{NI}$ or $\dfrac{1+QQ}{R}$.

And the resistance is as the product of the density of the medium and the square of the velocity; and therefore the density of the medium is directly as the resistance, and inversely as the square of the velocity; that is, directly as $\dfrac{3S\sqrt{(1+QQ)}}{4RR}$ and inversely as $\dfrac{1+QQ}{R}$; that is, as $\dfrac{S}{R\sqrt{(1+QQ)}}$. Q.E.I.

COR. I. If the tangent HN be produced both ways, so as to meet any ordinate AF in T, $\dfrac{HT}{AC}$ will be equal to $\sqrt{(1+QQ)}$, and therefore in what has gone before may be put for $\sqrt{(1+QQ)}$. By this means the resistance will be to the gravity as $3S \cdot HT$ to $4RR \cdot AC$; the velocity will be as $\dfrac{HT}{AC\sqrt{R}}$, and the density of the medium will be as $\dfrac{S \cdot AC}{R \cdot HT}$.

COR. II. And hence, if the curved line PFHQ be defined by the relation between the base or abscissa AC and the ordinate CH, as is usual, and the value of the ordinate be resolved into a converging series, the Problem will be expeditiously solved by the first terms of the series; as in the following Examples.

EXAM. I. Let the line PFHQ be a semicircle described upon the diameter PQ; to find the density of the medium that shall make a projectile move in that line.

Bisect the diameter PQ in A; and call AQ, n; AC, a; CH, e; and CD, o; then DI^2 or $AQ^2 - AD^2 = nn - aa - 2ao - oo$, or $ee - 2ao - oo$; and the root being extracted by our method, will give

$$DI = e - \frac{ao}{e} - \frac{oo}{2e} - \frac{aaoo}{2e^3} - \frac{ao^3}{2e^3} - \frac{a^3o^3}{2e^5} -, \&c.$$

Here put nn for $ee + aa$, and DI will become $= ee - \dfrac{ao}{e} - \dfrac{nnoo}{2e^3} - \dfrac{anno^3}{2e^5} -$, &c.

In such a series I distinguish the successive terms after this manner: I call that the first term in which the infinitely small quantity o is not found; the second, in which that quantity is of one dimension only; the third, in which it arises to two dimensions; the fourth, in which it is of three; and so *ad infinitum*. And the first term, which here is e, will always denote the length of the ordinate CH, erected at the starting point of the indefinite quantity o. The second term, which here is $\dfrac{ao}{e}$, will denote the difference between CH and DN; that is, the short line MN which is cut off by completing the parallelogram HCDM; and therefore always determines the position of the tangent HN; as, in this case, by taking $MN : HM = \dfrac{ao}{e} : o = a : e$. The third term, which here is $\dfrac{nnoo}{2e^3}$, will represent the short line IN, which lies

between the tangent and the curve; and therefore determines the angle of contact IHN, or the curvature which the curved line has in H. If that short line IN is of a finite magnitude, it will be expressed by the third term, together with those that follow *in infinitum.* But if that short line be diminished *in infinitum,* the terms following become infinitely less than the third term, and therefore may be neglected. The fourth term determines the variation of the curvature; the fifth, the variation of the variation; and so on.

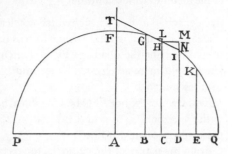

From this, by the way, appears the use, not to be disdained, which may be made of these series in the solution of problems that depend upon tangents, and the curvature of curves.

Now compare the series

$$e-\frac{ao}{e}-\frac{nnoo}{2e^3}-\frac{anno^3}{2e^5}-\&c.,$$

with the series

$$P-Qo-Roo-So^3-\&c.,$$

and for P, Q, R and S, put $e,\dfrac{a}{e},\dfrac{nn}{2e^3}$ and $\dfrac{ann}{2e^5}$, and for $\sqrt{(1+QQ)}$ put $\sqrt{\left(1+\dfrac{aa}{ee}\right)}$

or $\dfrac{n}{e}$; and the density of the medium will come out as $\dfrac{a}{ne}$; that is (because n is given), as $\dfrac{a}{e}$ or $\dfrac{AC}{CH}$, that is, as that length of the tangent HT, which is terminated at the semidiameter AF standing perpendicularly on PQ: and the resistance will be to the gravity as $3a$ to $2n$, that is, as 3AC to the diameter PQ of the circle; and the velocity will be as $\sqrt{CH}$. Therefore if the body goes from the place F, with a due velocity, in the direction of a line parallel to PQ, and the density of the medium in each of the places H is as the length of the tangent HT, and the resistance also in any place H is to the force of gravity as 3AC to PQ, that body will describe the quadrant FHQ of a circle. Q.E.I.

But if the same body should go from the place P, in the direction of a line perpendicular to PQ, and should begin to move in an arc of the semicircle PFQ, we must take AC or a on the contrary side of the centre A; and therefore its sign must be changed, and we must put $-a$ for $+a$. Then the density of the medium would come out as $-\dfrac{a}{e}$. But Nature does not admit of a negative density, that is, a density which accelerates the motion of bodies; and therefore it cannot naturally come to pass that a body by ascending from P should describe the quadrant PF of a circle. To produce such an effect, a body ought to be accelerated by an impelling medium, and not impeded by a resisting one.

EXAM. 2. Let the line PFQ be a parabola, having its axis AF perpendicular to the horizon PQ; to find the density of the medium, which will make a projectile move in that line.

From the nature of the parabola, the rectangle $-\mathrm{PD}\cdot\mathrm{DQ}$ is equal to the rectangle under the ordinate DI and some given right line; that is, if that right line be called b; PC, a; PQ, c; CH, e; and CD, o; the rectangle

$$(a+o)(c-a-o)=ac-aa-2ao+co-oo=b\cdot\mathrm{DI};$$

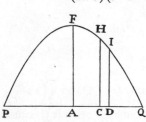

therefore $\mathrm{DI}=\dfrac{ac-aa}{b}+\dfrac{c-2a}{b}\cdot o-\dfrac{oo}{b}.$

Now the second term $\dfrac{c-2a}{b}\,o$ of this series is to be put for Qo, and the third term $\dfrac{oo}{b}$ for Roo. But since there are no more terms, the coefficient S of the fourth term will vanish; and therefore the quantity $\dfrac{\mathrm{S}}{\mathrm{R}\sqrt{(\mathrm{I}+\mathrm{QQ})}}$, to which the density of the medium is proportional, will be nothing. Therefore, where the medium is of no density, the projectile will move in a parabola; as *Galileo* hath heretofore demonstrated. Q.E.I.

EXAM. 3. Let the line AGK be an hyperbola, having its asymptote NX perpendicular to the horizontal plane AK; to find the density of the medium that will make a projectile move in that line.

Let MX be the other asymptote, meeting the ordinate DG produced in V; and from the nature of the hyperbola, the rectangle of XV into VG will be given. There is also given the ratio of DN to VX, and therefore the rectangle of DN into VG is given. Let that be bb; and, completing the parallelogram DNXZ, let BN be called a; BD, o; NX, c; and let the given ratio of VZ to ZX or DN be $\frac{m}{n}$. Then DN will be equal to $a-o$, VG equal to $\frac{bb}{a-o}$, VZ equal to $\frac{m}{n} \cdot (a-o)$ and GD or NX$-$VZ$-$VG equal to

$$c - \frac{m}{n} a + \frac{m}{n} o - \frac{bb}{a-o}.$$

Let the term $\frac{bb}{a-o}$ be resolved into the converging series

$$\frac{bb}{a} + \frac{bb}{aa} o + \frac{bb}{a^3} oo + \frac{bb}{a^4} o^3, \&c.,$$

and GD will become equal to

$$c - \frac{m}{n} a - \frac{bb}{a} + \frac{m}{n} o - \frac{bb}{aa} o - \frac{bb}{a^3} o^2 - \frac{bb}{a^4} o^3, \&c.$$

The second term $\frac{m}{n} o - \frac{bb}{aa} o$ of this series is to be used for Qo; the third $\frac{bb}{a^3} o^2$, with its sign changed for Ro^2; and the fourth $\frac{bb}{a^4} o^3$, with its sign changed also for So^3, and their coefficients $\frac{m}{n} - \frac{bb}{aa}$, $\frac{bb}{a^3}$, and $\frac{bb}{a^4}$ are to be put for Q, R, and S in the former rule. Which being done, the density of the medium will come out as

$$\frac{\dfrac{bb}{a^4}}{\dfrac{bb}{a^3} \sqrt{\left(1 + \dfrac{mm}{nn} - \dfrac{2mbb}{naa} + \dfrac{b^4}{a^4}\right)}}$$

or

$$\frac{1}{\sqrt{\left(aa + \dfrac{mm}{nn} aa - \dfrac{2mbb}{n} + \dfrac{b^4}{aa}\right)}},$$

that is, if in VZ you take VY equal to VG, as $\frac{1}{XY}$. For

$$aa \text{ and } \frac{m^2}{n^2} a^2 - \frac{2mbb}{n} + \frac{b^4}{aa}$$

are the squares of XZ and ZY. But the ratio of the resistance to gravity is found to be that of 3XY to 2YG; and the velocity is that with which the body would describe a parabola, whose vertex is G, diameter DG, latus rectum $\frac{XY^2}{VG}$. Suppose, therefore, that the densities of the medium in each of the places G are inversely as the distances XY, and that the resistance in any place G is to the gravity as 3XY to 2YG; and a body let go from the place A, with a due velocity, will describe that hyperbola AGK. Q.E.I.

EXAM. 4. Suppose, indefinitely, the line AGK to be an hyperbola described with the centre X, and the asymptotes MX, NX, so that, having constructed the rectangle XZDN, whose side ZD cuts the hyperbola in G and its asymptote in V, VG may be inversely as any power DN^n of the line

ZX or DN, whose index is the number n: to find the density of the medium in which a projected body will describe this curve.

For BN, BD, NX, put A, O, C, respectively, and let VZ be to XZ or DN as d to e, and VG be equal to $\frac{bb}{DN^n}$; then DN will be equal to $A - O$,

$$VG = \frac{bb}{(A-C)^n},$$

$$VZ = \frac{d}{e}(A-O)$$

and GD or NX $-$ VZ $-$ VG equal to

$$C - \frac{d}{e}A + \frac{d}{e}O - \frac{bb}{(A-O)^n}.$$

Let the term $\frac{bb}{(A-O)^n}$ be resolved into an infinite series

$$\frac{bb}{A^n} + \frac{nbb}{A^{n+1}} \cdot O + \frac{nn+n}{2A^{n+2}} \cdot bb\,O^2 + \frac{n^3+3nn+2n}{6A^{n+3}} \cdot bb\,O^3, \&c.,$$

and GD will be equal to

$$C - \frac{d}{e} A - \frac{bb}{A^n} + \frac{d}{e} O - \frac{nbb}{A^{n+1}} O - \frac{+nn+n}{2A^{n+2}} bbO^2 - \frac{+n^3+3nn+2n}{6A^{n+3}} bbO^3, \text{ &c.}$$

The second term $\frac{d}{e} O - \frac{nbb}{A^{n+1}} O$ of this series is to be used for Qo, the third

$\frac{nn+n}{2A^{n+2}} bbO^2$ for Roo, the fourth $\frac{n^3+3nn+2n}{6A^{n+3}} bbO$ for So^3. And thence the

density of the medium $\dfrac{S}{R\sqrt{(1+QQ)}}$, in any place G, will be

$$\frac{n+2}{3\sqrt{\left(A^2 + \dfrac{dd}{ee} A^2 - \dfrac{2dnbb}{eA^n} A + \dfrac{nnb^4}{A^{2n}}\right)}},$$

and therefore if in VZ you take VY equal to $n \cdot$ VG, that density is recipro-

cally as XY. For A^2 and $\dfrac{dd}{ee} A^2 - \dfrac{2dnbb}{eA^n} A + \dfrac{nnb^4}{A^{2n}}$ are the squares of XZ and

and ZY. But the resistance in the same place G is to the force of gravity as

$3S \cdot \dfrac{XY}{A}$ to 4RR, that is, as XY to $\dfrac{2nn+2n}{n+2}$ VG. And the velocity there is

the same wherewith the projected body would move in a parabola, whose

vertex is G, diameter GD, and latus rectum $\dfrac{1+QQ}{R}$ or $\dfrac{2XY^2}{(nn+n) \cdot VG}$. Q.E.I.

SCHOLIUM

In the same manner that the
density of the medium comes

out to be as $\dfrac{S \cdot AC}{R \cdot HT}$, in Cor. 1,

if the resistance is put as any
power V^n of the velocity V,
the density of the medium
will come out to be as

$$\frac{S}{R^{\frac{4-n}{2}}} \cdot \left(\frac{AC}{HT}\right)^{n-1}.$$

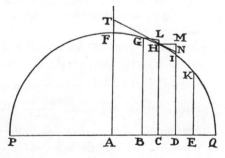

And therefore, if a curve can be found, such that the ratio of $\dfrac{S}{R^{\frac{4-n}{2}}}$ to

$\left(\dfrac{HT}{AC}\right)^{n-1}$, or of $\dfrac{S^2}{R^{4-n}}$ to $(1+QQ)^{n-1}$ may be given; the body, in an uniform medium, whose resistance is as the power V^n of the velocity V, will move in this curve. But let us return to more simple curves.

Since there can be no motion in a parabola except in a nonresisting medium, but in the hyperbolas here described it is produced by a continual resistance; it is evident that the line which a projectile describes in an uniformly resisting medium approaches nearer to these hyperbolas than to a parabola. That line is certainly of the hyperbolic kind, but about the vertex it is more distant from the asymptotes, and in the parts remote from the vertex draws nearer to them than these hyperbolas here described. The difference, however, is not so great between the one and the other but that these latter may be commodiously enough used in practice instead of the former. And perhaps these may prove more useful than an hyperbola that is more accurate, and at the same time more complex. They may be made use of, then, in this manner.

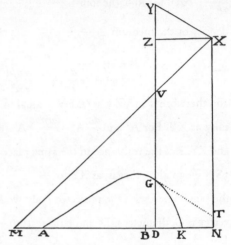

Complete the parallelogram XYGT, and the right line GT will touch the hyperbola in G, and therefore the density of the medium in G is inversely as the tangent GT, and the velocity there as $\sqrt{\dfrac{GT^2}{GV}}$; and the resistance is to the force of gravity as GT to $\dfrac{2nn + 2n}{n+2} \cdot GV$.

Therefore if a body projected from the place A, in the direction of the right line AH, describes the hyperbola AGK, and AH produced meets the asymptote NX in H, and AI drawn parallel to it meets the other asymptote MX in I; the density of the medium in A will be inversely as AH and the velocity of the body as $\sqrt{\dfrac{AH^2}{AI}}$, and the resistance there to the force of gravity as AH to $\dfrac{2nn+2n}{n+2}$ · AI. Hence the following Rules are deduced.

RULE 1. If the density of the medium at A, and the velocity with which the body is projected, remain the same, and the angle NAH be changed; the lengths AH, AI, HX will remain. Therefore if those lengths, in any one case, are found, the hyperbola may afterwards be easily determined from any given angle NAH.

RULE 2. If the angle NAH, and the density of the medium at A, remain the same, and the velocity with which the body is projected be changed, the length AH will continue the same; and AI will be changed inversely as the square of the velocity.

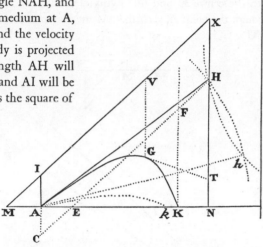

RULE 3. If the angle NAH, the velocity of the body at A, and the accelerative gravity remain the same, and the proportion of the resistance at A to the motive gravity be augmented in any ratio; the proportion of AH to AI will be augmented in the same ratio, the latus rectum of the above-mentioned parabola remaining the same, and also the length $\dfrac{AH^2}{AI}$ proportional to it; and therefore AH will be diminished in the same ratio, and AI will be diminished as the square of that ratio. But the proportion of the resistance to the weight is augmented, when either the specific gravity is made less, the magnitude remaining equal, or when the density of the medium is made greater, or when, by diminishing the magnitude, the resistance becomes diminished in a less ratio than the weight.

RULE 4. Because the density of the medium is greater near the vertex of the hyperbola than it is in the place A, that a mean density may be preserved, the ratio of the least of the tangents GT to the tangent AH ought to be found, and the density in A augmented in a ratio a little greater than that of half the sum of those tangents to the least of the tangents GT.

RULE 5. If the lengths AH, AI are given, and the figure AGK is to be described, produce HN to X, so that HX may be to AI as $n + 1$ to 1; and with the centre X, and the asymptotes MX, NX, describe an hyperbola through the point A, such that AI may be to any of the lines VG as XV^n to XI^n.

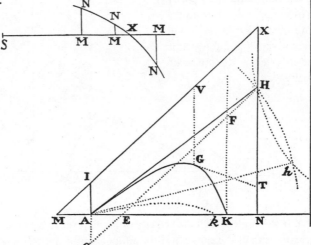

RULE 6. By how much the greater the number n is, so much the more accurate are these hyperbolas in the ascent of the body from A, and less accurate in its descent to K; and conversely. The conic hyperbola keeps a mean ratio between these, and is more simple than the rest. Therefore if the hyperbola be of this kind, and you are to find the point K, where the projected body falls upon any right line AN passing through the point A, let AN produced meet the asymptotes MX, NX in M and N, and take NK equal to AM.

RULE 7. And hence appears an expeditious method of determining this hyperbola from the phenomena. Let two similar and equal bodies be projected with the same velocity, in different angles HAK, hAk, and let them fall upon the plane of the horizon in K and k; and note the proportion of AK to Ak. Let it be as d to e. Then erecting a perpendicular AI of any length, assume any length AH or Ah, and thence graphically, or by scale and compass, collect the lengths AK, Ak (by Rule 6). If the ratio of AK to Ak be the same with that of d to e, the length of AH was rightly assumed. If not, take on the indefinite right line SM, the length SM equal to the assumed AH; and erect a perpendicular MN equal to the difference $\dfrac{AK}{A k} - \dfrac{d}{e}$ of the ratios multiplied by any given right line. By the like method, from several assumed lengths AH, you may find several points N; and

draw through them all a regular curve NNXN, cutting the right line SMMM in X. Lastly, assume AH equal to the abscissa SX, and thence find again the length AK; and the lengths, which are to the assumed length AI, and this last AH, as the length AK known by experiment, to the length AK last found, will be the true lengths AI and AH, which were to be found.

But these being given, there will be given also the resisting force of the medium in the place A, it being to the force of gravity as AH to ⅓AI. Let the density of the medium be increased by Rule 4, and if the resisting force just found be increased in the same ratio, it will become still more accurate.

RULE 8. The lengths AH, HX being found; let there be now required the position of the line AH, according to which a projectile thrown with that given velocity shall fall upon any point K. At the points A and K, erect the lines AC, KF perpendicular to the horizon; whereof let AC be drawn downwards, and be equal to AI or ½HX. With the asymptotes AK, KF, describe an hyperbola, whose conjugate shall pass through the point C; and from the centre A, with the interval AH, describe a circle cutting that hyperbola in the point H; then the projectile thrown in the direction of the right line AH will fall upon the point K. Q.E.I. For the point H, because of the given length AH, must be somewhere in the circumference of the described circle. Draw CH meeting AK and KF in E and F; and because CH, MX are parallel, and AC, AI equal, AE will be equal to AM, and therefore also equal to KN. But CE is to AE as FH to KN, and therefore CE and FH are equal. Therefore the point H falls upon the hyperbolic curve described with the asymptotes AK, KF whose conjugate passes through the point C; and is therefore found in the common intersection of this hyperbolic curve and the circumference of the described circle. Q.E.D. It is to be observed that this operation is the same, whether the right line AKN be parallel to the horizon, or inclined thereto in any angle; and that from two intersections H, *h,* there arise two angles NAH, NA*h*; and that in mechanical practice it is sufficient once to describe a circle, then to apply a ruler CH, of an indeterminate length, so to the point C, that its part FH, intercepted between the circle and the right line FK, may be equal to its part CE placed between the point C and the right line AK.

What has been said of hyperbolas may be easily applied to parabolas. For if a parabola be represented by XAGK, touched by a right line XV in the vertex X, and the ordinates IA, VG be as any powers XI^n, XV^n, of the abscissas XI, XV; draw XT, GT, AH, whereof let XT be parallel to VG, and let GT, AH touch the parabola in G and A: and a body projected from any place A, in the direction of the right line AH, with a due velocity, will describe this parabola, if the density of the medium in each of the places G be in-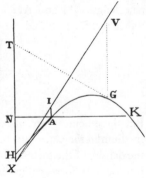
versely as the tangent GT. In that case the velocity in G will be the same as would cause a body, moving in a nonresisting space, to describe a conic

parabola, having G for its vertex, VG produced downwards for its diameter, and $\dfrac{2GT^2}{(nn-n) \cdot VG}$ for its latus rectum. And the resisting force in G will be to the force of gravity as GT to $\dfrac{2nn-2n}{n-2}$ VG. Therefore if NAK represent an horizontal line, and both the density of the medium at A, and the velocity with which the body is projected, remaining the same, the angle NAH be anyhow altered, the lengths AH, AI, HX will remain; and thence will be given the vertex X of the parabola, and the position of the right line XI; and by taking VG to IA as XV^n to XI^n, there will be given all the points G of the parabola, through which the projectile will pass.

SECTION III

The motion of bodies that are resisted partly in the ratio of the velocities,
and partly as the square of the same ratio.

PROPOSITION XI. THEOREM VIII

If a body be resisted partly in the ratio and partly as the square of the ratio
of its velocity, and moves in a similar medium by its innate force only; and
the times be taken in arithmetical progression: then quantities inversely
proportional to the velocities, increased by a certain given quantity, will be
in geometrical progression.

With the centre C, and the rectangular asymptotes CAD*d* and CH, describe an hyperbola BE*e,* and let AB, DE, *de* be parallel to the asymptote CH. In the asymptote CD let A, G be given points; and if the time be repre-

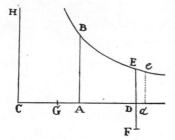

sented by the hyperbolic area ABED uniformly increasing, I say, that the velocity may be expressed by the length DF, whose reciprocal GD, together with the given line CG, compose the length CD increasing in a geometrical progression.

For let the small area DE*ed* be the least given increment of the time, and D*d* will be inversely as DE, and therefore directly as CD. Therefore the decrement of $\frac{1}{GD}$, which (by Lem. II, Book II) is $\frac{Dd}{GD^2}$, will be also as $\frac{CD}{GD^2}$ or $\frac{CG+GD}{GD^2}$, that is, as $\frac{1}{GD} + \frac{CG}{GD^2}$. Therefore, the time ABED uniformly increasing by the addition of the given intervals ED*de*, it follows that $\frac{1}{GD}$ decreases in the same ratio with the velocity. For the decrement of the velocity is as the resistance, that is (by the supposition), as the sum of two quantities, whereof one is as the velocity, and the other as the square of the velocity; and the decrement of $\frac{1}{GD}$ is as the sum of the quantities $\frac{1}{GD}$ and $\frac{CG}{GD^2}$, whereof the first is $\frac{1}{GD}$ itself, and the last

$\dfrac{CG}{GD^2}$ is as $\dfrac{1}{GD^2}$: therefore $\dfrac{1}{GD}$ is as the velocity, the decrements of both being analogous. And if the quantity GD inversely proportional to $\dfrac{1}{GD}$, be augmented by the given quantity CG; the sum CD, the time ABED uniformly increasing, will increase in a geometrical progression. Q.E.D.

Cor. i. Therefore, if, having the points A and G given, the time be represented by the hyperbolic area ABED, the velocity may be represented by $\dfrac{1}{GD}$ the reciprocal of GD.

Cor. ii. And by taking GA to GD as the reciprocal of the velocity at the beginning to the reciprocal of the velocity at the end of any time ABED, the point G will be found. And that point being found, the velocity may be found from any other time given.

PROPOSITION XII. THEOREM IX

The same things being supposed, I say, that if the spaces described are taken in arithmetical progression, the velocities augmented by a certain given quantity will be in geometrical progression.

In the asymptote CD let there be given the point R, and, erecting the perpendicular RS meeting the hyperbola in S, let the space described be represented by the hyperbolic area RSED; and the velocity will be as the length GD, which, together with the given line CG, composes a length CD

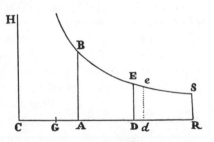

decreasing in a geometrical progression, while the space RSED increases in an arithmetical progression.

For, because the increment ED*de* of the space is given, the short line D*d*, which is the decrement of GD, will be reciprocally as ED, and therefore directly as CD; that is, as the sum of the same GD and the given length CG. But the decrement of the velocity, in a time inversely proportional thereto, in which the given interval of space D*de*E is described, is as the resistance and the time conjointly, that is, directly as the sum of two quantities, whereof one is as the velocity, the other as the square of the velocity, and inversely as the velocity; and therefore directly as the sum of two quantities, one of which is given, the other is as the velocity. Therefore the decrement both of the velocity and of the line GD is as a given quantity and a decreasing quantity conjointly; and, because the decrements are analogous, the decreasing quantities will always be analogous; viz., the velocity, and the line GD. Q.E.D.

Cor. I. If the velocity be represented by the length GD, the space described will be as the hyperbolic area DESR.

Cor. II. And if the point R be assumed anywhere, the point G will be found, by taking GR to GD as the velocity at the beginning to the velocity after any space RSED is described. The point G being given, the space is given from the given velocity; and conversely.

Cor. III. Whence since (by Prop. xI) the velocity is given from the given time, and (by this Proposition) the space is given from the given velocity, the space will be given from the given time; and conversely.

PROPOSITION XIII. THEOREM X

Supposing that a body attracted downwards by an uniform gravity ascends or descends in a right line; and that the same is resisted partly in the ratio of its velocity, and partly as the square of the ratio thereof: I say, that, if right lines parallel to the diameters of a circle and an hyperbola be drawn through the ends of the conjugate diameters, and the velocities be as some segments of those parallels drawn from a given point, the times will be as the sectors of the areas cut off by right lines drawn from the centre to the ends of the segments; and conversely.

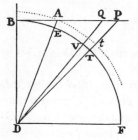

CASE I. Suppose first that the body is ascending, and from the centre D, with any semidiameter DB, describe a quadrant BETF of a circle, and through the end B of the semidiameter DB draw the indefinite line BAP, parallel to the semidiameter DF. In that line let there be given the point A, and take the segment AP proportional to the velocity. And since one part of the resistance is as the velocity, and another part as the square of the velocity, let the whole resistance be as $AP^2 + 2BA \cdot AP$. Join DA, DP, cutting the circle in E and T, and let the gravity be represented by DA^2, so that the gravity shall be to the resistance in P as DA^2 to $AP^2 + 2BA \cdot AP$; and the time of the whole ascent will be as the sector EDT of the circle.

For draw DVQ, cutting off the moment PQ of the velocity AP, and the moment DTV of the sector DET answering to a given moment of time; and that decrement PQ of the velocity will be as the sum of the forces of gravity DA^2 and of resistance $AP^2 + 2BA \cdot AP$; that is (by Prop. xii, Book ii, *Elem.* of *Euclid*), as DP^2. Then the area DPQ, proportional to PQ, is as DP^2, and the area DTV, which is to the area DPQ as DT^2 to DP^2, is as the given quantity DT^2. Therefore the area EDT decreases uniformly according to the rate of the future time, by subtraction of given intervals DTV, and is therefore proportional to the time of the whole ascent. Q.E.D.

CASE 2. If the velocity in the ascent of the body be represented by the length AP as before, and the resistance be made as $AP^2 + 2BA \cdot AP$; and if the force of gravity be less than can be expressed by DA^2; take BD of such a length, that $AB^2 - BD^2$ may be proportional to the gravity, and let DF be perpendicular and equal to DB, and through the vertex F describe the hyperbola FTVE, whose conjugate semidiameters

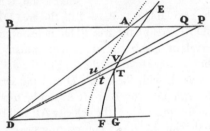

are DB and DF, and which cuts DA in E, and DP, DQ in T and V; and the time of the whole ascent will be as the hyperbolic sector TDE.

For the decrement PQ of the velocity, produced in a given interval of time, is as the sum of the resistance $AP^2 + 2BA \cdot AP$ and of the gravity $AB^2 - BD^2$, that is, as $BP^2 - BD^2$. But the area DTV is to the area DPQ as DT^2 to DP^2; and, therefore, if GT be drawn perpendicular to DF, as GT^2 or $GD^2 - DF^2$ to BD^2, and as GD^2 to BP^2, and, by subtraction, as DF^2 to $BP^2 - BD^2$. Therefore since the area DPQ is as PQ, that is, as $BP^2 - BD^2$, the area DTV will be as the given quantity DF^2. Therefore the area EDT decreases uniformly in each of the equal intervals of time, by the subtraction of so many given intervals DTV, and therefore is proportional to the time. Q.E.D.

CASE 3. Let AP be the velocity in the descent of the body, and $AP^2 + 2BA \cdot$ AP the force of resistance, and $BD^2 - AB^2$ the force of gravity, the angle DBA being a right one. And if with the centre D, and the principal vertex B, there be described a rectangular hyperbola BETV cutting DA, DP, and

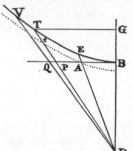

DQ produced in E, T, and V; the sector DET of this hyperbola will be as the whole time of descent.

For the increment PQ of the velocity, and the area DPQ proportional to it, is as the excess of the gravity above the resistance, that is, as

$$BD^2 - AB^2 - 2BA \cdot AP - AP^2$$

or $BD^2 - BP^2$. And the area DTV is to the area DPQ as DT^2 to DP^2; and therefore as GT^2 or $GD^2 - BD^2$ to BP^2; and as GD^2 to BD^2, and, by subtraction, as BD^2 to $BD^2 - BP^2$. Therefore since the area DPQ is as $BD^2 - BP^2$, the area DTV will be as the given quantity BD^2. Therefore the area EDT increases uniformly in the several equal intervals of time by the addition of as many given intervals DTV, and therefore is proportional to the time of the descent. Q.E.D.

COR. If with the centre D and the semidiameter DA there be drawn through the vertex A an arc A*t* similar to the arc ET, and similarly subtending the angle ADT, the velocity AP will be to the velocity which the body in the time EDT, in a nonresisting space, can lose in its ascent, or acquire in its descent, as the area of the triangle DAP to the area of the sector DA*t*; and therefore is given from the time given. For the velocity in a nonresisting medium is proportional to the time, and therefore to this sector; in a resisting medium, it is as the triangle; and in both mediums, where it is least, it approaches to the ratio of equality, as the sector and triangle do.

SCHOLIUM

One may demonstrate also that case in the ascent of the body, where the force of gravity is less than can be expressed by DA^2 or $AB^2 + BD^2$, and greater than can be expressed by $AB^2 - DB^2$, and must be expressed by AB^2. But I hasten to other things.

PROPOSITION XIV. THEOREM XI

The same things being supposed, I say, that the space described in the ascent or descent is as the difference of the area by which the time is expressed, and of some other area which is augmented or diminished in an arithmetical progression; if the forces compounded of the resistance and the gravity be taken in a geometrical progression.

Take AC (in these three figures) proportional to the gravity, and AK to the resistance; but take them on the same side of the point A, if the body is descending, otherwise on the contrary. Erect A*b*, which make to DB as

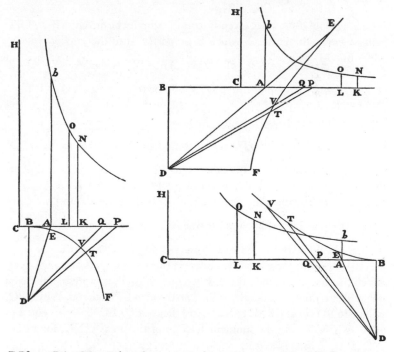

DB² to 4 BA · CA; and to the rectangular asymptotes CK, CH, describe the hyperbola *b*N; and, erecting KN perpendicular to CK, the area A*b*NK will be augmented or diminished in an arithmetical progression, while the

forces CK are taken in a geometrical progression. I say, therefore, that the distance of the body from its greatest altitude is as the excess of the area AbNK above the area DET.

For since AK is as the resistance, that is, as $AP^2 \cdot 2BA \cdot AP$; assume any given quantity Z, and put AK equal to $\dfrac{AP^2 + 2BA \cdot AP}{Z}$; then (by Lem. II of this Book) the moment KL of AK will be equal to $\dfrac{2PQ \cdot AP + 2BA \cdot PQ}{Z}$ or $\dfrac{2PQ \cdot BP}{Z}$, and the moment KLON of the area AbNK will be equal to $\dfrac{2PQ \cdot BP \cdot LO}{Z}$ or $\dfrac{PQ \cdot BP \cdot BD^3}{2Z \cdot CK \cdot AB}$.

CASE 1. Now if the body ascends, and the gravity be as $AB^2 + BD^2$, BET being a circle, the line AC, which is proportional to the gravity, will be $\dfrac{AB^2 + BD^2}{Z}$, and DP^2 or $AP^2 + 2BA \cdot AP + AB^2 + BD^2$ will be $AK \cdot Z + AC \cdot Z$ or $CK \cdot Z$; and therefore the area DTV will be to the area DPQ as DT^2 or DB^2 to $CK \cdot Z$.

CASE 2. If the body ascends, and the gravity be as $AB^2 - BD^2$, the line AC will be $\dfrac{AB^2 - BD^2}{Z}$, and DT^2 will be to DP^2 as DF^2 or DB^2 to $BP^2 - BD^2$ or $AP^2 + 2BA \cdot AP + AB^2 - BD^2$, that is, to $AK \cdot Z + AC \cdot Z$ or $CK \cdot Z$. And therefore the area DTV will be to the area DPQ as DB^2 to $CK \cdot Z$.

CASE 3. And by the same reasoning, if the body descends, and therefore the gravity is as $BD^2 - AB^2$, and the line AC becomes equal to $\dfrac{BD^2 - AB^2}{Z}$; the area DTV will be to the area DPQ as DB^2 to $CK \cdot Z$: as above.

Since, therefore, these areas are always in this ratio, if for the area DTV, by which the moment of the time, always equal to itself, is expressed, there be put any determinate rectangle, as $BD \cdot m$, the area DPQ, that is, $\frac{1}{2}BD \cdot PQ$, will be to $BD \cdot m$ as $CK \cdot Z$ to BD^2. And thence $PQ \cdot BD^3$ becomes equal to $2BD \cdot m \cdot CK \cdot Z$, and the moment KLON of the area AbNK, found before, becomes $\dfrac{BP \cdot BD \cdot m}{AB}$. From the area DET subtract its moment DTV or $BD \cdot m$, and there will remain $\dfrac{AP \cdot BD \cdot m}{AB}$. Therefore the difference of the moments, that is, the moment of the difference of the areas, is equal to $\dfrac{AP \cdot BD \cdot m}{AB}$; and therefore (because of the given quantity $\dfrac{BD \cdot m}{AB}$) as the velocity AP; that is, as the moment of the space which the body describes in its ascent or descent. And therefore the difference of the areas, and that space, increasing or decreasing by proportional moments, and beginning together or vanishing together, are proportional. Q.E.D.

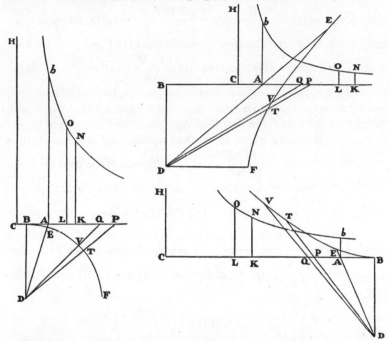

Cor. If the length, which arises by applying the area DET to the line BD, be called M; and another length V be taken in that ratio to the length M, which the line DA has to the line DE; the space which a body, in a resisting medium, describes in its whole ascent or descent, will be to the space which a body, in a nonresisting medium, falling from rest, can describe in the same time, as the difference of the aforesaid areas to $\dfrac{BD \cdot V^2}{AB}$; and therefore is given from the time given. For the space in a nonresisting medium is as the square of the time, or as V^2; and, because BD and AB are given, as $\dfrac{BD \cdot V^2}{AB}$. This area is equal to the area $\dfrac{DA^2 \cdot BD \cdot M^2}{DE^2 \cdot AB}$ and the moment of M is m; and therefore the moment of this area is $\dfrac{DA^2 \cdot BD \cdot 2M \cdot m}{DE^2 \cdot AB}$. But this moment is to the moment of the difference of the aforesaid areas DET and AbNK, viz., to $\dfrac{AP \cdot BD \cdot m}{AB}$, as $\dfrac{DA^2 \cdot BD \cdot M}{DE^2}$ to ½BD · AP, or as $\dfrac{DA^2}{DE^2}$ into DET to DAP; and, therefore, when the areas DET and DAP are least, in the ratio of equality. Therefore the area $\dfrac{BD \cdot V^2}{AB}$ and the difference of the areas DET and AbNK, when all these areas are least, have equal moments; and are therefore equal. Therefore since the velocities, and therefore also the spaces in both mediums described together, in the beginning of the descent, or the end of the ascent, approach to equality, and therefore are then one to another as the area $\dfrac{BD \cdot V^2}{AB}$, and the difference of the areas DET and AbNK; and moreover since the space, in a nonresisting medium, is continually as $\dfrac{BD \cdot V^2}{AB}$, and the space, in a resisting medium, is continually as the difference of the areas DET and AbNK; it necessarily follows, that the spaces, in both mediums, described in any equal times, are one to another as that area $\dfrac{BD \cdot V^2}{AB}$, and the difference of the areas DET and AbNK. Q.E.D.

SCHOLIUM

The resistance of spherical bodies in fluids arises partly from the tenacity, partly from the attrition, and partly from the density of the medium. And that part of the resistance which arises from the density of the fluid is, as I said, as the square of the velocity; the other part, which arises from the tenacity of the fluid, is uniform, or as the moment of the time; and, therefore, we might now proceed to the motion of bodies, which are resisted partly by an uniform force, or in the ratio of the moments of the time, and partly as the square of the velocity. But it is sufficient to have cleared the way to this speculation in Prop. VIII and IX foregoing, and their Corollaries. For in those Propositions, instead of the uniform resistance made to an ascending body arising from its gravity, one may substitute the uniform resistance which arises from the tenacity of the medium, when the body moves by its inertia alone; and when the body ascends in a right line, add this uniform resistance to the force of gravity, and subtract it when the body descends in a right line. One might also go on to the motion of bodies which are resisted in part uniformly, in part in the ratio of the velocity, and in part as the square of the same velocity. And I have opened a way to this in Prop. XIII and XIV foregoing, in which the uniform resistance arising from the tenacity of the medium may be substituted for the force of gravity, or be compounded with it as before. But I hasten to other things.

SECTION IV

The circular motion of bodies in resisting mediums.

LEMMA III

Let PQR *be a spiral cutting all the radii* SP, SQ, SR, &c., *in equal angles. Draw the right line* PT *touching the spiral in any point* P, *and cutting the radius* SQ *in* T; *draw* PO, QO *perpendicular to the spiral, and meeting in* O, *and join* SO: *I say, that if the points* P *and* Q *approach and coincide, the angle* PSO *will become a right angle, and the ultimate ratio of the rectangle* TQ · 2PS *to* PQ² *will be the ratio of equality.*

For, from the right angles OPQ, OQR, subtract the equal angles SPQ, SQR, and there will remain the equal angles OPS, OQS. Therefore a circle which passes through the points OSP will pass also through the point Q.

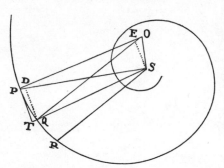

Let the points P and Q coincide, and this circle will touch the spiral in the place of coincidence PQ, and will therefore cut the right line OP perpendicularly. Therefore OP will become a diameter of this circle, and the angle OSP, being in a semicircle, becomes a right one. Q.E.D.

Draw QD, SE perpendicular to OP, and the ultimate ratios of the lines will be as follows:

TQ : PD = TS or PS : PE = 2PO : 2PS;

and   PD : PQ = PQ : 2PO;

multiplying together corresponding terms of equal ratios,

TQ : PQ = PQ : 2PS.

Whence PQ² becomes equal to TQ · 2PS. Q.E.D.

PROPOSITION XV. THEOREM XII

If the density of a medium in each place thereof be inversely as the distance of the places from an immovable centre, and the centripetal force be as the square of the density: I say, that a body may revolve in a spiral which cuts all the radii drawn from that centre in a given angle.

Suppose everything to be as in the foregoing Lemma, and produce SQ to V so that SV may be equal to SP. In any time let a body, in a resisting medium, describe the least arc PQ, and in double the time the least arc PR; and the decrements of those arcs arising from the resistance, or their differences from the arcs which would be described in a nonresisting medium in the same times, will be to each other as the squares of the times in which they are generated; therefore the decrement of the arc PQ is the fourth part of the decrement of the arc PR. Whence also if the area QSr be taken equal

to the area PSQ, the decrement of the arc PQ will be equal to half the short line Rr; and therefore the force of resistance and the centripetal force are to each other as the short line ½Rr and TQ which they generate in the same time. Because the centripetal force with which the body is urged in P is inversely as SP^2, and (by Lem. x, Book 1) the

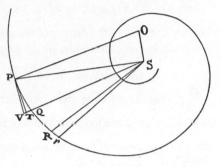

short line TQ, which is generated by that force, is in a ratio compounded of the ratio of this force and the squared ratio of the time in which the arc PQ is described (for in this case I neglect the resistance, as being infinitely less than the centripetal force), it follows that $TQ \cdot SP^2$, that is (by the last Lemma), $\frac{1}{2}PQ^2 \cdot SP$, will be as the square of the time, and therefore the time is as $PQ \cdot \sqrt{SP}$; and the velocity of the body, with which the arc PQ is described in that time, as $\dfrac{PQ}{PQ \cdot \sqrt{SP}}$ or $\dfrac{1}{\sqrt{SP}}$, that is, inversely as the square root of SP. And, by a like reasoning, the velocity with which the arc QR is described, is inversely as the square root of SQ. Now those arcs PQ and QR are as the describing velocities to each other; that is, as the square root of the ratio of SQ to SP, or as SQ to $\sqrt{(SP \cdot SQ)}$; and, because of the equal angles SPQ, SQr, and the equal areas PSQ, QSr, the arc PQ is to the arc Qr as SQ to SP. Take the differences of the proportional consequents, and the arc PQ will be to the arc Rr as SQ to $SP - \sqrt{(SP \cdot SQ)}$, or ½VQ. For, the points P and Q coinciding, the ultimate ratio of $SP - \sqrt{(SP \cdot SQ)}$ to

½VQ is the ratio of equality. Since the decrement of the arc PQ arising from the resistance, or its double Rr, is as the resistance and the square of the time conjointly, the resistance will be as $\frac{Rr}{PQ^2 \cdot SP}$. But PQ was to Rr as SQ to ½VQ, and thence $\frac{Rr}{PQ^2 \cdot SP}$ becomes as $\frac{\frac{1}{2}VQ}{PQ \cdot SP \cdot SQ}$, or as $\frac{\frac{1}{2}OS}{OP \cdot SP^2}$. For, the points P and Q coinciding, SP and SQ coincide also, and the angle PVQ becomes a right one; and, because of the similar triangles PVQ, PSO, PQ becomes to ½VQ as OP to ½OS. Therefore $\frac{OS}{OP \cdot SP^2}$ is as the resistance, that is, in the ratio of the density of the medium in P and the squared ratio of the velocity conjointly. Subtract the squared ratio of the velocity, namely, the ratio $\frac{1}{SP}$, and there will remain the density of the medium in P, as $\frac{OS}{OP \cdot SP}$. Let the spiral be given, and, because of the given ratio of OS to OP, the density of the medium in P will be as $\frac{1}{SP}$. Therefore in a medium whose density is inversely as SP the distance from the centre, a body will revolve in this spiral. Q.E.D.

Cor. i. The velocity in any place P, is always the same wherewith a body in a nonresisting medium with the same centripetal force would revolve in a circle, at the same distance SP from the centre.

Cor. ii. The density of the medium, if the distance SP be given, is as $\frac{OS}{OP}$, but if that distance is not given, as $\frac{OS}{OP \cdot SP}$. And thence a spiral may be fitted to any density of the medium.

Cor. iii. The force of the resistance in any place P is to the centripetal force in the same place as ½OS to OP. For those forces are to each other as ½Rr and TQ, or as $\frac{\frac{1}{4}VQ \cdot PQ}{SQ}$ and $\frac{\frac{1}{2}PQ^2}{SP}$, that is, as ½VQ and PQ, or ½OS and OP. The spiral therefore being given, there is given the proportion of the resistance to the centripetal force; and, conversely, from that proportion given the spiral is given.

Cor. iv. Therefore the body cannot revolve in this spiral, except where the force of resistance is less than half the centripetal force. Let the resistance be made equal to half the centripetal force, and the spiral will coincide with the right line PS, and in that right line the body will descend to the centre with a velocity that is to the velocity with which it was proved before in the case of the parabola (Theor. x, Book 1) that the descent would be made in a nonresisting medium, as the square root of the ratio of unity to the number 2. And the times of the descent will be here inversely as the velocities, and therefore given.

Cor. v. And because at equal distances from the centre the velocity is the same in the spiral PQR as it is in the right line SP, and the length of the spiral is to the length of the right line PS in a given ratio, namely, in the ratio of OP to OS; the time of the descent in the spiral will be to the time of the descent in the right line SP in the same given ratio, and therefore given.

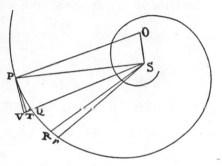

Cor. vi. If from the centre S, with any two given radii, two circles are described; and these circles remaining, the angle which the spiral makes with the radius PS be changed in any manner; the number of revolutions which the body can complete in the space between the circumferences of those circles, going round in the spiral from one circumference to another, will be as $\frac{PS}{OS}$, or as the tangent of the angle which the spiral makes with the radius PS; and the time of the same revolutions will be as $\frac{OP}{OS}$, that is, as the secant of the same angle, or inversely as the density of the medium.

Cor. vii. If a body, in a medium whose density is inversely as the distances of places from the centre, revolves in any curve AEB about that centre, and cuts the first radius AS in the same angle in B as it did before in A, and that with a velocity that shall be to its first velocity in A inversely as the square root of the distances from the centre (that is, as AS to a mean proportional between AS and BS), that body will continue to describe innumerable similar revolutions BFC, CGD, &c., and by its intersections will divide the radius AS into parts AS, BS, CS, DS, &c., that are continually proportional. But the times of the revolutions will be directly as the perimeters of the orbits AEB, BFC, CGD, &c., and inversely as the velocities at the beginnings A, B, C of those orbits; that is, as $AS^{3/2}$, $BS^{3/2}$, $CS^{3/2}$. And the whole time in which the body will arrive at the centre, will be to the time of the first revolution as the sum of all the continued proportionals $AS^{3/2}$, $BS^{3/2}$, $CS^{3/2}$, going on *ad infinitum,* is to the first term $AS^{3/2}$; that is, as the first term $AS^{3/2}$ is to the difference of the two first $AS^{3/2} - BS^{3/2}$, or as $\frac{2}{3}AS$ is to AB, very nearly. Whence the whole time may be easily found.

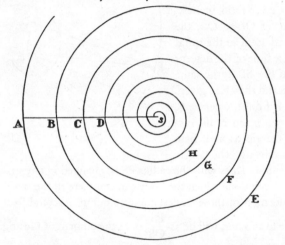

Cor. viii. From hence also may be deduced, near enough, the motions of bodies in mediums whose density is either uniform, or observes any other assigned law. From the centre S, with radii SA, SB, SC, &c., continually proportional, describe as many circles; and suppose the time of the revolutions between the perimeters of any two of those circles, in the medium whereof we treated, to be to the time of the revolutions between the same in the medium proposed as the mean density of the proposed medium between those circles is to the mean density of the medium whereof we treated, between the same circles, nearly; and that the secant of the angle in which the spiral above determined, in the medium whereof we treated, cuts the radius AS, is in the same ratio to the secant of the angle in which the new spiral, in the proposed medium, cuts the same radius; and also that the number of all the revolutions between the same two circles is nearly as the tangents of those angles. If this be done everywhere between every two circles, the motion will be continued through all the circles. And by this means one may without difficulty ascertain at what rate and in what time bodies ought to revolve in any regular medium.

COR. IX. And although these motions becoming eccentric should be performed in spirals approaching to an oval figure, yet, assuming the several revolutions of those spirals to be at the same distances from each other, and to approach to the centre by the same degrees as the spiral above described, we may also understand how the motions of bodies may be performed in spirals of that kind.

PROPOSITION XVI. THEOREM XIII

If the density of the medium in each of the places be inversely as the distance of the places from the immovable centre, and the centripetal force be inversely as any power of the same distance: I say, that the body may revolve in a spiral intersecting all the radii drawn from that centre in a given angle.

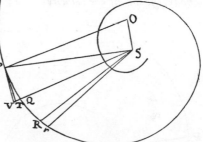

This is demonstrated in the same manner as the foregoing Proposition. For if the centripetal force in P be inversely as any power SP^{n+1} of the distance SP whose index is $n + 1$; it will be concluded, as above, that the time in which the body describes any arc PQ, will be as $PQ \cdot PS^{\frac{1}{2}n}$;

and the resistance in P as $\dfrac{Rr}{PQ^2 \cdot SP^n}$, or as $\dfrac{(1 - \frac{1}{2}n) \cdot VQ}{PQ \cdot SP^n \cdot SQ}$, and therefore as $\dfrac{(1 - \frac{1}{2}n) \cdot OS}{OP \cdot SP^{n+1}}$, that is (because $\dfrac{(1 - \frac{1}{2}n) \cdot OS}{OP}$ is a given quantity), inversely as SP^{n+1}. And therefore, since the velocity is inversely as $SP^{\frac{1}{2}n}$, the density in P will be reciprocally as SP.

Cor. I. The resistance is to the centripetal force as $(1 - \tfrac{1}{2}n) \cdot$ OS to OP.

Cor. II. If the centripetal force be inversely as SP^3, $1 - \tfrac{1}{2}n$ will be $= 0$; and therefore the resistance and density of the medium will be nothing, as in Prop. IX, Book I.

Cor. III. If the centripetal force be inversely as any power of the radius SP, whose index is greater than the number 3, the positive resistance will be changed into a negative.

SCHOLIUM

This Proposition and the former, which relate to mediums of unequal density, are to be understood as applying only to the motion of bodies that are so small, that the greater density of the medium on one side of the body above that on the other is not to be considered. I suppose also the resistance, other things being equal, to be proportional to its density. Hence, in mediums whose force of resistance is not as the density, the density must be so much augmented or diminished, that either the excess of the resistance may be taken away, or the defect supplied.

PROPOSITION XVII. PROBLEM IV

To find the centripetal force and the resisting force of the medium, by which a body, the law of the velocity being given, shall revolve in a given spiral.

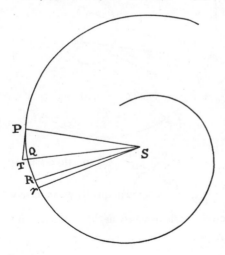

Let that spiral be PQR. From the velocity, with which the body goes over the very small arc PQ, the time will be given; and from the altitude TQ, which is as the centripetal force, and the square of the time, that force will be given. Then from the difference RS*r* of the areas PSQ and QSR described in equal intervals of time, the retardation of the body will be given; and from the retardation will be found the resisting force and density of the medium.

PROPOSITION XVIII. PROBLEM V

The law of centripetal force being given, to find the density of the medium in each of the places thereof, by which a body may describe a given spiral.

From the centripetal force the velocity in each place must be found; then from the retardation of the velocity the density of the medium is found, as in the foregoing Proposition.

But I have explained the method of managing these Problems in the tenth Proposition and second Lemma of this Book; and will no longer detain the reader in these complicated investigations. I shall now add some things relating to the forces of progressive bodies, and to the density and resistance of those mediums in which the motions hitherto discussed, and those akin to them, are performed.

SECTION V

The density and compression of fluids; hydrostatics.

THE DEFINITION OF A FLUID

A fluid is any body whose parts yield to any force impressed on it, and, by yielding, are easily moved among themselves.

PROPOSITION XIX. THEOREM XIV

All the parts of an homogeneous and unmoved fluid included in any un-moved vessel, and compressed on every side (setting aside the consideration of condensation, gravity, and all centripetal forces), will be equally pressed on every side, and remain in their places without any motion arising from that pressure.

CASE 1. Let a fluid be included in the spherical vessel ABC, and uniformly compressed on every side: I say, that no part of it will be moved by that pressure. For if any part, as D, be moved, all such parts at the same distance from the centre on every side must necessarily be moved at the same time

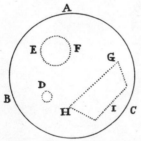

by a like motion; because the pressure of them all is similar and equal; and all other motion is excluded that does not arise from that pressure. But if these parts come all of them nearer to the centre, the fluid must be condensed towards the centre, contrary to the supposition. If they recede from it, the fluid must be condensed towards the circumference; which is also contrary to the supposition. Neither can they move in any one direction retaining their distance from the centre, because, for the same reason, they may move in a contrary direction; but the same part cannot be moved contrary ways at the same time. Therefore no part of the fluid will be moved from its place. Q.E.D.

CASE 2. I say now, that all the spherical parts of this fluid are equally pressed on every side. For let EF be a spherical part of the fluid; if this be not pressed equally on every side, augment the lesser pressure till it be pressed equally on every side; and its parts (by Case 1) will remain in their places. But before the increase of the pressure, they would remain in their places (by Case 1); and by the addition of a new pressure they will be moved, by the definition of a fluid, from those places. Now these two conclusions contradict each other. Therefore it was false to say that the sphere EF was not pressed equally on every side. Q.E.D.

CASE 3. I say besides, that different spherical parts have equal pressures. For the contiguous spherical parts press each other mutually and equally in the point of contact (by Law III). But (by Case 2) they are pressed on every side with the same force. Therefore any two spherical parts not contiguous, since an intermediate spherical part can touch both, will be pressed with the same force. Q.E.D.

CASE 4. I say now, that all the parts of the fluid are everywhere pressed equally. For any two parts may be touched by spherical parts in any points whatever; and there they will equally press those spherical parts (by Case 3), and are in reaction equally pressed by them (by Law III). Q.E.D.

CASE 5. Since, therefore, any part GHI of the fluid is inclosed by the rest of the fluid as in a vessel, and is equally pressed on every side; and also its parts equally press one another, and are at rest among themselves; it is manifest that all the parts of any fluid as GHI, which is pressed equally on every side, do press each other mutually and equally, and are at rest among themselves. Q.E.D.

CASE 6. Therefore if that fluid be included in a vessel of a yielding substance, or that is not rigid, and be not equally pressed on every side, the same will give way to a stronger pressure, by the definition of fluidity.

CASE 7. And therefore, in an inflexible or rigid vessel, a fluid will not sustain a stronger pressure on one side than on the other, but will give way to it, and that in a moment of time; because the rigid side of the vessel does not follow the yielding liquor. But the fluid, by thus yielding, will press against the opposite side, and so the pressure will tend on every side to equality. And because the fluid, as soon as it endeavors to recede from the part that is most pressed, is withstood by the resistance of the vessel on the opposite side, the pressure will on every side be reduced to equality, in a moment of time, without any local motion; and from thence the parts of the fluid (by Case 5) will press each other mutually and equally, and be at rest among themselves. Q.E.D.

Cor. Hence neither will a motion of the parts of the fluid among themselves be changed by a pressure communicated to the external surface, except so far as either the figure of the surface may be somewhere altered, or that all the parts of the fluid, by pressing one another more intensely or remissly, may slide with more or less difficulty among themselves.

PROPOSITION XX. THEOREM XV

If all the parts of a spherical fluid, homogeneous at equal distances from the centre, lying on a spherical concentric bottom, gravitate towards the centre of the whole, the bottom will sustain the weight of a cylinder, whose base is equal to the surface of the bottom, and whose altitude is the same with that of the incumbent fluid.

Let DHM be the surface of the bottom, and AEI the upper surface of the fluid. Let the fluid be divided into concentric orbs of equal thickness, by the innumerable spherical surfaces BFK, CGL; and conceive the force of gravity to act only in the upper surface of every orb, and the actions to be

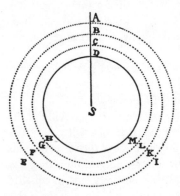

equal on the equal parts of the surfaces. Therefore the upper surface AE is pressed by the single force of its own gravity, by which all the parts of the upper orb, and the second surface BFK, will (by Prop. xix), according to its measure, be equally pressed. The second surface BFK is pressed likewise by the force of its own gravity, which, added to the former force, makes the pressure double. The third surface CGL is, according to its measure, acted on by this pressure and the force of its own gravity besides, which makes its pressure triple. And in like manner the fourth surface receives a quadruple pressure, the fifth surface a quintuple, and so on. Therefore the pressure acting on every surface is not as the solid quantity of the incumbent fluid, but as the number of the orbs reaching to the upper surface of the fluid; and is equal to the gravity of the lowest orb multiplied by the number of orbs; that is, to the gravity of a solid whose ultimate ratio to the cylinder above mentioned (when the number of the orbs is increased and their thickness diminished, *ad infinitum,* so that the action of gravity from the lowest surface to the uppermost may become continued) is the ratio of equality. Therefore the lowest surface sustains the weight of the cylinder above determined. Q.E.D. And by a like reasoning the Proposition will be evident, where the gravity of the fluid decreases in any assigned ratio of the distance from the centre, and also where the fluid is more rare above and denser below. Q.E.D.

Cor. i. Therefore the bottom is not pressed by the whole weight of the incumbent fluid, but only sustains that part of it which is described in the Proposition; the rest of the weight being sustained archwise by the spherical figure of the fluid.

Cor. ii. The quantity of the pressure is the same always at equal distances from the centre, whether the surface pressed be parallel to the horizon, or perpendicular, or oblique; or whether the fluid, continued upwards from the compressed surface, rises perpendicularly in a rectilinear direction, or creeps obliquely through crooked cavities and canals, whether those passages be regular or irregular, wide or narrow. That the pressure is not altered by any of these circumstances, may be inferred by applying the demonstration of this Theorem to the several cases of fluids.

Cor. iii. From the same demonstration it may also be concluded (by Prop. xix), that the parts of a heavy fluid acquire no motion among themselves by the pressure of the incumbent weight, except that motion which arises from condensation.

Cor. iv. And therefore if another body of the same specific gravity, incapable of condensation, be immersed in this fluid, it will acquire no motion by the pressure of the incumbent weight: it will neither descend nor ascend, nor change its figure. If it be spherical, it will remain so, notwithstanding the pressure; if it be square, it will remain square; and that, whether it be soft or fluid; whether it swims freely in the fluid, or lies at the bottom. For any internal part of a fluid is in the same state with the submersed body; and the case of all submersed bodies that have the same magnitude, figure, and specific gravity, is alike. If a submersed body, retaining its weight, should dissolve and put on the form of a fluid, this body, if before it should have ascended, descended, or from any pressure assumed a new figure, would now likewise ascend, descend, or put on a new figure; and that, because its gravity and the other causes of its motion remain. But (by Case 5, Prop. xix) it would now be at rest, and retain its figure. Therefore also in the former case.

Cor. v. Therefore a body that is specifically heavier than a fluid contiguous to it will sink; and that which is specifically lighter will ascend, and attain so much motion and change of figure as that excess or defect of gravity is able to produce. For that excess or defect is the same thing as an impulse, by which a body, otherwise in equilibrium with the parts of the fluid, is acted on; and may be compared with the excess or defect of a weight in one of the scales of a balance.

Cor. vi. Therefore bodies placed in fluids have a twofold gravity: the one true and absolute, the other apparent, common, and comparative. Absolute gravity is the whole force with which the body tends downwards; relative and common gravity is the excess of gravity with which the body tends downwards more than the ambient fluid. By the first kind of gravity the parts of all fluids and bodies gravitate in their proper places; and therefore their weights taken together compose the weight of the whole. For the whole taken together is heavy, as may be experienced in vessels full of liquor; and the weight of the whole is equal to the weights of all the parts, and is therefore composed of them. By the other kind of gravity bodies do not gravitate in their places; that is, compared with one another, they do not preponderate, but, hindering one another's endeavor to descend, remain in their proper places as if they were not heavy. Those things which are in the air, and do not preponderate, are commonly looked on as not heavy. Those which do preponderate are commonly reckoned heavy, inasmuch as they are not sustained by the weight of the air. The common weights are nothing else but the excess of the true weights above the weight of the air. Hence also, commonly, those things are called light which are less heavy, and, by yielding to the preponderating air, mount upwards. But these are only comparatively light, and not truly so, because they descend in a vacuum. Thus, in water, bodies which, by their greater or less gravity, descend or ascend, are comparatively and apparently heavy or light; and their comparative and apparent gravity or levity is the excess or defect by which their true gravity either exceeds the gravity of the water or is exceeded by it. But those things which neither by preponderating descend, nor, by yielding to the preponderating fluid, ascend, although by their true weight they do increase the weight of the whole, yet comparatively, and as commonly understood, they do not gravitate in the water. For these cases are alike demonstrated.

Cor. vii. These things which have been demonstrated concerning gravity take place in any other centripetal forces.

Cor. viii. Therefore if the medium in which any body moves be acted on either by its own gravity, or by any other centripetal force, and the body be urged more powerfully by the same force; the difference of the forces is that very motive force, which, in the foregoing Proposition, I have considered as a centripetal force. But if the body be more lightly urged by that force, the difference of the forces becomes a centrifugal force, and is to be considered as such.

Cor. ix. But since fluids by pressing the included bodies do not change their external figures, it appears also (by Cor., Prop. xix) that they will not change the situation of their internal parts in relation to one another; and therefore if animals were immersed therein, and if all sensation did arise from the motion of their parts, the fluid would neither hurt the immersed bodies, nor excite any sensation, unless so far as those bodies might be condensed by the compression. And the case is the same of any system of bodies encompassed with a compressing fluid. All the parts of the system will be agitated with the same motions as if they were placed in a vacuum, and would only retain their comparative gravity; unless so far as the fluid may somewhat resist their motions, or be requisite to unite them by compression.

PROPOSITION XXI. THEOREM XVI

Let the density of any fluid be proportional to the compression, and its parts be attracted downwards by a centripetal force inversely proportional to the distances from the centre: I say, that, if those distances be taken continually proportional, the densities of the fluid at the same distances will be also continually proportional.

Let ATV denote the spherical bottom of the fluid, S the centre, SA, SB, SC, SD, SE, SF, &c., distances continually proportional. Erect the perpendiculars AH, BI, CK, DL, EM, FN, &c., which shall be as the densities of the medium in the places A, B, C, D, E, F; and the specific gravities in those places will be as $\dfrac{AH}{AS}, \dfrac{BI}{BS}, \dfrac{CK}{CS}$, &c., or, which is all one, as $\dfrac{AH}{AB}, \dfrac{BI}{BC}, \dfrac{CK}{CD}$, &c. Suppose, first, these gravities to be uniformly continued from A to B, from B to C, from C to D, &c., the decrements in the points B, C, D, &c., being taken by steps. And these gravities multiplied by the altitudes AB, BC, CD, &c., will give the pressures AH, BI, CK, &c., by which the bottom ATV is acted on (by Theor. xv). Therefore the particle A sustains all the pressures

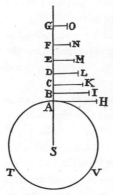

AH, BI, CK, DL, &c., proceeding *in infinitum;* and the particle B sustains the pressures of all but the first AH; and the particle C all but the two first AH, BI; and so on: and therefore the density AH of the first particle A is to the density BI of the second particle B as the sum of all AH + BI + CK + DL, *in infinitum,* to the sum of all BI + CK + DL, &c. And BI the density of the second particle B is to CK the density of the third C, as the sum of all BI + CK + DL, &c., to the sum of all CK + DL, &c. Therefore these sums are proportional to their differences AH, BI, CK, &c., and therefore continually proportional (by Lem. 1 of this Book); and therefore the differences AH, BI, CK, &c., proportional to the sums, are also continually proportional. Therefore since the densities in the places A, B, C, &c., are as AH, BI, CK, &c., they will also be continually proportional. Proceed intermissively, and, at the distances SA, SC, SE, continually proportional, the densities AH, CK, EM will be continually proportional. And by the same reasoning, at any distances SA, SD, SG, continually proportional, the densities AH, DL, GO will be continually proportional. Let now the points A, B, C, D, E, &c., coincide, so that the progression of the specific gravities from the bottom A to the top of the fluid may be made continual; and at any distances SA, SD, SG, continually proportional, the densities AH, DL, GO, being all along continually proportional, will still remain continually proportional. Q.E.D.

Cor. Hence if the density of the fluid in two places, as A and E, be given, its density in any other place Q may be obtained. With the centre S, and the rectangular asymptotes SQ, SX, describe an hyperbola cutting the perpendiculars AH, EM, QT in *a, e,* and *q,* as also the perpendiculars HX, MY, TZ, let fall upon the asymptote SX, in *h, m,* and *t.* Make the area Y*mt*Z to the given area Y*mh*X as the given area E*eq*Q to the given area E*ea*A; and the line Z*t* produced will cut off the line QT proportional to the density. For if the lines SA, SE, SQ are continually proportional, the areas E*eq*Q, E*ea*A will be equal, and thence the areas Y*mt*Z, X*hm*Y, proportional to them, will be also equal; and the lines SX, SY, SZ, that is, AH, EM, QT continually proportional, as they ought to be. And if the lines SA, SE, SQ obtain any other order in the series of continued proportionals, the lines AH, EM, QT, because of the

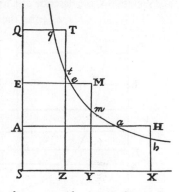

proportional hyperbolic areas, will obtain the same order in another series of quantities continually proportional.

PROPOSITION XXII. THEOREM XVII

Let the density of any fluid be proportional to the compression, and its parts be attracted downwards by a gravitation inversely proportional to the squares of the distances from the centre: I say, that if the distances be taken in harmonic progression, the densities of the fluid at those distances will be in a geometrical progression.

Let S denote the centre, and SA, SB, SC, SD, SE the distances in geometrical progression. Erect the perpendiculars AH, BI, CK, &c., which shall be as the densities of the fluid in the places A, B, C, D, E, &c., and the specific gravities thereof in those places will be as $\frac{AH}{SA^2}, \frac{BI}{SB^2}, \frac{CK}{SC^2}$, &c. Suppose these gravities to be uniformly continued, the first from A to B, the second from B to C, the third from C to D, &c. And these multiplied by the altitudes AB, BC, CD, DE, &c., or, which is the same thing, by the distances SA, SB, SC, &c., proportional to those altitudes, will give $\frac{AH}{SA}, \frac{BI}{SB}, \frac{CK}{SC}$, &c., representing the pressures. Therefore since the densities are as the sums of those pressures, the differences AH − BI, BI − CK, &c., of the densities will be as the differences of those sums $\frac{AH}{SA}$, $\frac{BI}{SB}$, $\frac{CK}{SC}$, &c. With the centre S, and the asymptotes SA, Sx, describe any hyperbola, cutting the perpendiculars AH, BI, CK, &c., in a, b, c, &c., and the perpendiculars Ht, Iu, Kw, let fall upon

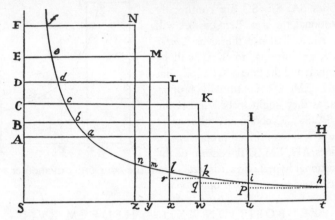

the asymptote Sx, in h, i, k; and the differences of the densities, tu, uw, &c., will be as $\frac{AH}{SA}$, $\frac{BI}{SB}$, &c. And the rectangles $tu \cdot th$, $uw \cdot ui$, &c., or tp, uq, &c., as $\frac{AH \cdot th}{SA}$, $\frac{BI \cdot ui}{SB}$, &c., that is, as Aa, Bb, &c. For, by the nature of the hyperbola, SA is to AH or St as th to Aa, and therefore $\frac{AH \cdot th}{SA}$ is equal to Aa. And, by a like reasoning, $\frac{BI \cdot ui}{SB}$ is equal to Bb, &c. But Aa, Bb, Cc, &c., are continually proportional, and therefore proportional to their differences Aa − Bb, Bb − Cc, &c., therefore the rectangles tp, uq, &c., are proportional to those differences; as also the sums of the rectangles $tp + uq$ or $tp + uq + wr$ to the sums of the differences Aa − Cc or Aa − Dd. Suppose several of these terms, and the sum of all the differences, as Aa − Ff, will be proportional to the sum of all the rectangles, as $zthn$. Increase the number of terms, and diminish the distances of the points A, B, C, &c., *in infinitum*, and those rectangles will become equal to the hyperbolic area *zthn*, and therefore the difference $Aa - Ff$ is proportional to this area. Take now any distances, as SA, SD, SF, in harmonic progression, and the differences $Aa - Dd$, $Dd - Ff$ will be equal; and therefore the areas *thlx*, *xlnz*, proportional to those differences, will be equal among themselves, and the densities S*t*, S*x*, S*z*, that is, AH, DL, FN, continually proportional. Q.E.D.

Cor. Hence if any two densities of the fluid, as AH and BI, be given, the area *thiu*, answering to their difference *tu*, will be given; and thence the density FN will be found at any height SF, by taking the area *thnz* to that given area *thiu* as the difference $Aa - Ff$ to the difference $Aa - Bb$.

SCHOLIUM

By a like reasoning it may be proved, that if the gravity of the particles of a fluid diminishes as the cube of the distances from the centre, and the reciprocals of the squares of the distances SA, SB, SC, &c., (namely, $\frac{SA^3}{SA^2}$, $\frac{SA^3}{SB^2}$, $\frac{SA^3}{SC^2}$) be taken in an arithmetical progression, the densities AH, BI, CK, &c., will be in a geometrical progression. And if the gravity be diminished as the fourth power of the distances, and the reciprocals of the cubes of the distances (as $\frac{SA^4}{SA^3}$, $\frac{SA^4}{SB^3}$, $\frac{SA^4}{SC^3}$, &c.) be taken in arithmetical progression, the densities AH, BI, CK, &c., will be in geometrical progression. And so *in infinitum*. Again; if the gravity of the particles of the fluid be the same at all distances, and the distances be in arithmetical progression, the densities will be in a geometrical progression, as Dr. *Halley* hath found. If the gravity be as the distance, and the squares of the distances be in arithmetical progression, the densities will be in geometrical progression. And so *in infinitum*. These things will be so, when the density of the fluid condensed by compression is as the force of compression; or, which is the same thing, when the space possessed by the fluid is inversely as this force. Other laws of condensation may be supposed, as that the cube of the compressing force may be as the fourth power of the density, or the cube of the ratio of the force the same with the fourth power of the ratio of the density: in which case, if the gravity be inversely as the square of the distance from the centre, the density will be inversely as the cube of the distance. Suppose that the cube of the compressing force be as the fifth power of the density; and if the gravity be inversely as the square of the distance, the density will be inversely as the ³⁄₂th power of the distance. Suppose the compressing force to be as the square of the density, and the gravity inversely as the square of the distance, then the density will be inversely as the distance. To run over all the cases that might be offered would be tedious. But as to our own air, this is certain from experiment, that its density is either accurately, or very nearly at least, as the compressing force; and therefore the density of the air in the atmosphere of the earth is as the weight of the whole incumbent air, that is, as the height of the mercury in the barometer.

PROPOSITION XXIII. THEOREM XVIII

If a fluid be composed of particles fleeing from each other, and the density be as the compression, the centrifugal forces of the particles will be inversely proportional to the distances of their centres. And, conversely, particles fleeing from each other, with forces that are inversely proportional to the distances of their centres, compose an elastic fluid, whose density is as the compression.

Let the fluid be supposed to be included in a cubic space ACE, and then to be reduced by compression into a lesser cubic space *ace*; and the distances of the particles retaining a like situation with respect to each other in both the spaces, will be as the sides AB, *ab* of the cubes; and the densities of the

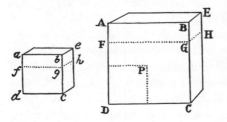

mediums will be inversely as the containing spaces AB³, *ab*³. In the plane side of the greater cube ABCD take the square DP equal to the plane side *db* of the lesser cube; and, by the supposition, the pressure with which the square DP urges the inclosed fluid will be to the pressure with which that square *db* urges the inclosed fluid as the densities of the mediums are to each other, that is, as *ab*³ to AB³. But the pressure with which the square DB urges the included fluid is to the pressure with which the square DP urges the same fluid as the square DB to the square DP, that is, as AB^2 to ab^2. Therefore, multiplying together corresponding terms of the proportions, the pressure with which the square DB urges the fluid is to the pressure with which the square db urges the fluid as ab to AB. Let the planes FGH, fgh be drawn through the interior of the two cubes, and divide the fluid into two parts. These parts will press each other with the same forces with which they are themselves pressed by the planes AC, $ac,$ that is, in the proportion of ab to AB: and therefore the centrifugal forces by which these pressures are sustained are in the same ratio. The number of the particles being equal, and the situation alike, in both cubes, the forces which all the particles exert, according to the planes FGH, $fgh,$ upon all, are as the forces which each exerts on each. Therefore the forces which each exerts on each, according to the plane FGH in the greater cube, are to the forces which each exerts on each, according to the plane fgh in the lesser cube, as ab to AB, that is, inversely as the distances of the particles from each other. Q.E.D.

And, conversely, if the forces of the single particles are inversely as the distances, that is, inversely as the sides of the cubes AB, ab; the sums of the forces will be in the same ratio, and the pressures of the sides DB, db as the sums of the forces; and the pressure of the square DP to the pressure of the side DB as ab^2 to AB^2. And, multiplying together corresponding terms of the proportions, one obtains the pressure of the square DP to the pressure of the side db as ab^3 to AB^3; that is, the force of compression in the one is to the force of compression in the other as the density in the former to the density in the latter. Q.E.D.

SCHOLIUM

By a like reasoning, if the centrifugal forces of the particles are inversely as the square of the distances between the centres, the cubes of the compressing forces will be as the fourth power of the densities. If the centrifugal forces be inversely as the third or fourth power of the distances, the cubes of the compressing forces will be as the fifth or sixth power of the densities. And universally, if D be put for the distance, and E for the density of the compressed fluid, and the centrifugal forces be inversely as any power D^n of the distance, whose index is the number n, the compressing forces will be as the cube roots of the power E^{n+2}, whose index is the number $n + 2$; and conversely. All these things are to be understood of particles whose centrifugal forces terminate in those particles that are next them, or are diffused not much farther. We have an example of this in magnetic bodies. Their attractive force is terminated nearly in bodies of their own kind that are next them. The force of the magnet is reduced by the interposition of an iron plate, and is almost terminated at it: for bodies farther off are not attracted by the magnet so much as by the iron plate. If in this manner particles repel others of their own kind that lie next them, but do not exert their force on the more remote, particles of this kind will compose such fluids as are treated of in this Proposition. If the force of any particle diffuse itself every way *in infinitum,* there will be required a greater force to produce an equal condensation of a greater quantity of the fluid. But whether elastic fluids do really consist of particles so repelling each other, is a physical question. We have here demonstrated mathematically the property of fluids consisting of particles of this kind, that hence philosophers may take occasion to discuss that question.

SECTION VI

The motion and resistance of pendulous bodies.

PROPOSITION XXIV. THEOREM XIX

The quantities of matter in pendulous bodies, whose centres of oscillation are equally distant from the centre of suspension, are in a ratio compounded of the ratio of the weights and the squared ratio of the times of the oscillations in a vacuum.

For the velocity which a given force can generate in a given matter in a given time is directly as the force and the time, and inversely as the matter. The greater the force or the time is, or the less the matter, the greater the velocity generated. This is manifest from the second Law of Motion. Now if pendulums are of the same length, the motive forces in places equally distant from the perpendicular are as the weights: and therefore if two bodies by oscillating describe equal arcs, and those arcs are divided into equal parts; since the times in which the bodies describe each of the correspondent parts of the arcs are as the times of the whole oscillations, the velocities in the correspondent parts of the oscillations will be to each other directly as the motive forces and the whole times of the oscillations, and inversely as the quantities of matter: and therefore the quantities of matter are directly as the forces and the times of the oscillations, and inversely as the velocities. But the velocities are inversely as the times, and therefore the times are directly and the velocities inversely as the squares of the times; and therefore the quantities of matter are as the motive forces and the squares of the times, that is, as the weights and the squares of the times. Q.E.D.

Cor. I. Therefore if the times are equal, the quantities of matter in each of the bodies are as the weights.

Cor. II. If the weights are equal, the quantities of matter will be as the squares of the times.

Cor. III. If the quantities of matter are equal, the weights will be inversely as the squares of the times.

Cor. IV. Since the squares of the times, other things being equal, are as the lengths of the pendulums, therefore if both the times and the quantities of matter are equal, the weights will be as the lengths of the pendulums.

Cor. v. And, in general, the quantity of matter in the pendulous body is directly as the weight and the square of the time, and inversely as the length of the pendulum.

Cor. vi. But in a nonresisting medium, the quantity of matter in the pendulous body is directly as the comparative weight and the square of the time, and inversely as the length of the pendulum. For the comparative weight is the motive force of the body in any heavy medium, as was shown above; and therefore does the same thing in such a nonresisting medium as the absolute weight does in a vacuum.

Cor. vii. And hence appears a method both of comparing bodies one with another, as to the quantity of matter in each; and of comparing the weights of the same body in different places, to know the variation of its gravity. And by experiments made with the greatest accuracy, I have always found the quantity of matter in bodies to be proportional to their weight.

PROPOSITION XXV. THEOREM XX

Pendulous bodies that are, in any medium, resisted in the ratio of the moments of time, and pendulous bodies that move in a nonresisting medium of the same specific gravity, perform their oscillations in a cycloid in the same time, and describe proportional parts of arcs together.

Let AB be an arc of a cycloid, which a body D, by vibrating in a nonresisting medium, shall describe in any time. Bisect that arc in C, so that C may be the lowest point thereof; and the accelerative force with which the body is urged in any place D, or *d,* or E, will be as the length of the arc CD, or C*d,* or CE. Let that force be expressed by that same arc; and since the resistance is as the moment of the time, and therefore given, let it be expressed by the given part CO of the cycloidal arc, and take the arc O*d* in the same ratio to the arc CD that the arc OB has to the arc CB: and the force with which the body in *d* is urged in a resisting medium, being the excess of the force C*d* above the resistance CO, will be expressed by the arc O*d,* and will therefore be to the force with which the body D is urged in a nonresisting medium in the place D, as the arc O*d* to the arc CD; and therefore also in the place B, as the arc OB to the arc CB. Therefore if two bodies D, *d* go from the place B, and are urged by these forces; since the forces at the begin-

ning are as the arcs CB and OB, the first velocities and arcs first described
will be in the same ratio. Let those arcs be BD and B*d,* and the remaining
arcs CD, O*d* will be in the same ratio. Therefore the forces, being proportional to those arcs CD, O*d,* will remain in the same ratio as at the beginning, and therefore the bodies will continue describing together arcs in the
same ratio. Therefore the forces and velocities and the remaining arcs CD,
O*d,* will be always as the whole arcs CB, OB, and therefore those remaining
arcs will be described together. Therefore the two bodies D and *d* will

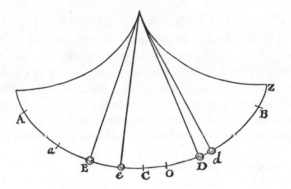

arrive together at the places C and O; that which moves in the nonresisting
medium, at the place C, and the other, in the resisting medium, at the place
O. Now since the velocities in C and O are as the arcs CB, OB, the arcs
which the bodies describe when they go farther will be in the same ratio.
Let those arcs be CE and O*e.* The force with which the body D in a nonresisting medium is retarded in E is as CE, and the force with which the
body *d* in the resisting medium is retarded in *e,* is as the sum of the force C*e*
and the resistance CO, that is, as O*e;* and therefore the forces with which
the bodies are retarded are as the arcs CB, OB, proportional to the arcs CE,
O*e;* and therefore the velocities, retarded in that given ratio, remain in the
same given ratio. Therefore the velocities and the arcs described with those
velocities are always to each other in that given ratio of the arcs CB and
OB; and therefore if the entire arcs AB, *a*B are taken in the same ratio, the
bodies D and *d* will describe those arcs together, and in the places A and *a*
will lose all their motion together. Therefore the whole oscillations are

isochronal, or are performed in equal times; and any parts of the arcs, as BD, B*d,* or BE, B*e,* that are described together, are proportional to the whole arcs BA, B*a.* Q.E.D.

Cor. Therefore the swiftest motion in a resisting medium does not fall upon the lowest point C, but is found in that point O, in which the whole arc described B*a* is bisected. And the body, proceeding from thence to *a,* is retarded at the same rate with which it was accelerated before in its descent from B to O.

PROPOSITION XXVI. THEOREM XXI

Pendulous bodies, that are resisted in the ratio of the velocity, have their oscillations in a cycloid isochronal.

For if two bodies, equally distant from their centres of suspension, describe, in oscillating, unequal arcs, and the velocities in the correspondent parts of the arcs be to each other as the whole arcs; the resistances, proportional to the velocities, will be also to each other as the same arcs. Therefore if these resistances be subtracted from or added to the motive forces arising from gravity which are as the same arcs, the differences or sums will be to each other in the same ratio of the arcs; and since the increments and decrements of the velocities are as these differences or sums, the velocities will be always as the whole arcs; therefore if the velocities are in any one case as the whole arcs, they will remain always in the same ratio. But at the beginning of the motion, when the bodies begin to descend and describe those arcs, the forces, which at that time are proportional to the arcs, will generate velocities proportional to the arcs. Therefore the velocities will be always as the whole arcs to be described, and therefore those arcs will be described in the same time. Q.E.D.

PROPOSITION XXVII. THEOREM XXII

If pendulous bodies are resisted as the square of their velocities, the differences between the times of the oscillations in a resisting medium, and the times of the oscillations in a nonresisting medium of the same specific gravity, will be proportional to the arcs described in oscillating, nearly.

For let equal pendulums in a resisting medium describe the unequal arcs A, B; and the resistance of the body in the arc A will be to the resistance of the body in the correspondent part of the arc B as the square of the velocities, that is, as AA to BB, nearly. If the resistance in the arc B were to the resistance in the arc A as AB to AA, the times in the arcs A and B would be equal (by the last Proposition). Therefore the resistance AA in the arc A,

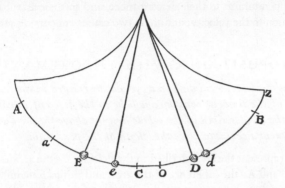

or AB in the arc B, causes the excess of the time in the arc A above the time in a nonresisting medium; and the resistance BB causes the excess of the time in the arc B above the time in a nonresisting medium. But those excesses are as the efficient forces AB and BB nearly, that is, as the arcs A and B. Q.E.D.

COR. I. Hence from the times of the oscillations in unequal arcs in a resisting medium, may be known the times of the oscillations in a nonresisting medium of the same specific gravity. For the difference of the times will be to the excess of the time in the shorter arc above the time in a nonresisting medium as the difference of the arcs is to the shorter arc.

COR. II. The shorter oscillations are more isochronal, and very short ones are performed nearly in the same times as in a nonresisting medium. But the times of those which are performed in greater arcs are a little greater, because the resistance in the descent of the body, by which the time is prolonged, is greater, in proportion to the length described in the descent than the resistance in the subsequent ascent, by which the time is contracted. But the time of the oscillations, both short and long, seems to be prolonged in some measure by the motion of the medium. For retarded bodies are resisted somewhat less in proportion to the velocity, and accelerated bodies somewhat more than those that proceed uniformly forwards; because the medium, by the motion it has received from the bodies, going forwards the same way with them, is more agitated in the former case, and less in the latter; and so conspires more or less with the bodies moved. Therefore it resists the pendulums in their descent more, and in their ascent less, than in proportion to the velocity; and these two causes concurring prolong the time.

PROPOSITION XXVIII. THEOREM XXIII

If a pendulous body, oscillating in a cycloid, be resisted in the ratio of the moments of the time, its resistance will be to the force of gravity, as the excess of the arc described in the whole descent above the arc described in the subsequent ascent is to twice the length of the pendulum.

Let BC represent the arc described in the descent, C*a* the arc described in the ascent, and A*a* the difference of the arcs: and things remaining as they were constructed and demonstrated in Prop. xxv, the force with which the oscillating body is urged in any place D will be to the force of resistance as

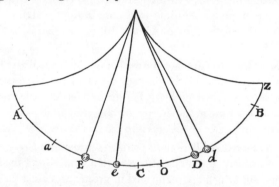

the arc CD to the arc CO, which is half of that difference A*a*. Therefore the force with which the oscillating body is urged at the beginning or the highest point of the cycloid, that is, the force of gravity, will be to the resistance as the arc of the cycloid, between that highest point and the lowest point C, is to the arc CO; that is (doubling those arcs), as the whole cycloidal arc, or twice the length of the pendulum, is to the arc A*a*. Q.E.D.

PROPOSITION XXIX. PROBLEM VI

Supposing that a body oscillating in a cycloid is resisted as the square of the velocity; to find the resistance in each place.

Let B*a* be an arc described in one entire oscillation, C the lowest point of the cycloid, and CZ half the whole cycloidal arc, equal to the length of the pendulum; and let it be required to find the resistance of the body in any place D. Cut the indefinite right line OQ in the points O, S, P, Q, so that (erecting the perpendiculars OK, ST, PI, QE, and with the centre O, and the asymptotes OK, OQ, describing the hyperbola TIGE cutting the perpendiculars ST, PI, QE in T, I, and E, and through the point I drawing KF, parallel to the asymptote OQ, meeting the asymptote OK in K, and

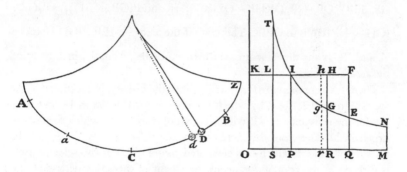

the perpendiculars ST and QE in L and F) the hyperbolic area PIEQ may be to the hyperbolic area PITS as the arc BC, described in the descent of the body, is to the arc C*a* described in the ascent; and that the area IEF may be to the area ILT as OQ to OS. Then with the perpendicular MN cut off the hyperbolic area PINM, and let that area be to the hyperbolic area PIEQ as the arc CZ to the arc BC described in the descent. And if the perpendicular RG cuts off the hyperbolic area PIGR, which shall be to the area PIEQ as any arc CD is to the arc BC described in the whole descent, the resistance in any place D will be to the force of gravity as the area $\frac{OR}{OQ}$ IEF − IGH is to the area PINM.

For since the forces arising from gravity with which the body is urged in the places Z, B, D, a are as the arcs CZ, CB, CD, Ca, and those arcs are as the areas PINM, PIEQ, PIGR, PITS; let those areas represent both the arcs and the forces respectively. Let Dd be a very small space described by the body in its descent; and let it be expressed by the very small area RGgr, comprehended between the parallels RG, rg; and produce rg to h, so that GHhg and RGgr may be the contemporaneous decrements of the areas IGH, PIGR. And the increment GH$hg - \dfrac{Rr}{OQ}$ IEF, or R$r \cdot$ HG $- \dfrac{Rr}{OQ}$ IEF, of the area $\dfrac{OR}{OQ}$ IEF $-$ IGH will be to the decrement RGgr, or R$r \cdot$ RG, of the area PIGR, as HG $- \dfrac{IEF}{OQ}$ is to RG; and therefore as OR $\cdot$ HG $- \dfrac{OR}{OQ}$ IEF is to OR $\cdot$ GR or OP $\cdot$ PI, that is (because of the equal quantities OR $\cdot$ HG, OR $\cdot$ HR $-$ OR $\cdot$ GR, ORHK $-$ OPIK, PIHR and PIGR $+$ IGH), as PIGR $+$ IGH $- \dfrac{OR}{OQ}$ IEF is to OPIK. Therefore if the area $\dfrac{OR}{OQ}$ IEF $-$ IGH be called Y, and RGgr the decrement of the area PIGR be given, the increment of the area Y will be as PIGR $-$ Y.

Then if V represent the force arising from the gravity, proportional to the arc CD to be described, by which the body is acted upon in D, and R be put for the resistance, V $-$ R will be the whole force with which the body is urged in D. Therefore the increment of the velocity is as V $-$ R and the interval of time in which it is generated conjointly. But the velocity itself is directly as the contemporaneous increment of the space described and inversely as the same interval of time. Therefore, since the resistance is, by the supposition, as the square of the velocity, the increment of the resistance will (by Lem. II) be as the velocity and the increment of the velocity conjointly, that is, as the moment of the space and V $-$ R conjointly; and, therefore, if the moment of the space be given, as V $-$ R; that is, if for the force V we put its expression PIGR, and the resistance R be expressed by any other area Z, as PIGR $-$ Z.

Therefore the area PIGR uniformly decreasing by the subtraction of given moments, the area Y increases in proportion of PIGR $-$ Y, and the area Z in proportion of PIGR $-$ Z. And therefore if the areas Y and Z begin together, and at the beginning are equal, these, by the addition of equal moments, will continue to be equal; and in like manner decreasing by equal moments, will vanish together. And, conversely, if they together begin and vanish, they will have equal moments and be always equal. For, if the resistance Z be augmented, then the velocity together with the arc Ca, described in the ascent of the body, will be diminished; and, the point in which all the motion together with the resistance ceases, coming nearer to the point C, then the resistance vanishes sooner than the area Y. And the contrary will happen when the resistance is diminished.

Now the area Z begins and ends where the resistance is nothing, that is, at the beginning of the motion where the arc CD is equal to the arc CB, and the right line RG falls upon the right line QE; and at the end of the motion where the arc CD is equal to the arc Ca, and RG falls upon the right line ST. And the area Y or $\frac{OR}{OQ}$ IEF – IGH begins and ends also where the resistance is nothing, and therefore where $\frac{OR}{OQ}$ IEF and IGH are equal; that is (by the construction), where the right line RG falls successively upon the

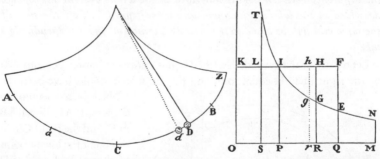

right lines QE and ST. Therefore those areas begin and vanish together, and are therefore always equal. Hence, the area $\frac{OR}{OQ}$ IEF – IGH is equal to the area Z, by which the resistance is expressed, and therefore is to the area PINM, by which the gravity is expressed, as the resistance is to the gravity. Q.E.D.

Cor. I. Therefore the resistance in the lowest place C is to the force of gravity as the area $\frac{OP}{OQ}$ IEF is to the area PINM.

Cor. II. But it becomes greatest where the area PIHR is to the area IEF as OR is to OQ. For in that case its moment (that is, PIGR – Y) becomes nothing.

Cor. III. Hence also may be known the velocity in each place, as varying as the square root of the resistance, and at the beginning of the motion being equal to the velocity of the body oscillating in the same cycloid without any resistance.

However, by reason of the difficulty of the calculation by which the resistance and the velocity are found by this Proposition, we have thought fit to subjoin the Proposition following.

PROPOSITION XXX. THEOREM XXIV

If a right line aB be equal to the arc of a cycloid which an oscillating body describes, and at each of its points D the perpendiculars DK be erected, which shall be to the length of the pendulum as the resistance of the body in the corresponding points of the arc is to the force of gravity: I say, that the difference between the arc described in the whole descent and the arc described in the whole subsequent ascent multiplied by half the sum of the same arcs will be equal to the area BKa which all those perpendiculars take up.

Let the arc of the cycloid, described in one entire oscillation, be expressed by the right line aB, equal to it, and the arc which would have been de-

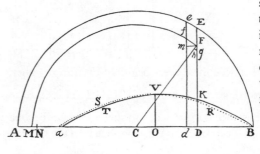

scribed in a vacuum by the length AB. Bisect AB in C, and the point C will represent the lowest point of the cycloid, and CD will be as the force arising from gravity, with which the body in D is urged in the direction of the tangent of the cycloid, and will have the same ratio to the length of the pendulum as the force in D has to the force of gravity. Let that force, therefore, be expressed by that length CD, and the force of gravity by the length of the pendulum; and if in DE you take DK in the same ratio to the length of the pendulum as the resistance is to the gravity, DK will be the exponent of the resistance. From the centre C with the interval CA or CB describe a semicircle BE*e*A. Let the body describe, in the least time, the space D*d*; and, erecting the perpendiculars DE, *de,* meeting the circumference in E and *e,* they will be as the velocities which the body descending in a vacuum from the point B would acquire in the places D and *d.* This appears by Prop. LII, Book I. Let, therefore, these velocities be expressed by those perpendiculars DE, *de*; and let DF be the velocity which it acquires in D by falling from B in the resisting medium. And if from the centre C with the interval CF we describe the circle F*f*M meeting the right lines *de* and AB in *f* and M, then M will be the place to which it would thenceforward, without further resistance, ascend, and *df* the velocity it would acquire in *d.* Hence, also, if F*g* represent the moment of the velocity which the body D, in describing the least space D*d,* loses by the resistance of the medium; and CN be taken equal to C*g;* then will N be the place to which the body, if it met no further resistance, would thenceforward ascend, and MN will be the decrement of the ascent arising from the loss of that velocity. Draw F*m* perpendicular to *df,* and the decrement F*g* of the velocity DF generated by the resistance DK will be to the increment *fm* of the same velocity, generated by the force CD, as the generating force DK to the generating force CD. But because of the similar triangles F*mf,* F*hg,* FDC, *fm* is to F*m* or D*d* as CD to DF; and, by multiplication of corresponding terms, F*g* to D*d* as DK to DF. Also F*h* is to F*g* as DF to CF; and, again by multiplication of corresponding terms, F*h* or MN to D*d* as DK to CF or CM; and therefore the sum of all the MN · CM will be equal to the sum of all the D*d* · DK. At the movable point M suppose always a rectangular ordinate erected equal to the indeterminate CM, which by a continual motion is multiplied by the whole length A*a*; and the trapezium described by that motion, or its equal, the rectangle A*a* · ½*a*B, will be equal to the sum of all the MN · CM, and therefore to the sum of all the D*d* · DK, that is, to the area BKVT*a.* Q.E.D.

COR. Hence from the law of resistance, and the difference A*a* of the arcs C*a,* CB, may be derived the proportion of the resistance to the gravity, nearly.

For if the resistance DK be uniform, the figure BKTa will be a rectangle under Ba and DK; and hence the rectangle under ½Ba and Aa will be equal to the rectangle under Ba and DK, and DK will be equal to ½Aa. Therefore since DK represents the resistance, and the length of the pendulum represents the gravity, the resistance will be to the gravity as ½Aa is to the length of the pendulum; altogether as in Prop. xxviii is demonstrated.

If the resistance be as the velocity, the figure BKTa will be nearly an ellipse. For if a body, in a nonresisting medium, by one entire oscillation, should describe the length BA, the velocity in any place D would be as the ordinate DE of the circle described on the diameter AB. Therefore since Ba in the resisting medium, and BA in the nonresisting one, are described nearly in the same times; and therefore the velocities in each of the points of Ba are to the velocities in the corresponding points of the length BA nearly as Ba is to BA, the velocity in the point D in the resisting medium will be as the ordinate of the circle or ellipse described upon the diameter Ba; and therefore the figure BKVTa will be nearly an ellipse. Since the resistance is supposed proportional to the velocity, let OV represent the resistance in the middle point O; and an ellipse BRVSa described with the centre O, and the semiaxes OB, OV, will be nearly equal to the figure BKVTa, and to its equal the rectangle A$a \cdot$ BO. Therefore A$a \cdot$ BO is to OV$\cdot$ BO as the area of this ellipse to OV$\cdot$ BO; that is, Aa is to OV as the area of the semicircle is to the square of the radius, or as 11 to 7 nearly; and, therefore, ⁷⁄₁₁ Aa is to the length of the pendulum as the resistance of the oscillating body in O is to its gravity.

Now if the resistance DK varies as the square of the velocity, the figure BKVTa will be almost a parabola having V for its vertex and OV for its axis, and therefore will be nearly equal to the rectangle under ⅔Ba and OV. Therefore the rectangle under ½Ba and Aa is equal to the rectangle ⅔B$a \cdot$ OV, and therefore OV is equal to ¾Aa; and therefore the resistance in O made to the oscillating body is to its gravity as ¾Aa is to the length of the pendulum.

And I take these conclusions to be accurate enough for practical uses. For since an ellipse or parabola BRVSa falls in with the figure BKVTa in the middle point V, that figure, if greater towards the part BRV or VSa, is less towards the contrary part, and is therefore nearly equal to it.

PROPOSITION XXXI. THEOREM XXV

If the resistance made to an oscillating body in each of the proportional parts of the arcs described be augmented or diminished in a given ratio, the difference between the arc described in the descent and the arc described in the subsequent ascent will be augmented or diminished in the same ratio.

For that difference arises from the retardation of the pendulum by the resistance of the medium, and therefore is as the whole retardation and the retarding resistance proportional thereto. In the foregoing Proposition the rectangle under the right line ½aB and the difference A*a* of the arcs CB, C*a,* was equal to the area BKT*a.* And that area, if the length *a*B remains, is augmented or diminished in the ratio of the ordinates DK; that is, in the ratio of the resistance, and is therefore as the length *a*B and the resistance conjointly. And therefore the rectangle under A*a* and ½*a*B is as *a*B and the resistance conjointly, and therefore A*a* is as the resistance. Q.E.D.

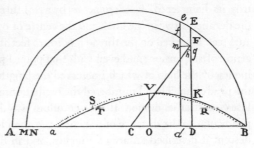

Cor. i. Hence if the resistance be as the velocity, the difference of the arcs in the same medium will be as the whole arc described; and conversely.

Cor. ii. If the resistance varies as the square of the velocity, that difference will vary as the square of the whole arc; and conversely.

Cor. iii. And generally, if the resistance varies as the third or any other power of the velocity, the difference will vary as the same power of the whole arc; and conversely.

Cor. iv. If the resistance varies partly as the first power of the velocity and partly as the square of the same, the difference will vary partly as the first power and partly as the square of the whole arc; and conversely. So that the law and ratio of the resistance will be the same for the velocity as the law and ratio of that difference for the length of the arc.

Cor. v. And therefore if a pendulum describe successively unequal arcs, and we can find the ratio of the increment or decrement of this difference for the length of the arc described, there will be had also the ratio of the increment or decrement of the resistance for a greater or less velocity.

GENERAL SCHOLIUM

From these Propositions we may find the resistance of mediums by pendulums oscillating therein. I found the resistance of the air by the following experiments. I suspended a wooden globe or ball weighing 57 7/22 ounces troy, its diameter 6⅞ *London* inches, by a fine thread on a firm hook, so that the distance between the hook and the centre of oscillation of the globe was 10½ feet. I marked on the thread a point 10 feet and 1 inch distant from the centre of suspension; and even with that point I placed a ruler divided into inches, by the help of which I observed the lengths of the arcs described by the pendulum. Then I numbered the oscillations in which the globe would lose ⅛ part of its motion. If the pendulum was drawn aside from the perpendicular to the distance of 2 inches, and then let go, so that in its whole descent it described an arc of 2 inches, and in the first whole oscillation, compounded of the descent and subsequent ascent, an arc of almost 4 inches, the pendulum in 164 oscillations lost ⅛ part of its motion, so as in its last ascent to describe an arc of 1¾ inches. If in the first descent it described an arc of 4 inches, it lost ⅛ part of its motion in 121 oscillations, so as in its last ascent to describe an arc of 3½ inches. If in the first descent it described an arc of 8, 16, 32, or 64 inches, it lost ⅛ part of its motion in 69, 35½, 18½, 9⅔ oscillations, respectively. Therefore the difference between the arcs described in the first descent and the last ascent was in the 1st, 2d, 3d, 4th, 5th, 6th cases, ¼, ½, 1, 2, 4, 8 inches, respectively. Divide those differences by the number of oscillations in each case, and in one mean oscillation, in which an arc of 3¾, 7½, 15, 30, 60, 120 inches was described, the difference of the arcs described in the descent and subsequent ascent will be 1/656, 1/242, 1/69, 4/71, 8/37, 24/29 parts of an inch, respectively. But these differences in the greater oscillations are as the square of the arcs described, nearly, but in lesser oscillations somewhat greater than in that ratio; and therefore (by Cor. II, Prop. XXXI of this Book) the resistance of the globe, when it moves very swiftly, varies as the square of the velocity, nearly; and when it moves slowly, in a somewhat greater ratio.

Now let V represent the greatest velocity in any oscillation, and let A, B, and C be given quantities, and let us suppose the difference of the arcs to be $AV + BV^{3/2} + CV^2$. Since the greatest velocities are in the cycloid as ½ the arcs described in oscillating, and in the circle as ½ the chords of those arcs; and therefore in equal arcs are greater in the cycloid than in the circle in the ratio of ½ the arcs to their chords; but the times in the circle are greater than in the cycloid, in a ratio inversely as the velocity; it is plain that the differences of the arcs (which are as the resistance and the square of the time conjointly) are nearly the same in both curves: for in the cycloid those differences must be on the one hand augmented, with the resistance, in about the squared ratio of the arc to the chord, because of the velocity augmented in the simple ratio of the same; and on the other hand diminished, with the square of the time, in the same squared ratio. Therefore to reduce these observations to the cycloid, we must take the same differences of the arcs as were observed in the circle, and suppose the greatest velocities analogous to the half, or the whole arcs, that is, to the numbers ½, 1, 2, 4, 8, 16. Therefore in the 2d, 4th, and 6th cases put 1, 4, and 16 for V; and the difference of the arcs in the 2d case will become $\frac{1/2}{121} = A + B + C$; in the 4th case, $\frac{2}{35\frac{1}{2}} = 4A + 8B + 16C$; in the 6th case, $\frac{8}{9\frac{2}{3}} = 16A + 64B + 256C$. These equations reduced give A = 0.0000916, B = 0.0010847, and C = 0.0029558. Therefore the difference of the arcs is as $0.0000916V + 0.0010847V^{3/2} + 0.0029558V^2$; and therefore since (by Cor., Prop. xxx, applied to this case) the resistance of the globe in the middle of the arc described in oscillating, where the velocity is V, is to its weight as $\frac{7}{11}AV + \frac{7}{10}BV^{3/2} + \frac{3}{4}CV^2$ is to the length of the pendulum, if for A, B, and C you put the numbers found, the resistance of the globe will be to its weight as $0.0000583V + 0.0007593V^{3/2} + 0.0022169V^2$ is to the length of the pendulum between the centre of suspension and the ruler, that is, to 121 inches. Therefore since V in the second case represents 1, in the 4th case 4, and in the 6th case 16, the resistance will be to the weight of the globe, in the 2d case, as 0.0030345 is to 121; in the 4th, as 0.041748 is to 121; in the 6th, as 0.61705 is to 121.

The arc, which the point marked in the thread described in the 6th case, was $120 - \frac{8}{9\frac{2}{3}}$, or $119\frac{5}{29}$ inches. And therefore since the radius was 121 inches, and the length of the pendulum between the point of suspension and the centre of the globe was 126 inches, the arc which the centre of the globe described was 124³/₃₁ inches. Because the greatest velocity of the oscillating body, by reason of the resistance of the air, does not fall on the lowest point of the arc described, but near the middle place of the whole arc, this velocity will be nearly the same as if the globe in its whole descent in a nonresisting medium should describe 62³¹/₆₂ inches, the half of that arc, and that in a cycloid, to which we have above reduced the motion of the pendulum; and therefore that velocity will be equal to that which the globe would acquire by falling perpendicularly from a height equal to the versed sine of that arc. But that versed sine in the cycloid is to that arc 62³¹/₆₂ as the same arc to twice the length of the pendulum 252, and therefore equal to 15.278 inches. Therefore the velocity of the pendulum is the same which a body would acquire by falling, and in its fall describing a space of 15.278 inches. Therefore with such a velocity the globe meets with a resistance which is to its weight as 0.61705 is to 121, or (if we take that part only of the resistance which is in the squared ratio of the velocity) as 0.56752 to 121.

I found, by an hydrostatical experiment, that the weight of this wooden globe was to the weight of a globe of water of the same magnitude as 55 to 97; and therefore since 121 is to 213.4 in the same ratio, the resistance made to this globe of water, moving forwards with the above-mentioned velocity, will be to its weight as 0.56752 to 213.4, that is, as 1 to 376¹/₅₀. Since the weight of a globe of water, in the time in which the globe with a velocity uniformly continued describes a length of 30.556 inches, will generate all that velocity in the falling globe, it is manifest that the force of resistance uniformly continued in the same time will take away a velocity, which will be less than the other in the ratio of 1 to 376¹/₅₀, that is, the $\frac{1}{376\frac{1}{50}}$ part of the whole velocity. And therefore in the time that the globe, with the same velocity uniformly continued, would describe the length of its semidiameter, or 3⁷/₁₆ inches, it would lose the $\frac{1}{3342}$ part of its motion.

I also counted the oscillations in which the pendulum lost ¼ part of its motion. In the following table the upper numbers denote the length of the arc described in the first descent, expressed in inches and parts of an inch; the middle numbers denote the length of the arc described in the last ascent; and in the lowest place are the numbers of the oscillations. I give an account of this experiment, as being more accurate than that in which only ⅛ part of the motion was lost. I leave the calculation to such as are disposed to make it.

First descent	2	4	8	16	32	64
Last ascent	1½	3	6	12	24	48
No. of oscillations . .	374	272	162½	83⅓	41⅔	22⅔

I afterwards suspended a leaden globe of 2 inches in diameter, weighing 26¼ ounces troy by the same thread, so that between the centre of the globe and the point of suspension there was an interval of 10½ feet, and I counted the oscillations in which a given part of the motion was lost. The first of the following tables exhibits the number of oscillations in which ⅛ part of the whole motion was lost; the second the number of oscillations in which there was lost ¼ part of the same.

First descent . .	1	2	4	8	16	32	64
Last ascent . . .	⅞	7/4	3½	7	14	28	56
No. of oscillations	226	228	193	140	90½	53	30
First descent . .	1	2	4	8	16	32	64
Last ascent . . .	¾	1½	3	6	12	24	48
No. of oscillations	510	518	420	318	204	121	70

Selecting in the first table the 3d, 5th, and 7th observations, and expressing the greatest velocities in these observations particularly by the numbers 1, 4, 16, respectively, and generally by the quantity V as above, there will come out in the 3d observation $\frac{1/2}{193} = A + B + C$, in the 5th observation $\frac{2}{90\frac{1}{2}} = 4A + 8B + 16C$, in the 7th observation $\frac{8}{30} = 16A + 64B + 256C$. These equations reduced give $A = 0.001414$, $B = 0.000297$, $C = 0.000879$. And thence the resistance of the globe moving with the velocity V will be to its weight 26¼ ounces in the same ratio as $0.0009V + 0.000208V^{3/2} + 0.000659V^2$ to 121 inches, the length of the pendulum. And if we regard that part only of the resistance which is as the square of the velocity, it will be to the weight of the globe as $0.000659V^2$ to 121 inches. But this part of the resistance in the first experiment was to the weight of the wooden globe of 57 7/22 ounces as $0.002217V^2$ to 121; hence the resistance of the wooden globe is to the resistance of the leaden one (their velocities being equal) as 57 7/22 into 0.002217 to 26¼ into 0.000659, that is, as 7⅓ to 1. The diameters of the two globes were 6⅞ and 2 inches, and the squares of these are to each other as 47¼ and 4, or 11¹³⁄₁₆ and 1, nearly. Therefore the resistances of these equally swift globes were in less than a squared ratio of the diameters. But we have not yet considered the resistance of the thread, which was certainly very considerable, and ought to be subtracted from the resistance of the pendulums here found. I could not determine this accurately, but I found it greater than ⅓ part of the whole resistance of the lesser pendulum; hence I gathered that the resistances of the globes, when the resistance of the thread is subtracted, are nearly in the squared ratio of their diameters. For the ratio of 7½ − ⅓ to 1 − ⅓, or 10½ to 1 is not very different from the squared ratio of the diameters 11¹³⁄₁₆ to 1.

Since the resistance of the thread is of less moment in greater globes, I tried the experiment also with a globe whose diameter was 18¾ inches. The length of the pendulum between the point of suspension and the centre of oscillation was 122½ inches, and between the point of suspension and the knot in the thread 109½ inches. The arc described by the knot at the first descent of the pendulum was 32 inches. The arc described by the same knot in the last ascent after five oscillations was 28 inches. The sum of the arcs, or the whole arc described in one mean oscillation, was 60 inches; the difference of the arcs, 4 inches. The ¹⁄₁₀ part of this, or the difference between the descent and ascent in one mean oscillation, is ⅖ of an inch. Then as the radius 109½ is to the radius 122½, so is the whole arc of 60 inches described by the knot in one mean oscillation to the whole arc of 67⅛ inches described by the centre of the globe in one mean oscillation; and so is the difference ⅖ to a new difference 0.4475. If the length of the arc described were to remain, and the length of the pendulum should be augmented in the ratio of 126 to 122½, the time of the oscillation would be augmented, and the velocity of the pendulum would be diminished as the square root of that ratio; so that the difference 0.4475 of the arcs described in the descent and subsequent ascent would remain. Then if the arc described be augmented in the ratio of 124³⁄₃₁ to 67⅛, that difference 0.4475 would be augmented as the square of that ratio, and so would become 1.5295. These things would be so upon the supposition that the resistance of the pendulum were as the square of the velocity. Therefore if the pendulum describe the whole arc of 124$\frac{3}{31}$ inches, and its length between the point of suspension and the centre of oscillation be 126 inches, the difference of the arcs described in the descent and subsequent ascent would be 1.5295 inches. And this difference multiplied by the weight of the pendulous globe, which was 208 ounces, produces 318.136. Again, in the pendulum above mentioned, made of a wooden globe, when its centre of oscillation, being 126 inches from the point of suspension, described the whole arc of 124$\frac{3}{31}$ inches, the difference of the arcs described in the descent and ascent was $\frac{126}{121}$ into $\frac{8}{9\frac{2}{3}}$. This multiplied by the weight of the globe, which was 57$\frac{7}{22}$ ounces, produces 49.396. But I multiply these differences by the weights of the globes, in order to find their resistances. For the differences arise from the resistances, and are as the resistances directly and the weights inversely. Therefore the resistances are as the numbers 318.136 and 49.396. But that part of the resistance of the lesser globe, which is as the square of the velocity, was to the whole resistance as 0.56752 to 0.61675, that is, as 45.453 to 49.396, whereas that part of the resistance of the greater globe is almost equal to its whole resistance, and so those parts are nearly as 318.136 and 45.453, that is, as 7 and 1. But the diameters of the globes are 18$\frac{3}{4}$ and 6$\frac{7}{8}$; and their squares 351$\frac{9}{16}$ and 47$\frac{17}{64}$ are as 7.438 and 1, that is, nearly as the resistances of the globes 7 and 1. The difference of these ratios is barely greater than may arise from the resistance of the thread. Therefore those parts of the resistances which are, when the globes are equal, as the squares of the velocities, are also, when the velocities are equal, as the squares of the diameters of the globes.

But the greatest of the globes I used in these experiments was not perfectly spherical, and therefore in this calculation I have, for brevity's sake, neglected some little niceties; being not very solicitous for an accurate calculus in an experiment that was not very accurate. So that I could wish that these experiments were tried again with other globes, of a larger size, more in number, and more accurately formed; since the demonstration of a vacuum depends thereon. If the globes be taken in a geometrical proportion, whose diameters, let us suppose, are 4, 8, 16, 32 inches; one may infer from the progression observed in the experiments what would happen if the globes were still larger.

In order to compare the resistances of different fluids with each other, I made the following trials. I procured a wooden vessel 4 feet long, 1 foot broad, and 1 foot high. This vessel, being uncovered, I filled with spring water, and, having immersed pendulums therein, I made them oscillate in the water. And I found that a leaden globe weighing 166⅙ ounces, and in diameter 3⅝ inches, moved therein as it is set down in the following table; the length of the pendulum from the point of suspension to a certain point marked in the thread being 126 inches, and to the centre of oscillation 134⅜ inches.

The arc described in the first descent, by a point marked in the thread was inches	64	32	16	8	4	2	1	½	¼
The arc described in the last ascent was inches	48	24	12	6	3	1½	¾	⅜	3/16
The difference of the arcs, proportional to the motion lost, was inches	16	8	4	2	1	½	¼	⅛	1/16
The number of the oscillations in water			29/60	1⅕	3	7	11¼	12⅔	13⅓
The number of the oscillations in air	85½	287	535						

In the experiments of the 4th column there were equal motions lost in 535 oscillations made in the air, and 1⅕ in water. The oscillations in the air were indeed a little swifter than those in the water. But if the oscillations in the water were accelerated in such a ratio that the motions of the pendulums might be equally swift in both mediums, there would be still the same number of 1⅕ oscillations in the water, and by these the same quantity of motion would be lost as before; because the resistance is increased, and the square of the time diminished in the same squared ratio. The pendulums, therefore, being of equal velocities, there were equal motions lost in 535 oscillations in the air, and 1⅕ in the water; and therefore the resistance of the pendulum in the water is to its resistance in the air as 535 to 1⅕. This is the proportion of the whole resistances in the case of the 4th column.

Now let $AV + CV^2$ represent the difference of the arcs described in the descent and subsequent ascent by the globe moving in air with the greatest velocity V; and since the greatest velocity is in the case of the 4th column to the greatest velocity in the case of the 1st column as 1 is to 8; and that difference of the arcs in the case of the 4th column to the difference in the case of the 1st column as $\frac{2}{535}$ to $\frac{16}{85\frac{1}{2}}$, or as $85\frac{1}{2}$ to 4280; put in these cases 1 and 8 for the velocities, and $85\frac{1}{2}$ and 4280 for the differences of the arcs, and $A + C$ will be $= 85\frac{1}{2}$, and $8A + 64C = 4280$ or $A + 8C = 535$; and then, by reducing these equations, there will come out $7C = 449\frac{1}{2}$ and $C = 64\frac{3}{14}$ and $A = 21\frac{2}{7}$; and therefore the resistance, which is as $\frac{7}{11}AV + \frac{3}{4}CV^2$, will become as $13\frac{6}{11}V + 48\frac{5}{56}V^2$. Therefore in the case of the 4th column, where the velocity was 1, the whole resistance is to its part proportional to the square of the velocity as $13\frac{6}{11} + 48\frac{5}{56}$ or $61\frac{12}{17}$ to $48\frac{5}{56}$; and therefore the resistance of the pendulum in water is to that part of the resistance in air, which is proportional to the square of the velocity, and which in swift motions is the only part that deserves consideration, as $61\frac{12}{17}$ to $48\frac{5}{56}$ and 535 to $1\frac{1}{5}$ conjointly, that is, as 571 to 1. If the whole thread of the pendulum oscillating in the water had been immersed, its resistance would have been still greater; so that the resistance of the pendulum oscillating in the water, that is, that part which is proportional to the square of the velocity, and which only needs to be considered in swift bodies, is to the resistance of the same whole pendulum, oscillating in air with the same velocity, as about 850 to 1, that is, as the density of water is to the density of air, nearly.

In this calculation we ought also to have taken in that part of the resistance of the pendulum in the water which was as the square of the velocity; but I found (which will perhaps seem strange) that the resistance in the water was augmented in more than a squared ratio of the velocity. In searching after the cause, I thought upon this, that the vessel was too narrow for the magnitude of the pendulous globe, and by its narrowness obstructed the motion of the water as it yielded to the oscillating globe. For when I immersed a pendulous globe, whose diameter was one inch only, the resistance was augmented nearly as the square of the velocity. I tried this by making a pendulum of two globes, of which the lesser and lower oscillated in the water, and the greater and higher was fastened to the thread just above the water, and, by oscillating in the air, assisted the motion of the pendulum, and continued it longer. The experiments made by this contrivance resulted as shown in the following table (p. 324).

Arc described in first descent 16 . 8 . 4 . 2 . 1 . ½ . ¼

Arc described in last ascent___ 12 . 6 . 3 . 1½ . ¾ . ⅜ . ³⁄₁₆

Difference of arcs, propor-⎱ 4 . 2 . 1 . ½ . ¼ . ⅛ . ¹⁄₁₆
tional to motion lost_____⎰

Number of oscillations_____ 3⅜ . 6½ . 12¹⁄₁₂ . 21⅕ . 34 . 53 . 62⅕

In comparing the resistances of the mediums with each other, I also caused iron pendulums to oscillate in quicksilver. The length of the iron wire was about 3 feet, and the diameter of the pendulous globe about ⅓ of an inch. To the wire, just above the quicksilver, there was fixed another leaden globe of a bigness sufficient to continue the motion of the pendulum for some time. Then a vessel, that would hold about 3 pounds of quicksilver, was filled by turns with quicksilver and common water, so that, by making the pendulum oscillate successively in these two different fluids, I might find the proportion of their resistances; and the resistance of the quicksilver proved to be to the resistance of water as about 13 or 14 to 1; that is, as the density of quicksilver to the density of water. When I made use of a pendulous globe something bigger, as of one whose diameter was about ½ or ⅔ of an inch, the resistance of the quicksilver proved to be to the resistance of the water as about 12 or 10 to 1. But the former experiment is more to be relied on, because in the latter the vessel was too narrow in proportion to the magnitude of the immersed globe; for the vessel ought to have been enlarged together with the globe. I intended to repeat these experiments with larger vessels, and in melted metals, and other liquors both cold and hot; but I had not leisure to try all; and besides, from what is already described, it appears sufficiently that the resistance of bodies moving swiftly is nearly proportional to the densities of the fluids in which they move. I do not say accurately; for more tenacious fluids, of equal density, will undoubtedly resist more than those that are more liquid; as cold oil more than warm, warm oil more than rain water, and water more than spirit of wine. But in liquors, which are sensibly fluid enough, as in air, in salt and fresh water, in spirit of wine, of turpentine, and salts, in oil cleared of its feces by distillation and warmed, in oil of vitriol, and in mercury, and melted metals, and any other such like, that are fluid enough to retain for some time the motion impressed upon them by the agitation of the vessel, and which being poured out are easily resolved into drops, I doubt not that the rule already laid down may be accurate enough, especially if the experiments be made with larger pendulous bodies and more swiftly moved.

Lastly, since it is the opinion of some that there is a certain ethereal medium extremely rare and subtile, which freely pervades the pores of all bodies; and from such a medium, so pervading the pores of bodies, some resistance must needs arise; in order to try whether the resistance, which we experience in bodies in motion, be made upon their outward surfaces only, or whether their internal parts meet with any considerable resistance upon their surfaces, I thought of the following experiment. I suspended a round deal box by a thread 11 feet long, on a steel hook, by means of a ring of the same metal, so as to make a pendulum of the aforesaid length. The hook had a sharp hollow edge on its upper part, so that the upper arc of the ring pressing on the edge might move the more freely; and the thread was fastened to the lower arc of the ring. The pendulum being thus prepared, I drew it aside from the perpendicular to the distance of about 6 feet, and that in a plane perpendicular to the edge of the hook, lest the ring, while the pendulum oscillated, should slide to and fro on the edge of the hook; for the point of suspension, in which the ring touches the hook, ought to remain immovable. I therefore accurately noted the place to which the pendulum was brought, and letting it go, I marked three other places, to which it returned at the end of the 1st, 2d, and 3d oscillation. This I often repeated, that I might find those places as accurately as possible. Then I filled the box with lead and other heavy metals that were near at hand. But, first, I weighed the box when empty, and that part of the thread that went round it, and half the remaining part, extended between the hook and the suspended box; for the thread so extended always acts upon the pendulum, when drawn aside from the perpendicular, with half its weight. To this weight I added the weight of the air contained in the box. And this whole weight was about $\frac{1}{78}$ of the weight of the box when filled with the metals. Then because the box when full of the metals, by extending the thread with its weight, increased the length of the pendulum, I shortened the thread so as to make the length of the pendulum, when oscillating, the same as before. Then drawing aside the pendulum to the place first marked, and letting it go, I reckoned about 77 oscillations before the box returned to the second mark, and as many afterwards before it came to the third mark, and as many after that before it came to the fourth mark. From this I conclude that the whole resistance of the box, when full, had not a greater proportion to the resistance of the box, when empty, than 78 to 77. For if their resistances were equal, the box, when full, by reason of its inertia, which was 78 times greater than the inertia of the same when empty, ought to have continued its oscillating motion so much the longer, and therefore to have returned to those marks at the end of 78 oscillations. But it returned to them at the end of 77 oscillations.

Let, therefore, A represent the resistance of the box upon its external surface, and B the resistance of the empty box on its internal surface, and if the resistances to the internal parts of bodies equally swift be as the matter, or the number of particles that are resisted, then 78B will be the resistance made to the internal parts of the box, when full; and therefore the whole resistance $A + B$ of the empty box will be to the whole resistance $A + 78B$ of the full box as 77 to 78, and, by subtraction, $A + B$ to 77B as 77 to 1; and thence $A + B$ to B as $77 \cdot 77$ to 1, and, by subtraction, again, A to B as 5928 to 1. Therefore the resistance of the empty box in its internal parts will be above 5000 times less than the resistance on its external surface. This reasoning depends upon the supposition that the greater resistance of the full box arises not from any other latent cause, but only from the action of some subtile fluid upon the included metal.

This experiment is related by memory, the paper being lost in which I had described it; so that I have been obliged to omit some fractional parts, which are slipped out of my memory; and I have no leisure to try it again. The first time I made it, the hook being weak, the full box was retarded sooner. The cause I found to be, that the hook was not strong enough to bear the weight of the box; so that, as it oscillated to and fro, the hook was bent sometimes this and sometimes that way. I therefore procured a hook of sufficient strength, so that the point of suspension might remain unmoved, and then all things happened as is above described.

SECTION VII

The motion of fluids, and the resistance made to projected bodies.

PROPOSITION XXXII. THEOREM XXVI

Suppose two similar systems of bodies consisting of an equal number of particles, and let the correspondent particles be similar and proportional, each in one system to each in the other, and have a like situation among themselves, and the same given ratio of density to each other; and let them begin to move among themselves in proportional times, and with like motions (that is, those in one system among one another, and those in the other among one another). And if the particles that are in the same system do not touch one another, except in the moments of reflection; nor attract, nor repel each other, except with accelerative forces that are inversely as the diameters of the correspondent particles, and directly as the squares of the velocities: I say, that the particles of those systems will continue to move among themselves with like motions and in proportional times.

Like bodies in like situations are said to be moved among themselves with like motions and in proportional times, when their situations at the end of those times are always found alike in respect of each other; as suppose we compare the particles in one system with the correspondent particles in the other. Hence the times will be proportional, in which similar and proportional parts of similar figures will be described by correspondent particles. Therefore if we suppose two systems of this kind, the correspondent particles, by reason of the similitude of the motions at their beginning, will continue to be moved with like motions, so long as they move without meeting one another; for if they are acted on by no forces, they will go on uniformly in right lines, by the first Law. But if they agitate one another with some certain forces, and those forces are inversely as the diameters of the correspondent particles and directly as the squares of the velocities, then, because the particles are in like situations, and their forces are proportional, the whole forces with which correspondent particles are agitated, and which are compounded of each of the agitating forces (by Cor. II of the Laws), will have like directions, and have the same effect as if they respected centres placed alike among the particles; and those whole forces will be to each other as the several forces which compose them, that is, inversely as the diameters of the correspondent particles and directly as the squares of the velocities: and therefore will cause correspondent particles to continue to describe like figures. These things will be so (by Cor. 1 and VIII, Prop. IV, Book I), if those centres are at rest; but if they are moved, yet, by reason of the similitude of the translations, their situations among the particles of the system will remain similar, so that the changes introduced into the figures described by the particles will still be similar. So that the motions of correspondent and similar particles will continue similar till their first meeting with each other; and thence will arise similar collisions, and similar reflections; which will again beget similar motions of the particles among themselves (by what was just now shown), till they mutually fall upon one another again, and so on *ad infinitum*. Q.E.D.

Cor. I. Hence if any two bodies, which are similar and in like situations to the correspondent particles of the systems, begin to move amongst them in like manner and in proportional times, and their magnitudes and densities be to each other as the magnitudes and densities of the corresponding particles, these bodies will continue to be moved in like manner and in proportional times; for the case of the greater parts of both systems and of the particles is the very same.

Cor. II. And if all the similar and similarly situated parts of both systems be at rest among themselves; and two of them, which are greater than the rest, and mutually correspondent in both systems, begin to move in lines alike posited, with any similar motion whatsoever, they will excite similar motions in the rest of the parts of the systems, and will continue to move among those parts in like manner and in proportional times; and will therefore describe spaces proportional to their diameters.

PROPOSITION XXXIII. THEOREM XXVII

The same things being supposed, I say, that the greater parts of the systems are resisted in a ratio compounded of the squared ratio of their velocities, and the squared ratio of their diameters, and the simple ratio of the density of the parts of the systems.

For the resistance arises partly from the centripetal or centrifugal forces with which the particles of the system act on each other, partly from the collisions and reflections of the particles and the greater parts. The resistances of the first kind are to each other as the whole motive forces from which they arise, that is, as the whole accelerative forces and the quantities of matter in corresponding parts; that is (by the supposition), directly as the squares of the velocities and inversely as the distances of the corresponding particles, and directly as the quantities of matter in the correspondent parts: and therefore since the distances of the particles in one system are to the correspondent distances of the particles in the other, as the diameter of one particle or part in the former system to the diameter of the correspondent particle or part in the other, and since the quantities of matter are as the densities of the parts and the cubes of the diameters, the resistances are to each other as the squares of the velocities and the squares of the diameters and the densities of the parts of the systems. Q.E.D. The resistances of the latter sort are as the number of correspondent reflections and the forces of those reflections conjointly; but the number of the reflections are to each other directly as the velocities of the corresponding parts and inversely as the spaces between their reflections. And the forces of the reflections are as the velocities and the magnitudes and the densities of the corresponding parts conjointly; that is, as the velocities and the cubes of the diameters and the densities of the parts. And, joining all these ratios, the resistances of the corresponding parts are to each other as the squares of the velocities and the squares of the diameters and the densities of the parts conjointly. Q.E.D.

Cor. I. Therefore if those systems are two elastic fluids, like our air, and their parts are at rest among themselves; and two similar bodies proportional in magnitude and density to the parts of the fluids, and similarly situated among those parts, be in any manner projected in the direction of lines similarly posited; and the accelerative forces with which the particles of the fluids act upon each other are inversely as the diameters of the bodies projected and directly as the squares of their velocities; those bodies will excite similar motions in the fluids in proportional times, and will describe similar spaces and proportional to their diameters.

Cor. II. Therefore in the same fluid a projected body that moves swiftly meets with a resistance that is as the square of its velocity, nearly. For if the forces with which distant particles act upon one another should be augmented as the square of the velocity, the projected body would be resisted in the same squared ratio accurately; and therefore in a medium, whose parts when at a distance do not act with any force on one another, the resistance is as the square of the velocity, accurately. Let there be, therefore, three mediums A, B, C, consisting of similar and equal parts regularly disposed at equal distances. Let the parts of the mediums A and B recede from each other with forces that are among themselves as T and V; and let the parts of the medium C be entirely destitute of any such forces. And if four equal bodies D, E, F, G move in these mediums, the two first D and E in the two first A and B, and the other two F and G in the third C; and if the velocity of the body D be to the velocity of the body E, and the velocity of the body F to the velocity of the body G, as the square root of the ratio of the force T to the force V; then the resistance of the body D to the resistance of the body E, and the resistance of the body F to the resistance of the body G, will be as the square of the velocities; and therefore the resistance of the body D will be to the resistance of the body F as the resistance of the body E to the resistance of the body G. Let the bodies D and F be equally swift, as also the bodies E and G; and, augmenting the velocities of the bodies D and F in any ratio, and diminishing the forces of the particles of the medium B as the square of the same ratio, the medium B will approach to the form and condition of the medium C at pleasure; and therefore the resistances of the equal and equally swift bodies E and G in these mediums will continually approach to equality, so that their difference will at last become less than any given. Therefore since the resistances of the bodies D and F are to each other as the resistances of the bodies E and G, those will also in like manner approach to the ratio of equality. Therefore the bodies D and F, when they move with very great swiftness, meet with resistances very nearly equal; and therefore since the resistance of the body F is in a squared ratio of the velocity, the resistance of the body D will be nearly in the same ratio.

Cor. iii. Hence the resistance of a body moving very swiftly in an elastic fluid is almost the same as if the parts of the fluid were destitute of their centrifugal forces, and did not fly from each other; provided only that the elasticity of the fluid arise from the centrifugal forces of the particles, and the velocity be so great as not to allow the particles time enough to act.

Cor. iv. Since the resistances of similar and equally swift bodies, in a medium whose distant parts do not fly from each other, are as the squares of the diameters, therefore the resistances made to bodies moving with very great and equal velocities in an elastic fluid will be as the squares of the diameters, nearly.

Cor. v. And since similar, equal, and equally swift bodies, moving through mediums of the same density, whose particles do not fly from each other, will strike against an equal quantity of matter in equal times, whether the particles of which the medium consists be more and smaller, or fewer and greater, and therefore impress on that matter an equal quantity of motion, and in return (by the third Law of Motion) suffer an equal reaction from the same, that is, are equally resisted; it is manifest, also, that in elastic fluids of the same density, when the bodies move with extreme swiftness, their resistances are nearly equal, whether the fluids consist of gross parts, or of parts ever so subtile. For the resistance of projectiles moving with exceedingly great celerities is not much diminished by the subtilty of the medium.

Cor. vi. All these things are so in fluids whose elastic force takes its rise from the centrifugal forces of the particles. But if that force arise from some other cause, as from the expansion of the particles after the manner of wool, or the boughs of trees, or any other cause, by which the particles are hindered from moving freely among themselves, the resistance, by reason of the lesser fluidity of the medium, will be greater than in the Corollaries above.

PROPOSITION XXXIV. THEOREM XXVIII

If in a rare medium, consisting of equal particles freely disposed at equal distances from each other, a globe and a cylinder described on equal diameters move with equal velocities in the direction of the axis of the cylinder, the resistance of the globe will be but half as great as that of the cylinder.

For since the action of the medium upon the body is the same (by Cor. v of the Laws) whether the body move in a quiescent medium, or whether the particles of the medium impinge with the same velocity upon the quiescent body, let us consider the body as if it were quiescent, and see with what force it would be impelled by the moving medium. Let, therefore, ABKI represent a spherical body described from the centre C with the semidiameter CA, and let the particles of the medium impinge with a given velocity upon that spherical body in the directions of right lines parallel to

AC; and let FB be one of those right lines. In FB take LB equal to the semi-diameter CB, and draw BD touching the sphere in B. Upon KC and BD let fall the perpendiculars BE, LD; and the force with which a particle of the medium, impinging on the globe obliquely in the direction FB, would strike the globe in B, will be to the force with which the same particle, meeting the cylinder ONGQ described about the globe with the axis ACI, would

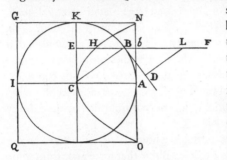

strike it perpendicularly in b, as LD is to LB, or BE to BC. Again; the efficacy of this force to move the globe, according to the direction of its incidence FB or AC, is to the efficacy of the same to move the globe, according to the direction of its determination, that is, in the direction of the right line BC in which it impels the globe directly, as BE to BC. And, joining these ratios, the efficacy of a particle, falling upon the globe obliquely in the direction of the right line FB, to move the globe in the direction of its incidence, is to the efficacy of the same particle falling in the same line perpendicularly on the cylinder, to move it in the same direction, as BE^2 to BC^2. Therefore if in bE, which is perpendicular to the circular base of the cylinder NAO, and equal to the radius AC, we take bH equal to $\dfrac{BE^2}{CB}$; then bH will be to bE as the effect of the particle upon the globe to the effect of the particle upon the cylinder. And therefore the solid which is formed by all the right lines bH will be to the solid formed by all the right lines bE as the effect of all the particles upon the globe to the effect of all the particles upon the cylinder. But the former of these solids is a paraboloid whose vertex is C, its axis CA, and latus rectum CA, and the latter solid is a cylinder circumscribing the paraboloid; and it is known that a paraboloid is half its circumscribed cylinder. Therefore the whole force of the medium upon the globe is half the entire force of the same upon the cylinder. And therefore if the particles of the medium are at rest, and the cylinder and globe move with equal velocities, the resistance of the globe will be half the resistance of the cylinder. Q.E.D.

SCHOLIUM[1]

By the same method other figures may be compared together as to their resistance; and those may be found which are most apt to continue their motions in resisting mediums. As if upon the circular base CEBH from the centre O, with the radius OC, and the altitude OD, one would construct a frustum CBGF of a cone, which should meet with less resistance than any other frustum constructed with the same base and altitude, and going forwards towards D in the direction of its axis: bisect the altitude OD in Q, and produce OQ to S, so that QS may be equal to QC, and S will be the vertex of the cone whose frustum is sought.

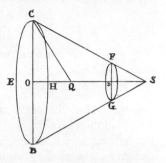

Incidentally, since the angle CSB is always acute, it follows from the above that, if the solid ADBE be generated by the convolution of an elliptical or oval figure ADBE about its axis AB, and the generating figure be touched by three right lines FG, GH, HI, in the points F, B, and I, so that

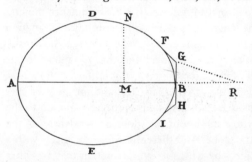

GH shall be perpendicular to the axis in the point of contact B, and FG, HI may be inclined to GH in the angles FGB, BHI of 135 degrees: the solid arising from the convolution of the figure ADFGHIE about the same axis AB will be less resisted than the former solid, provided that both move forwards in the direction of their axis AB, and that the extremity B of each go foremost. This Proposition I conceive may be of use in the building of ships.

If the figure DNFG be such a curve, that if, from any point thereof, as N, the perpendicular NM be let fall on the axis AB, and from the given point G there be drawn the right line GR parallel to a right line touching the figure in N, and cutting the axis produced in R, MN becomes to GR as GR^3 to 4BR · GB^2, the solid described by the revolution of this figure about its axis AB, moving in the before-mentioned rare medium from A towards B, will be less resisted than any other circular solid whatsoever, described of the same length and breadth.

PROPOSITION XXXV. PROBLEM VII

If a rare medium consist of very small quiescent particles of equal magnitudes, and freely disposed at equal distances from one another: to find the resistance of a globe moving uniformly forwards in this medium.

CASE 1. Let a cylinder described with the same diameter and altitude be conceived to go forwards with the same velocity in the direction of its axis through the same medium; and let us suppose that the particles of the medium, on which the globe or cylinder falls, fly back with as great a force of reflection as possible. Then since the resistance of the globe (by the last Proposition) is but half the resistance of the cylinder, and since the globe is to the cylinder as 2 to 3, and since the cylinder by falling perpendicularly on the particles, and reflecting them with the utmost force, communicates to them a velocity double to its own: it follows that the cylinder in moving forwards uniformly half the length of its axis, will communicate a motion to the particles which is to the whole motion of the cylinder as the density of the medium to the density of the cylinder; and that the globe, in the time it describes one length of its diameter in moving uniformly forwards, will communicate the same motion to the particles; and, in the time that it describes two-thirds of its diameter, will communicate a motion to the particles which is to the whole motion of the globe as the density of the medium to the density of the globe. And therefore the globe meets with a resistance, which is to the force by which its whole motion may be either taken away or generated in the time in which it describes two-thirds of its diameter moving uniformly forwards, as the density of the medium is to the density of the globe.

CASE 2. Let us suppose that the particles of the medium incident on the globe or cylinder are not reflected; and then the cylinder falling perpendicularly on the particles will communicate its own simple velocity to them, and therefore meets a resistance but half so great as in the former case, and the globe also meets with a resistance but half so great.

Case 3. Let us suppose the particles of the medium to fly back from the globe with a force which is neither the greatest, nor yet none at all, but with a certain mean force; then the resistance of the globe will be in the same mean ratio between the resistance in the first case and the resistance in the second. Q.E.I.

Cor. i. Hence if the globe and the particles are infinitely hard, and destitute of all elastic force, and therefore of all force of reflection, the resistance of the globe will be to the force by which its whole motion may be destroyed or generated, in the time that the globe describes four third parts of its diameter, as the density of the medium is to the density of the globe.

Cor. ii. The resistance of the globe, other things being equal, varies as the square of the velocity.

Cor. iii. The resistance of the globe, other things being equal, varies as the square of the diameter.

Cor. iv. The resistance of the globe, other things being equal, varies as the density of the medium.

Cor. v. The resistance of the globe varies jointly as the square of the velocity, as the square of the diameter, and as the density of the medium.

Cor. vi. The motion of the globe and its resistance may be thus represented. Let AB be the time in which the globe may, by its resistance uniformly continued, lose its whole motion. Erect AD, BC perpendicular to AB. Let BC be that whole motion, and through the point C, the asymptotes being AD, AB, describe the hyperbola CF. Produce AB to any point E. Erect the perpendicular EF meeting the hyperbola in F. Complete the parallelogram CBEG, and draw AF meeting BC in H. Then if the 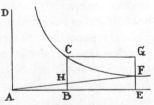 globe in any time BE, with its first motion BC uniformly continued, describes in a nonresisting medium the space CBEG represented by the area of the parallelogram, the same in a resisting medium will describe the space CBEF, represented by the area of the hyperbola; and its motion at the end of that time will be represented by EF, the ordinate of the hyperbola, there being lost of its motion the part FG. And its resistance at the end of the same time will be represented by the length BH, there being lost of its resistance the part CH. All these things appear by Cor. I and III, Prop. v, Book II.

Cor. VII. Hence if the globe in the time T by the resistance R uniformly continued lose its whole motion M, the same globe in the time t in a resisting medium, wherein the resistance R decreases as the square of the velocity, will lose out of its motion M the part $\dfrac{tM}{T+t}$, the part $\dfrac{TM}{T+t}$ remaining; and will describe a space which is to the space described in the same time t, with the uniform motion M, as the logarithm of the number $\dfrac{T+t}{T}$ multiplied by the number 2.302585092994 is to the number $\dfrac{t}{T}$, because the hyperbolic area BCFE is to the rectangle BCGE in that proportion.

SCHOLIUM

I have exhibited in this Proposition the resistance and retardation of spherical projectiles in mediums that are not continued, and shown that this resistance is to the force by which the whole motion of the globe may be destroyed or produced in the time in which the globe can describe two-thirds of its diameter, with a velocity uniformly continued, as the density of the medium is to the density of the globe, provided the globe and the particles of the medium be perfectly elastic, and are endued with the utmost force of reflection; and that this force, where the globe and particles of the medium are infinitely hard and void of any reflecting force, is diminished one-half. But in continued mediums, as water, hot oil, and quicksilver, the globe as it passes through them does not immediately strike against all the particles of the fluid that generate the resistance made to it, but presses only the particles that lie next to it, which press the particles beyond, which press other particles, and so on; and in these mediums the resistance is diminished one other half. A globe in these extremely fluid mediums meets with a resistance that is to the force by which its whole motion may be destroyed or generated in the time wherein it can describe, with that motion uniformly continued, eight third parts of its diameter, as the density of the medium is to the density of the globe. This I shall endeavor to show in what follows.

PROPOSITION XXXVI. PROBLEM VIII

To find the motion of water running out of a cylindrical vessel through a hole made at the bottom.[1]

Let ACDB be a cylindrical vessel, AB the mouth of it, CD the bottom parallel to the horizon, EF a circular hole in the middle of the bottom, G the centre of the hole, and GH the axis of the cylinder perpendicular to the horizon. And suppose a cylinder of ice APQB to be of the same breadth with the cavity of the vessel, and to have the same axis, and to descend continually with an uniform motion, and that its parts, as soon as they touch the surface AB, dissolve into water, and flow down by their weight into the vessel, and in their fall compose the cataract or column of water ABNFEM, passing through the hole EF, and filling up the same exactly. Let the uniform velocity of the descending ice and of the contiguous water in the circle AB be that which the water would acquire by falling through the space IH; and let IH and HG lie in the same right line; and through the point I let there be drawn the right line KL parallel to the horizon, and meeting the ice on both the sides thereof in K and L. Then the velocity of the water running out at the hole EF will be the same that it would acquire by falling from I through the space IG. Therefore, by *Galileo's* Theorems, IG will be to IH as the square of the velocity of the water that runs out at the hole to the velocity of the water in the circle AB, that is, as the square of the ratio of the circle AB to the circle EF; those circles being inversely as the velocities of the water which in the same time and in equal quantities passes through each of them, and completely fills them both. We are now considering the velocity with which the water tends to the plane of the horizon. But the motion parallel to the same, by which the parts of the falling water approach to each other, is not here taken notice of; since it is neither produced by gravity, nor at all changes the motion perpendicular to the horizon which the gravity produces. We suppose, indeed, that the parts of the

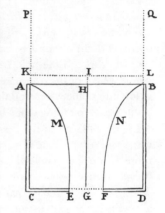

[1 Appendix, Note 36.]

water cohere a little, that by their cohesion they may in falling approach to each other with motions parallel to the horizon in order to form one single cataract, and to prevent their being divided into several; but the motion parallel to the horizon arising from this cohesion does not come under our present consideration.

CASE I. Conceive now the whole cavity in the vessel, which surrounds the falling water ABNFEM, to be full of ice, so that the water may pass through the ice as through a funnel. Then if the water pass very near to the ice only, without touching it; or, which is the same thing, if by reason of the perfect smoothness of the surface of the ice, the water, though touching it, glides over it with the utmost freedom, and without the least resistance; the water will run through the hole EF with the same velocity as before, and the whole weight of the column of water ABNFEM will be taken up as before in forcing out the water, and the bottom of the vessel will sustain the weight of the ice surrounding that column.

Let now the ice in the vessel dissolve into water; but the efflux of the water will remain, as to its velocity, the same as before. It will not be less, because the ice now dissolved will endeavor to descend; it will not be greater, because the ice, now become water, cannot descend without hindering the descent of other water equal to its own descent. The same force ought always to generate the same velocity in the effluent water.

But the hole at the bottom of the vessel, by reason of the oblique motions of the particles of the effluent water, must be a little greater than before. For now the particles of the water do not all of them pass through the hole perpendicularly, but, flowing down on all parts from the sides of the vessel, and converging towards the hole, pass through it with oblique motions; and in tending downwards they meet in a stream whose diameter is a little smaller below the hole than at the hole itself; its diameter being to the diameter of the hole as 5 to 6, or as 5½ to 6½, very nearly, if I measured those diameters rightly. I procured a thin flat plate, having a hole pierced in the middle, the diameter of the circular hole being five eighth parts of an inch. And that the stream of running water might not be accelerated in falling, and by that acceleration become narrower, I fixed this plate not to the bottom, but to the side of the vessel, so as to make the water go out in the direction of a line parallel to the horizon. Then, when the vessel was full of water, I opened the hole to let it run out; and the diameter of the stream, measured with great accuracy at the distance of about half an inch from the hole, was $^{21}\!/_{40}$ of an inch. Therefore the diameter of this circular hole was to the diameter of the stream very nearly as 25 to 21. So that the water in passing through the hole converges on all sides, and, after it has run out of the vessel, becomes smaller by converging in that manner, and by becoming smaller is accelerated till it comes to the distance of half an inch from the hole, and at that distance flows in a smaller stream and with greater celerity than in the hole itself, and this in the ratio of 25·25 to 21·21, or 17 to 12, very nearly; that is, in about the ratio of $\sqrt{2}$ to 1. Now it is certain from experiments, that the quantity of water running out in a given time through a circular hole made in the bottom of a vessel is equal to the quantity, which, flowing freely with the aforesaid velocity, would run out in the same time through another circular hole, whose diameter is to the diameter of the former as 21 to 25. And therefore this running water in passing through the hole itself has a velocity downwards nearly equal to that which a heavy body would acquire in falling through half the height of the stagnant water in the vessel. But then, after it has run out, it is still accelerated by converging, till it arrives at a distance from the hole that is nearly equal to its diameter, and acquires a velocity greater than the other in about the ratio of $\sqrt{2}$ to 1; this velocity a heavy body would nearly acquire by falling freely through the whole height of the stagnant water in the vessel.

Therefore in what follows let the diameter of the stream be represented by that lesser hole which we shall call EF. And imagine another plane VW above the hole EF, and parallel to the plane thereof, to be placed at a distance equal to the diameter of the same hole, and to be pierced through with a greater hole ST, of such a magnitude that a stream which will exactly fill the lower hole EF may pass through it; the diameter of this hole will therefore be to the diameter of the lower hole nearly as 25 to 21. By this means the water will run perpendicularly out at

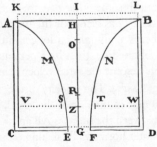

the lower hole; and the quantity of the water running out will be, according to the magnitude of this last hole, very nearly the same as that which the solution of the Problem requires. The space included between the two planes and the falling stream may be considered as the bottom of the vessel. But to make the solution more simple and mathematical, it is better to take the lower plane alone for the bottom of the vessel, and to suppose that the water which flowed through the ice as through a funnel, and ran out of the vessel through the hole EF made in the lower plane, preserves its motion continually, and that the ice continues at rest. Therefore in what follows let ST be the diameter of a circular hole described from the centre Z, and let the stream run out of the vessel through that hole, when the water in the vessel is all fluid. And let EF be the diameter of the hole, which the stream, in falling through, exactly fills up, whether the water runs out of the vessel by that upper hole ST, or flows through the middle of the ice in the vessel, as through a funnel. And let the diameter of the upper hole ST be to the diameter of the lower EF as about 25 to 21, and let the perpendicular distance between the planes of the holes be equal to the diameter of the lesser hole EF. Then the velocity of the water downwards, in running out of the vessel through the hole ST, will be in that hole the same that a body may acquire by falling freely from half the height IZ; and the velocity of both the falling streams will be in the hole EF, the same which a body would acquire by falling freely from the whole height IG.

CASE 2. If the hole EF be not in the middle of the bottom of the vessel, but in some other part thereof, the water will still run out with the same velocity as before, if the magnitude of the hole be the same. For though a heavy body takes a longer time in descending to the same depth, by an oblique line, than by a perpendicular line, yet in both cases it acquires in its descent the same velocity; as *Galileo* hath demonstrated.

CASE 3. The velocity of the water is the same when it runs out through a hole in the side of the vessel. For if the hole be small, so that the interval between the surfaces AB and KL may vanish as to sense, and the stream of water horizontally issuing out may form a parabolic figure; from the latus rectum of this parabola one may see, that the velocity of the effluent water is that which a body may acquire by falling the height IG or HG of the stagnant water in the vessel. For, by making an experiment, I found that if the height of the stagnant water above the hole were 20 inches, and the height of the hole above a plane parallel to the horizon were also 20 inches, a stream of water springing out from thence would fall upon the plane, at the distance of very nearly 37 inches, from a perpendicular let fall upon that plane from the hole. For without resistance the stream would have fallen upon the plane at the distance of 40 inches, the latus rectum of the parabolic stream being 80 inches.

CASE 4. If the effluent water tend upwards, it will still issue forth with the same velocity. For the small stream of water springing upwards, ascends with a perpendicular motion to GH or GI, the height of the stagnant water in the vessel; except so far as its ascent is hindered a little by the resistance of the air; and therefore it springs out with the same velocity that it would acquire in falling from that height. Every particle of the stagnant water is equally pressed on all sides (by Prop. XIX, Book II), and, yielding to the pressure, tends always with an equal force, whether it descends through the hole in the bottom of the vessel, or gushes out in an horizontal direction through a hole in the side, or passes into a canal, and springs up from thence through a little hole made in the upper part of the canal. And it may not only be inferred from reasoning, but is manifest also from the well-known experiments just mentioned, that the velocity with which the water runs out is the very same that is assigned in this Proposition.

CASE 5. The velocity of the effluent water is the same, whether the figure of the hole be circular, or square, or triangular, or of any other figure whatever equal to the circular; for the velocity of the effluent water does not depend upon the figure of the hole, but arises from such depth of the hole as it may have below the plane KL.

CASE 6. If the lower part of the vessel ABDC be immersed into stagnant water, and the height of the stagnant water above the bottom of the vessel be GR, the velocity with which the water that is in the vessel will run out at the hole EF into the stagnant water will be the same which the water would acquire by falling from the height IR; for the weight of all the water

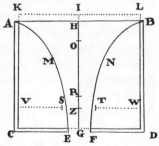

in the vessel that is below the surface of the stagnant water will be sustained in equilibrium by the weight of the stagnant water, and therefore does not at all accelerate the motion of the descending water in the vessel. This case will also become evident from experiments, measuring the times in which the water will run out.

Cor. i. Hence if CA, the depth of the water, be produced to K, so that AK may be to CK as the square of the ratio of the area of a hole made in any part of the bottom to the area of the circle AB, the velocity of the effluent water will be equal to the velocity which the water would acquire by falling freely from the height KC.

Cor. ii. And the force with which the whole motion of the effluent water may be generated is equal to the weight of a cylindric column of water, whose base is the hole EF, and its altitude 2GI or 2CK. For the effluent water, in the time it becomes equal to this column, may acquire, by falling by its own weight from the height GI, a velocity equal to that with which it runs out.

Cor. iii. The weight of all the water in the vessel ABDC is to that part of the weight which is employed in forcing out the water as the sum of the circles AB and EF is to twice the circle EF. For let IO be a mean proportional between IH and IG, and the water running out at the hole EF will, in the time that a drop falling from I would describe the altitude IG, become equal to a cylinder whose base is the circle EF and its altitude 2IG, that is, to a cylinder whose base is the circle AB, and whose altitude is 2IO. For the circle EF is to the circle AB as the square root of the ratio of the altitude IH to the altitude IG; that is, in the simple ratio of the mean proportional IO to the altitude IG. Moreover, in the time that a drop falling from I can describe the altitude IH, the water that runs out will have become equal to a cylinder whose base is the circle AB, and its altitude 2IH; and in the time that a drop falling from I through H to G describes HG, the difference of the altitudes, the effluent water, that is, the water contained within the solid ABNFEM, will be equal to the difference of the cylinders, that is, to a cylinder whose base is AB, and its altitude 2HO. And therefore all the water contained in the vessel ABDC is to the whole falling water contained in the said solid ABNFEM as HG is to 2HO, that is, as HO + OG to 2HO, or IH + IO to 2IH. But the weight of all the water in the solid ABNFEM is employed in forcing out the water; and therefore the weight of all the water in the vessel is to that part of the weight that is employed in forcing out the water as IH + IO is to 2IH, and therefore as the sum of the circles EF and AB is to twice the circle EF.

Cor. iv. And hence the weight of all the water in the vessel ABDC is to the other part of the weight which is sustained by the bottom of the vessel as the sum of the circles AB and EF is to the difference of the same circles.

Cor. v. And that part of the weight which the bottom of the vessel sustains is to the other part of the weight employed in forcing out the water as the difference of the circles AB and EF is to twice the lesser circle EF, or as the area of the bottom to twice the hole.

Cor. vi. That part of the weight which presses upon the bottom is to the whole weight of the water perpendicularly incumbent thereon as the circle AB is to the sum of the circles AB and EF, or as the circle AB is to the excess of twice the circle AB above the area of the bottom. For that part of the weight which presses upon the bottom is to the weight of the whole water in the vessel as the difference of the circles AB and EF is to the sum of the same circles (by Cor. iv); and the weight of the whole water in the vessel is to the weight of the whole water perpendicularly incumbent on the bottom as the circle AB is to the difference of the circles AB and EF. Therefore, multiplying together corresponding terms of the two proportions, that part of the weight which presses upon the bottom is to the weight of the whole water perpendicularly incumbent thereon as the circle AB to the sum of the circles AB and EF, or the excess of twice the circle AB above the bottom.

Cor. vii. If in the middle of the hole EF there be placed the little circle PQ described about the centre G, and parallel to the horizon, the weight of water which that little circle sustains is greater than the weight of a third part of a cylinder of water whose base is that little circle and its height GH. For let ABNFEM be the cataract or column of falling water whose axis is GH, as above, and let all the water, whose fluidity is not requisite for the ready and quick descent of the water, be supposed to be congealed, as well round about the cataract, as above the little circle. And let PHQ be the column of water congealed above the little circle, whose vertex is H, and its altitude GH. And suppose this cataract to fall with its whole weight downwards, and not in the least to lie against or to press PHQ, but to glide freely by it without any friction, unless, perhaps, just at the very vertex of the ice, where the cataract at the beginning of its fall may tend to a concave figure. And as the congealed water AMEC, BNFD, lying

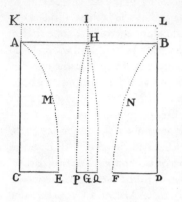

round the cataract, is convex in its internal surfaces AME, BNF towards the falling cataract, so this column PHQ will be convex towards the cataract also, and will therefore be greater than a cone whose base is that little circle PQ and its altitude GH; that is, greater than a third part of a cylinder described with the same base and altitude. Now that little circle sustains the weight of this column, that is, a weight greater than the weight of the cone, or a third part of the cylinder.

COR. VIII. The weight of water which the circle PQ, when very small, sustains, seems to be less than the weight of two-thirds of a cylinder of water whose base is that little circle, and its altitude HG. For, things standing as above supposed, imagine the half of a spheroid described whose base is that little circle, and its semiaxis or altitude HG. This figure will be equal to two-thirds of that cylinder, and will comprehend within it the column of congealed water PHQ, the weight of which is sustained by that little circle. For though the motion of the water tends directly downwards, the external surfaces of that column must yet meet the base PQ in an angle somewhat acute, because the water in its fall is continually accelerated, and by reason of that acceleration becomes narrower. Therefore, since that angle is less than a right one, this column in the lower parts thereof will lie within the hemispheroid. In the upper parts also it will be acute or pointed; because to make it otherwise, the horizontal motion of the water must be at the vertex infinitely more swift than its motion towards the horizon. And the less this circle PQ is, the more acute will the vertex of this column be; and the circle being diminished *in infinitum,* the angle PHQ will be diminished *in infinitum,* and therefore the column will lie within the hemispheroid.

Therefore that column is less than that hemispheroid, or than two third parts of the cylinder whose base is that little circle, and its altitude GH. Now the little circle sustains a force of water equal to the weight of this column, the weight of the ambient water being employed in causing its efflux out at the hole.

Cor. ix. The weight of water which the little circle PQ sustains, when it is very small, is very nearly equal to the weight of a cylinder of water whose base is that little circle, and its altitude ½GH; for this weight is an arithmetical mean between the weights of the cone and the hemispheroid above mentioned. But if that little circle be not very small, but on the contrary increased till it be equal to the hole EF, it will sustain the weight of all the water lying perpendicularly above it, that is, the weight of a cylinder of water whose base is that little circle, and its altitude GH.

Cor. x. And (as far as I can judge) the weight which this little circle sustains is always to the weight of a cylinder of water whose base is that little circle, and its altitude ½GH, as EF^2 is to $EF^2 - \frac{1}{2}PQ^2$, or as the circle EF is to the excess of this circle above half the little circle PQ, very nearly.

LEMMA IV

If a cylinder moves uniformly forwards in the direction of its length, the resistance made thereto is not at all changed by augmenting or diminishing that length; and is therefore the same with the resistance of a circle, described with the same diameter, and moving forwards with the same velocity in the direction of a right line perpendicular to its plane.

For the sides are not at all opposed to the motion; and a cylinder becomes a circle when its length is diminished *in infinitum.*

PROPOSITION XXXVII. THEOREM XXIX

If a cylinder moves uniformly forwards in a compressed, infinite, and nonelastic fluid, in the direction of its length, the resistance arising from the magnitude of its transverse section is to the force by which its whole motion may be destroyed or generated, in the time that it moves four times its length, as the density of the medium is to the density of the cylinder, nearly.

For let the vessel ABDC touch the surface of stagnant water with its bottom CD, and let the water run out of this vessel into the stagnant water through the cylindric canal EFTS perpendicular to the horizon; and let the little circle PQ be placed parallel to the horizon anywhere in the middle of the canal; and produce CA to K, so that AK may be to CK as the square of the ratio, which the excess of the orifice of the canal EF above

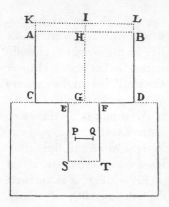

the little circle PQ bears to the circle AB. Then it is manifest (by Case 5, Case 6, and Cor. I, Prop. xxxvi) that the velocity of the water passing through the annular space between the little circle and the sides of the vessel will be the very same as that which the water would acquire by falling, and in its fall describing the altitude KC or IG.

And (by Cor. x, Prop. xxxvi) if the breadth of the vessel be infinite, so that the short line HI may vanish, and the altitudes IG, HG become equal; the force of the water that flows down and presses upon the circle will be to the weight of a cylinder whose base is that little circle, and the altitude ½IG, as EF^2 is to $EF^2 - \frac{1}{2}PQ^2$, very nearly. For the force of the water flowing downwards uniformly through the whole canal will be the same upon the little circle PQ in whatsoever part of the canal it be placed.

Let now the orifices of the canal EF, ST be closed, and let the little circle ascend in the fluid compressed on every side, and by its ascent let it oblige the water that lies above it to descend through the annular space between the little circle and the sides of the canal. Then will the velocity of the ascending little circle be to the velocity of the descending water as the difference of the circles EF and PQ is to the circle PQ; and the velocity of the ascending little circle will be to the sum of the velocities, that is, to the relative velocity of the descending water with which it passes by the little circle in its ascent, as the difference of the circles EF and PQ is to the circle EF, or as $EF^2 - PQ^2$ to EF^2. Let that relative velocity be equal to the velocity with which it was shown above that the water would pass through the annular space, if the circle were to remain unmoved, that is, to the velocity which the water would acquire by falling, and in its fall describing the altitude IG; and the force of the water upon the ascending circle will be the same as before (by Cor. v of the Laws of Motion); that is, the resistance of the ascending little circle will be to the weight of a cylinder of water whose base is that little circle, and its altitude ½IG, as EF^2 is to $EF^2 - ½PQ^2$, nearly. But the velocity of the little circle will be to the velocity which the water acquires by falling, and in its fall describing the altitude IG, as $EF^2 - PQ^2$ is to EF^2.

Let the breadth of the canal be increased *in infinitum*; and the ratios between $EF^2 - PQ^2$ and EF^2, and between EF^2 and $EF^2 - ½PQ^2$, will become at last ratios of equality. And therefore the velocity of the little circle will now be the same as that which the water would acquire in falling, and in its fall describing the altitude IG; and the resistance will become equal to the weight of a cylinder whose base is that little circle, and its altitude half the altitude IG, from which the cylinder must fall to acquire the velocity of the ascending circle; and with this velocity the cylinder in the time of its fall will describe four times its length. But the resistance of the cylinder moving forwards with this velocity in the direction of its length is the same with the resistance of the little circle (by Lem. iv), and is therefore nearly equal to the force by which its motion may be generated while it describes four times its length.

If the length of the cylinder be augmented or diminished, its motion, and the time in which it describes four times its length, will be augmented or diminished in the same ratio, and therefore the force by which the motion, so increased or diminished, may be destroyed or generated, will continue the same; because the time is increased or diminished in the same proportion; and therefore that force remains still equal to the resistance of the cylinder, because (by Lem. iv) that resistance will also remain the same.

If the density of the cylinder be augmented or diminished, its motion, and the force by which its motion may be generated or destroyed in the same time, will be augmented or diminished in the same ratio. Therefore the resistance of any cylinder whatsoever will be to the force by which its whole motion may be generated or destroyed, in the time during which it moves four times its length, as the density of the medium is to the density of the cylinder, nearly. Q.E.D.

A fluid must be compressed to become continued; it must be continued and nonelastic, that all the pressure arising from its compression may be propagated in an instant; and so, acting equally upon all parts of the body moved, may produce no change of the resistance. The pressure arising from the motion of the body is spent in generating a motion in the parts of the fluid, and this creates the resistance. But the pressure arising from the compression of the fluid, be it ever so forcible, if it be propagated in an instant, generates no motion in the parts of a continued fluid, produces no change at all of motion therein; and therefore neither augments nor lessens the resistance. This is certain, that the action of the fluid arising from the compression cannot be stronger on the hinder parts of the body moved than on its fore parts, and therefore cannot lessen the resistance described in this Proposition. And if its propagation be infinitely swifter than the motion of the body pressed, it will not be stronger on the fore parts than on the hinder parts. But that action will be infinitely swifter, and propagated in an instant, if the fluid be continued and nonelastic.

COR. I. The resistances, made to cylinders going uniformly forwards in the direction of their lengths through continued infinite mediums, are in a ratio compounded of the square of the ratio of the velocities and the square of the ratio of the diameters, and the ratio of the density of the mediums.

COR. II. If the breadth of the canal be not infinitely increased, but the cylinder go forwards in the direction of its length through an included quiescent medium, its axis all the while coinciding with the axis of the canal, its resistance will be to the force by which its whole motion, in the time in which it describes four times its length, may be generated or destroyed, in a ratio compounded of the ratio of EF^2 to $EF^2 - \frac{1}{2}PQ^2$, and the square of the ratio of EF^2 to $EF^2 - PQ^2$, and the ratio of the density of the medium to the density of the cylinder.

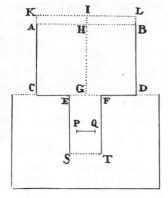

COR. III. The same thing supposed, and that a length L is to four times the length of the cylinder in a ratio compounded of the ratio $EF^2 - \frac{1}{2}PQ^2$ to EF^2, and the square of the ratio of $EF^2 - PQ^2$ to EF^2:

the resistance of the cylinder will be to the force by which its whole motion, in the time during which it describes the length L, may be destroyed or generated, as the density of the medium is to the density of the cylinder.

SCHOLIUM

In this Proposition we have investigated that resistance alone which arises from the magnitude of the transverse section of the cylinder, neglecting that part of the same which may arise from the obliquity of the motions. For as, in Case 1 of Prop. XXXVI, the obliquity of the motions with which the parts of the water in the vessel converged on every side to the hole EF hindered the efflux of the water through the hole, so, in this Proposition, the obliquity of the motions, with which the parts of the water, pressed by the antecedent extremity of the cylinder, yield to the pressure, and diverge on all sides, retards their passage through the places that lie round that antecedent extremity, towards the hinder parts of the cylinder, and causes the fluid to be moved to a greater distance; which increases the resistance, and that in the same ratio almost in which it diminished the efflux of the water out of the vessel, that is, in the squared ratio of 25 to 21, nearly. And as, in Case 1 of that Proposition, we made the parts of the water pass through the hole EF perpendicularly and in the greatest plenty, by supposing all the water in the vessel lying round the cataract to be frozen, and that part of the water whose motion was oblique and useless to remain without motion, so in this Proposition, that the obliquity of the motions may be taken away, and the parts of the water may give the freest passage to the cylinder, by yielding to it with the most direct and quick motion possible, so that only so much resistance may remain as arises from the magnitude of the transverse section, and as is incapable of diminution, unless by diminishing the diameter of the cylinder; we must conceive those parts of the fluid whose motions are oblique and useless, and produce resistance, to be at rest among themselves at both extremities of the cylinder, and there to cohere, and be joined to the cylinder.

Let ABCD be a rectangle, and let AE and BE be two parabolic arcs, described with the axis AB, and with a latus rectum that is to the space HG, which must be

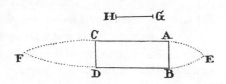

described by the cylinder in falling, in order to acquire the velocity with which it moves, as HG to ½AB. Let CF and DF be two other parabolic arcs described with the axis CD, and a latus rectum four times the former; and by the revolution of the figure about the axis EF let there be generated a solid, whose middle part ABDC is the cylinder we are here speaking of, and whose extreme parts ABE and CDF contain the parts of the fluid at rest among themselves, and concreted into two hard bodies, adhering to the cylinder at each end like a head and tail. Then if this solid EACFDB move in the direction of the length of its axis FE towards the parts beyond E, the resistance will be nearly the same as that which we have here determined in this Proposition; that is, it will have the same ratio to the force with which the whole motion of the cylinder may be destroyed or generated, in the time that it is describing the length 4AC with that motion uniformly continued, as the density of the fluid has to the density of the cylinder, nearly. And (by Cor. vii, Prop. xxxvi) the resistance must be to this force in the ratio of 2 to 3, at the least.

LEMMA V

If a cylinder, a sphere, and a spheroid, of equal breadths be placed successively in the middle of a cylindric canal, so that their axes may coincide with the axis of the canal, these bodies will equally hinder the passage of the water through the canal.

For the spaces lying between the sides of the canal, and the cylinder, sphere, and spheroid, through which the water passes, are equal; and the water will pass equally through equal spaces.

This is true, upon the supposition that all the water above the cylinder, sphere, or spheroid, whose fluidity is not necessary to make the passage of the water the quickest possible, is congealed, as was explained above in Cor. vii, Prop. xxxvi.

LEMMA VI

The same supposition remaining, the fore-mentioned bodies are equally acted on by the water flowing through the canal.

This appears by Lem. v and the third Law. For the water and the bodies act upon each other mutually and equally.

LEMMA VII

If the water be at rest in the canal, and these bodies move with equal velocity and in opposite directions through the canal, their resistances will be equal among themselves.

This appears from the last Lemma, for the relative motions remain the same among themselves.

SCHOLIUM

The case is the same for all convex and round bodies, whose axes coincide with the axis of the canal. Some difference may arise from a greater or less friction; but in these Lemmas we suppose the bodies to be perfectly smooth, and the medium to be void of all tenacity and friction; and that those parts of the fluid which by their oblique and superfluous motions may disturb, hinder, and retard the flux of the water through the canal, are at rest among themselves; being fixed like water by frost, and adhering to the fore and hinder parts of the bodies in the manner explained in the Scholium of the last Proposition; for in what follows we consider the very least resistance that round bodies described with the greatest given transverse sections can possibly meet with.

Bodies swimming upon fluids, when they move straight forwards, cause the fluid to ascend at their fore parts and subside at their hinder parts, especially if they are of an obtuse figure; and hence they meet with a little more resistance than if they were acute at the head and tail. And bodies moving in elastic fluids, if they are obtuse behind and before, condense the fluid a little more at their fore parts, and relax the same at their hinder parts; and therefore meet also with a little more resistance than if they were acute at the head and tail. But in these Lemmas and Propositions we are not treating of elastic but nonelastic fluids; not of bodies floating on the surface of the fluid, but deeply immersed therein. And when the resistance of bodies in nonelastic fluids is once known, we may then augment this resistance a little in elastic fluids, as our air; and in the surfaces of stagnating fluids, as lakes and seas.

PROPOSITION XXXVIII. THEOREM XXX

If a globe move uniformly forwards in a compressed, infinite, and non-elastic fluid, its resistance is to the force by which its whole motion may be destroyed or generated, in the time that it describes eight third parts of its diameter, as the density of the fluid is to the density of the globe, very nearly.

For the globe is to its circumscribed cylinder as 2 to 3; and therefore the force which can destroy all the motion of the cylinder, while the same cylinder is describing the length of four of its diameters, will destroy all the motion of the globe, while the globe is describing two-thirds of this length, that is, eight third parts of its own diameter. Now the resistance of the cylinder is to this force very nearly as the density of the fluid is to the density of the cylinder or globe (by Prop. xxxvii), and the resistance of the globe is equal to the resistance of the cylinder (by Lem. v, vi, vii). Q.E.D.

Cor. i. The resistances of globes in infinite compressed mediums are in a ratio compounded of the squared ratio of the velocity, and the squared ratio of the diameter, and the ratio of the density of the mediums.

Cor. ii. The greatest velocity, with which a globe can descend by its comparative weight through a resisting fluid, is the same as that which it may acquire by falling with the same weight, and without any resistance, and in its fall describing a space that is to four third parts of its diameter as the density of the globe is to the density of the fluid. For the globe in the time of its fall, moving with the velocity acquired in falling, will describe a space that will be to eight third parts of its diameter as the density of the globe is to the density of the fluid; and the force of its weight which generates this motion will be to the force that can generate the same motion, in the time that the globe describes eight third parts of its diameter, with the same velocity as the density of the fluid is to the density of the globe; and therefore (by this Proposition) the force of weight will be equal to the force of resistance, and therefore cannot accelerate the globe.

Cor. iii. If there be given both the density of the globe and its velocity at the beginning of the motion, and the density of the compressed quiescent fluid in which the globe moves, there is given at any time both the velocity of the globe and its resistance, and the space described by it (by Cor. vii, Prop. xxxv).

Cor. iv. A globe moving in a compressed quiescent fluid of the same density with itself will lose half its motion before it can describe the length of two of its diameters (by the same Cor. vii).

PROPOSITION XXXIX. THEOREM XXXI

If a globe move uniformly forwards through a fluid inclosed and compressed in a cylindric canal, its resistance is to the force by which its whole motion may be generated or destroyed, in the time in which it describes eight third parts of its diameter, in a ratio compounded of the ratio of the orifice of the canal to the excess of that orifice above half the greatest circle of the globe; and the squared ratio of the orifice of the canal to the excess of that orifice above the greatest circle of the globe; and the ratio of the density of the fluid to the density of the globe, nearly.

This appears by Cor. ii, Prop. xxxvii, and the demonstration proceeds in the same manner as in the foregoing Proposition.

SCHOLIUM

In the last two Propositions we suppose (as was done before in Lem. v) that all the water which precedes the globe, and whose fluidity increases the resistance of the same, is congealed. Now if that water becomes fluid, it will somewhat increase the resistance. But in these Propositions that increase is so small, that it may be neglected, because the convex surface of the globe produces the very same effect almost as the congelation of the water.

PROPOSITION XL. PROBLEM IX

To find by experiment the resistance of a globe moving through a perfectly fluid compressed medium.

Let A be the weight of the globe in a vacuum, B its weight in the resisting medium, D the diameter of the globe, F a space which is to ⅓D as the density of the globe is to the density of the medium, that is, as A is to A − B, G the time in which the globe falling with the weight B without resistance describes the space F, and H the velocity which the body acquires by that fall. Then H will be the greatest velocity with which the globe can possibly descend with the weight B in the resisting medium, by Cor. ii, Prop. xxxviii; and the resistance which the globe meets with, when descending with that velocity, will be equal to its weight B; and the resistance it meets with in any other velocity will be to the weight B as the square of the ratio of that velocity to the greatest velocity H, by Cor. I, Prop. XXXVIII.

This is the resistance that arises from the inactivity of the matter of the fluid. That resistance which arises from the elasticity, tenacity, and friction of its parts, may be thus investigated.

Let the globe be let fall so that it may descend in the fluid by the weight B; and let P be the time of falling, and let that time be expressed in seconds, if the time G be given in seconds. Find the absolute number N agreeing to the logarithm 0.4342944819 $\frac{2P}{G}$, and let L be the logarithm of the number $\frac{N+I}{N}$; and the velocity acquired in falling will be $\frac{N-I}{N+I}$ H, and the height described will be $\frac{2PF}{G}$ − 1.3862943611F + 4.605170186LF. If the fluid be of a sufficient depth, we may neglect the term 4.605170186LF; and $\frac{2PF}{G}$ − 1.3862943611F will be the altitude described, nearly. These things appear by Prop. IX, Book II, and its Corollaries, and are true upon this supposition, that the globe meets with no other resistance but that which arises from the inactivity of matter. Now if it really meet with any resistance of another kind, the descent will be slower, and from the amount of that retardation will be known the amount of this new resistance.

That the velocity and descent of a body falling in a fluid might more easily be known, I have composed the following table (p. 355), the first column of which denotes the times of descent; the second shows the velocities acquired in falling, the greatest velocity being 100000000; the third exhibits the spaces described by falling in those times, 2F being the space which the body describes in the time G with the greatest velocity; and the fourth gives the spaces described with the greatest velocity in the same times. The numbers in the fourth column are $\frac{2P}{G}$, and by subtracting the number 1.3862944 − 4.6051702L, are found the numbers in the third column; and these numbers must be multiplied by the space F to obtain the spaces described in falling. A fifth column is added to all these, containing the spaces described in the same times by a body falling in a vacuum with the force of B its comparative weight.

The Times P	Velocities of the body falling in the fluid	The spaces described in falling in the fluid	The spaces described with the greatest motion	The spaces described by falling in a vacuum
0.001G	$99999^{29}/_{30}$	0.000001F	0.002F	0.000001F
0.01G	999967	0.0001F	0.02F	0.0001F
0.1G	9966799	0.0099834F	0.2F	0.01F
0.2G	19737532	0.0397361F	0.4F	0.04F
0.3G	29131261	0.0886815F	0.6F	0.09F
0.4G	37994896	0.1559070F	0.8F	0.16F
0.5G	46211716	0.2402290F	1.0F	0.25F
0.6G	53704957	0.3402706F	1.2F	0.36F
0.7G	60436778	0.4545405F	1.4F	0.49F
0.8G	66403677	0.5815071F	1.6F	0.64F
0.9G	71629787	0.7196609F	1.8F	0.81F
1G	76159416	0.8675617F	2F	1F
2G	96402758	2.6500055F	4F	4F
3G	99505475	4.6186570F	6F	9F
4G	99932930	6.6143765F	8F	16F
5G	99990920	8.6137964F	10F	25F
6G	99998771	10.6137179F	12F	36F
7G	99999834	12.6137073F	14F	49F
8G	99999980	14.6137059F	16F	64F
9G	99999997	16.6137057F	18F	81F
10G	$99999999^{3}/_{5}$	18.6137056F	20F	100F

SCHOLIUM

In order to investigate the resistances of fluids from experiments, I procured a square wooden vessel, whose length and breadth on the inside was 9 inches *English* measure, and its depth 9½ feet; this I filled with rain water; and having provided globes made up of wax, and lead included therein, I noted the times of the descents of these globes, the height through which they descended being 112 inches. A solid cubic foot of *English* measure contains 76 pounds troy weight of rain water; and a solid inch contains $^{19}/_{36}$ ounces troy weight, or 253⅓ grains; and a globe of water of one inch in diameter contains 132.645 grains in air, or 132.8 grains in a vacuum; and any other globe will be as the excess of its weight in a vacuum above its weight in water.

EXPER. 1. A globe whose weight was 156¼ grains in air, and 77 grains in water, described the whole height of 112 inches in 4 seconds. And, upon repeating the experiment, the globe spent again the very same time of 4 seconds in falling.

The weight of this globe in a vacuum is 156¹³⁄₃₈ grains; and the excess of this weight above the weight of the globe in water is 79¹³⁄₃₈ grains. Hence the diameter of the globe appears to be 0.84224 parts of an inch. Then it will be, as that excess to the weight of the globe in a vacuum, so is the density of the water to the density of the globe; and so is ⅔ parts of the diameter of the globe (viz., 2.24597 inches) to the space 2F, which will be therefore 4.4256 inches. Now a globe falling in a vacuum with its whole weight of 156¹³⁄₃₈ grains in one second of time will describe 193⅓ inches; and falling in water in the same time with the weight of 77 grains without resistance, will describe 95.219 inches; and in the time G, which is to one second of time as the square root of the ratio of the space F, or of 2.2128 inches to 95.219 inches, will describe 2.2128 inches, and will acquire the greatest velocity H with which it is capable of descending in water. Therefore the time G is 0.15244 sec. And in this time G, with that greatest velocity H, the globe will describe the space 2F, which is 4.4256 inches; and therefore in 4 seconds will describe a space of 116.1245 inches. Subtract the space 1.3862944 · F, or 3.0676 inches, and there will remain a space of 113.0569 inches, which the globe falling through water in a very wide vessel will describe in 4 seconds. But this space, by reason of the narrowness of the wooden vessel before mentioned, ought to be diminished in a ratio compounded of the square root of the ratio of the orifice of the vessel to the excess of this orifice above half a great circle of the globe, and of the simple ratio of the same orifice to its excess above a great circle of the globe, that is, in a ratio of 1 to 0.9914. This done, we have a space of 112.08 inches, which a globe falling through the water in this wooden vessel in 4 seconds of time ought nearly to describe by this theory; but it described 112 inches by the experiment.

EXPER. 2. Three equal globes, whose weights were severally 76⅓ grains in air, and 5¹⁄₁₆ grains in water, were let fall successively; and every one fell through the water in 15 seconds of time, describing in its fall a height of 112 inches.

By computation, the weight of each globe in a vacuum is 76$\frac{5}{12}$ grains; the excess of this weight above the weight in water is 71$\frac{17}{48}$ grains; the diameter of the globe 0.81296 of an inch; $\frac{2}{3}$ parts of this diameter 2.16789 inches; the space 2F is 2.3217 inches; the space which a globe of 5$\frac{1}{16}$ grains in weight would describe in one second without resistance, 12.808 inches, and the time G 0.301056 sec. Therefore the globe, with the greatest velocity it is capable of receiving from a weight of 5$\frac{1}{16}$ grains in its descent through water, will describe in the time 0.301056 seconds the space 2.3217 inches; and in 15 seconds the space 115.678 inches. Subtract the space 1.3862944F, or 1.609 inches, and there remains the space 114.069 inches; which therefore the falling globe ought to describe in the same time, if the vessel were very wide. But because our vessel was narrow, the space ought to be diminished by about 0.895 of an inch. And so the space will remain 113.174 inches, which a globe falling in this vessel ought nearly to describe in 15 seconds. But by the experiment it described 112 inches. The difference is not sensible.

EXPER. 3. Three equal globes, whose weights were severally 121 grains in air, and 1 grain in water, were successively let fall; and they fell through the water in the times 46 seconds, 47 seconds, and 50 seconds, describing a height of 112 inches.

By the theory, these globes ought to have fallen in about 40 sec. Now whether their falling more slowly were occasioned from the consideration that in slow motions the resistance arising from the force of inactivity does really bear a less proportion to the resistance arising from other causes; or whether it is to be attributed to little bubbles that might chance to stick to the globes, or to the rarefaction of the wax by the warmth of the weather, or of the hand that let them fall; or, lastly, whether it proceeded from some insensible errors in weighing the globes in the water, I am not certain. Therefore the weight of the globe in water should be of several grains, that the experiment may be certain, and to be depended on.

EXPER. 4. I began the foregoing Experiments to investigate the resistances of fluids, before I was acquainted with the theory laid down in the Propositions immediately preceding. Afterwards, in order to examine the theory after it was discovered, I procured a wooden vessel, whose breadth on the inside was 8$\frac{2}{3}$ inches, and its depth 15$\frac{1}{3}$ feet. Then I made four globes of wax, with lead included, each of which weighed 139$\frac{1}{4}$ grains in air, and

7⅛ grains in water. These I let fall, measuring the times of their falling in the water with a pendulum oscillating to half-seconds. The globes were cold, and had remained so some time, both when they were weighed and when they were let fall; because warmth rarefies the wax, and by rarefying it diminishes the weight of the globe in the water; and wax, when rarefied, is not instantly reduced by cold to its former density. Before they were let fall, they were totally immersed under water, lest, by the weight of any part of them that might chance to be above the water, their descent should be accelerated in its beginning. Then, when after their immersion they were perfectly at rest, they were let go with the greatest care, that they might not receive any impulse from the hand that let them down. And they fell successively in the times of 47½, 48½, 50, and 51 oscillations, describing a height of 15 feet and 2 inches. But the weather was now a little colder than when the globes were weighed, and therefore I repeated the experiment another day; and then the globes fell in the times of 49, 49½, 50, and 53; and at a third trial in the times of 49½, 50, 51, and 53 oscillations. And by making the experiment several times over, I found that the globes fell mostly in the times of 49½ and 50 oscillations. When they fell slower, I suspect them to have been retarded by striking against the sides of the vessel.

Now, computing from the theory, the weight of the globe in a vacuum is 139⅖ grains; the excess of this weight above the weight of the globe in water 132¹¹⁄₄₀ grains; the diameter of the globe 0.99868 of an inch; ⅔ parts of the diameter 2.66315 inches; the space 2F 2.8066 inches; the space which a globe weighing 7⅛ grains falling without resistance describes in a second of time 9.88164 inches; and the time G 0.376843 sec. Therefore the globe with the greatest velocity with which it is capable of descending through the water by the force of a weight of 7⅛ grains, will in the time 0.376843 sec. describe a space of 2.8066 inches, and in one second of time a space of 7.44766 inches, and in the time 25 sec., or in 50 oscillations, the space 186.1915 inches. Subtract the space 1.386294F, or 1.9454 inches, and there will remain the space 184.2461 inches which the globe will describe in that time in a very wide vessel. Because our vessel was narrow, let this space be diminished in a ratio compounded of the square root of the ratio of the orifice of the vessel to the excess of this orifice above half a great circle of the globe, and of the simple ratio of the same orifice to its excess above a great circle of the globe; and we shall have the space of 181.86 inches, which the globe ought by the theory to describe in this vessel in the time of 50 oscillations, nearly. But it described the space of 182 inches, by experiment, in 49½ or 50 oscillations.

EXPER. 5. Four globes weighing 154⅜ grains in air, and 21½ grains in water, being let fall several times, fell in the times of 28½, 29, 29½, and 30, and sometimes of 31, 32, and 33 oscillations, describing a height of 15 feet and 2 inches.

They ought by the theory to have fallen in the time of 29 oscillations, nearly.

EXPER. 6. Five globes, weighing 212⅜ grains in air, and 79½ in water, being several times let fall, fell in the times of 15, 15½, 16, 17, and 18 oscillations, describing a height of 15 feet and 2 inches.

By the theory they ought to have fallen in the time of 15 oscillations, nearly.

EXPER. 7. Four globes, weighing 293⅜ grains in air, and 35⅞ grains in water, being let fall several times, fell in the times of 29½, 30, 30½, 31, 32, and 33 oscillations, describing a height of 15 feet and 1½ inches.

By the theory they ought to have fallen in the time of 28 oscillations, nearly.

In searching for the cause that occasioned these globes of the same weight and magnitude to fall, some swifter and some slower, I hit upon this: that the globes, when they were first let go and began to fall, oscillated about their centres; that side which chanced to be the heavier descending first, and producing an oscillating motion. Now by oscillating thus, the globe communicates a greater motion to the water than if it descended without any oscillations; and by this communication loses part of its own motion with which it should descend; and therefore as this oscillation is greater or less, it will be more or less retarded. Besides, the globe always recedes from that side of itself which is descending in the oscillation, and by so receding comes nearer to the sides of the vessel, so as even to strike against them sometimes. And the heavier the globes are, the stronger this oscillation is; and the greater they are, the more is the water agitated by it. Therefore to diminish this oscillation of the globes, I made new ones of lead and wax, sticking the lead in one side of the globe very near its surface; and I let fall the globe in such a manner, that, as near as possible, the heavier side might be lowest at the beginning of the descent. By this means the oscillations became much less than before, and the times in which the globes fell were not so unequal: as in the following Experiments.

EXPER. 8. Four globes weighing 139 grains in air, and 6½ in water, were let fall several times, and fell mostly in the time of 51 oscillations, never in more than 52, or in fewer than 50, describing a height of 182 inches.

By the theory they ought to fall in about the time of 52 oscillations.

EXPER. 9. Four globes weighing 273¼ grains in air, and 140¾ in water, being several times let fall, fell in never fewer than 12, and never more than 13 oscillations, describing a height of 182 inches.

These globes by the theory ought to have fallen in the time of 11⅓ oscillations, nearly.

EXPER. 10. Four globes, weighing 384 grains in air, and 119½ in water, being let fall several times, fell in the times of 17¾, 18, 18½, and 19 oscillations, describing a height of 181½ inches. And when they fell in the time of 19 oscillations, I sometimes heard them hit against the sides of the vessel before they reached the bottom.

By the theory they ought to have fallen in the time of 15⁵⁄₉ oscillations, nearly.

EXPER. 11. Three equal globes, weighing 48 grains in air, and 3²⁹⁄₃₂ in water, being several times let fall, fell in the times of 43½, 44, 44½, 45, and 46 oscillations, and mostly in 44 and 45, describing a height of 182½ inches, nearly.

By the theory they ought to have fallen in the time of 46⁵⁄₉ oscillations, nearly.

EXPER. 12. Three equal globes, weighing 141 grains in air, and 4⅜ in water, being let fall several times, fell in the times of 61, 62, 63, 64, and 65 oscillations, describing a space of 182 inches.

And by the theory they ought to have fallen in 64½ oscillations, nearly.

From these Experiments it is manifest, that when the globes fell slowly, as in the second, fourth, fifth, eighth, eleventh, and twelfth Experiments, the times of falling are rightly exhibited by the theory; but when the globes fell more swiftly, as in the sixth, ninth, and tenth Experiments, the resistance was somewhat greater than the square of the velocity. For the globes in falling oscillate a little; and this oscillation, in those globes that are light and fall slowly, soon ceases by the weakness of the motion; but in greater and heavier globes, the motion being strong, it continues longer, and is not to be checked by the ambient water till after several oscillations. Besides, the more swiftly the globes move, the less are they pressed by the fluid at their hinder parts; and if the velocity be continually increased, they will at last leave an empty space behind them, unless the compression of the fluid be increased at the same time. For the compression of the fluid ought to be increased (by Prop. XXXII and XXXIII) as the square of the velocity, in order to maintain the resistance in the same squared ratio. But because this is not done, the globes that move swiftly are not so much pressed at their hinder parts as the others; and by the defect of this pressure it comes to pass that their resistance is a little greater than the square of their velocity.

So that the theory agrees with the experiments on bodies falling in water. It remains that we examine the observations of bodies falling in air.

EXPER. 13. From the top of St. *Paul's* Church in *London,* in *June,* 1710, there were let fall together two glass globes, one full of quicksilver, the other of air; and in their fall they described a height of 220 *English* feet. A wooden table was suspended upon iron hinges on one side, and the other side of the table was supported by a wooden pin. The two globes lying upon this table were let fall together by pulling out the pin by means of an iron wire reaching thence down to the ground; so that, the pin being removed, the table, which had then no support but the iron hinges, fell downwards, and turning round upon the hinges, gave leave to the globes to drop off from it. At the same instant, with the same pull of the iron wire that took out the pin, a pendulum oscillating to seconds was let go, and began to oscillate. The diameters and weights of the globes, and their times of falling, are exhibited in the accompanying table (p. 362).

But the times observed must be corrected; for the globes of mercury (by *Galileo's* theory), in 4 seconds of time, will describe 257 *English* feet, and 220 feet in only 3 sec. 42 thirds.[1] So that the wooden table, when the pin was taken out, did not turn upon its hinges so quickly as it ought to have done; and the slowness of that revolution hindered the descent of the globes at the beginning. For the globes lay about the middle of the table, and indeed were rather nearer to the axis upon which it turned than to the pin. And

[1 Appendix, Note 37.]

hence the times of falling were prolonged about 18 thirds;[1] and therefore ought to be corrected by subtracting that excess, especially in the larger globes, which, by reason of the largeness of their diameters, lay longer upon the revolving table than the others. This being done, the times in which the six larger globes fell will come forth 8 sec. 12 thirds, 7 sec. 42 thirds, 7 sec. 42 thirds, 7 sec. 57 thirds, 8 sec. 12 thirds, and 7 sec. 42 thirds.

The globes filled with mercury			The globes full of air		
Weights	Diameters	Times in falling	Weights	Diameters	Times in falling
grains	inches	seconds	grains	inches	seconds
908	0.8	4	510	5.1	$8\frac{1}{2}$
983	0.8	4 −	642	5.2	8
866	0.8	4	599	5.1	8
747	0.75	4 +	515	5.0	$8\frac{1}{4}$
808	0.75	4	483	5.0	$8\frac{1}{2}$
784	0.75	4 +	641	5.2	8

Therefore the fifth in order among the globes that were full of air being 5 inches in diameter, and 483 grains in weight, fell in 8 sec. 12 thirds, describing a space of 220 feet. The weight of a bulk of water equal to this globe is 16600 grains; and the weight of an equal bulk of air is $\frac{16600}{860}$ grains, or 19⁹⁄₁₀ grains; and therefore the weight of the globe in a vacuum is 502³⁄₁₀ grains; and this weight is to the weight of a bulk of air equal to the globe as 502³⁄₁₀ is to 19⁹⁄₁₀; and so is 2F to ⅔ of the diameter of the globe, that is, to 13⅓ inches. Hence 2F becomes 28 feet 11 inches. A globe, falling in a vacuum with its whole weight of 502³⁄₁₀ grains, will in one second of time describe 193⅛ inches as above; and with the weight 483 grains will describe 185.905 inches; and with that weight 483 grains in a vacuum will describe the space F, or 14 feet 5½ inches, in the time of 57 thirds and 58 fourths, and acquire the greatest velocity it is capable of descending with in the air. With this velocity the globe in 8 sec. 12 thirds of time will describe 245 feet and 5⅓ inches. Subtract 1.3863·F, or 20 feet and ½ an inch, and there remain 225 feet 5 inches. This space, therefore, the falling globe ought by the theory to describe in 8 sec. 12 thirds. But by the experiment it described a space of 220 feet. The difference is inappreciable.

[1 Appendix, Note 37.]

By like calculations applied to the other globes full of air, I composed the following table.

The weights of the globes	The diameters	The times of falling from a height of 220 feet		The spaces which they would describe by the theory		The excesses	
grains	inches	seconds	thirds	feet	inches	feet	inches
510	5.1	8	12	226	11	6	11
642	5.2	7	42	230	9	10	9
599	5.1	7	42	227	10	7	0
515	5	7	57	224	5	4	5
483	5	8	12	225	5	5	5
641	5.2	7	42	230	7	10	7

EXPER. 14. In the year 1719, in the month of *July*, Dr. *Desaguliers* made some experiments of this kind again, by forming hogs' bladders into spherical orbs; which was done by means of a concave wooden sphere, which the bladders, being wetted well first, were put into. After that, being blown full of air, they were obliged to fill up the spherical cavity that contained them; and then, when dry, were taken out. These were let fall from the lantern on the top of the cupola of the same church, namely, from a height of 272 feet; and at the same moment of time there was let fall a leaden globe, whose weight was about 2 pounds troy weight. And in the meantime some persons standing in the upper part of the church where the globes were let fall observed the whole times of falling; and others standing on the ground observed the differences of the times between the fall of the leaden weight and the fall of the bladder. The times were measured by pendulums oscillating to half-seconds. And one of those that stood upon the ground had a machine vibrating four times in one second; and another had another machine accurately made with a pendulum vibrating four times in a second also. One of those also who stood at the top of the church had a like machine; and these instruments were so contrived, that their motions could be stopped or renewed at pleasure. Now the leaden globe fell in about 4¼ seconds of time; and from the addition of this time to the difference of time above spoken of, was obtained the whole time in which the bladder was falling. The times which the five bladders spent in falling, after the leaden globe had reached the ground, were, the first time, 14¾ sec., 12¾

sec., 14⅝ sec., 17¾ sec., and 16⅞ sec.; and the second time, 14½ sec., 14¼ sec., 14 sec., 19 sec., and 16¾ sec. Add to these 4¼ sec., the time in which the leaden globe was falling, and the whole times in which the five bladders fell were, the first time, 19 sec., 17 sec., 18⅞ sec., 22 sec., and 21⅛ sec.; and the second time, 18¾ sec., 18½ sec., 18¼ sec., 23¼ sec., and 21 sec. The times observed at the top of the church were, the first time, 19⅜ sec., 17¼ sec., 18¾ sec., 22⅛ sec., and 21⅝ sec.; and the second time, 19 sec., 18⅝ sec., 18⅜ sec., 24 sec., and 21¼ sec. But the bladders did not always fall directly down, but sometimes fluttered a little in the air, and waved to and fro, as they were descending. And by these motions the times of their falling were prolonged, and increased by half a second sometimes, and sometimes by a whole second. The second and fourth bladders fell most directly the first time, and the first and third the second time. The fifth bladder was wrinkled, and by its wrinkles was a little retarded. I found their diameters by their circumferences measured with a very fine thread wound about them twice. In the following table I have compared the experiments with the theory; making the density of air to be to the density of rain water as 1 to 860, and computing the spaces which by the theory the globes ought to describe in falling.

The weights of the bladders	The diameters	The times of falling from a height of 272 feet	The spaces which by the theory ought to have been described in those times		The difference between the theory and the experiments	
grains	inches	seconds	feet	inches	feet	inches
128	5.28	19	271	11	− 0	1
156	5.19	17	272	0½	+ 0	0½
137½	5.3	18	272	7	+ 0	7
97½	5.26	22	277	4	+ 5	4
99⅛	5	21⅛	282	0	+10	0

Our theory, therefore, exhibits rightly, within a very little, all the resistance that globes moving either in air or in water meet with; which appears to be proportional to the densities of the fluids in globes of equal velocities and magnitudes.

In the Scholium subjoined to the sixth Section, we showed, by experiments of pendulums, that the resistances of equal and equally swift globes moving in air, water, and quicksilver, are as the densities of the fluids. We here prove the same more accurately by experiments of bodies falling in air and water. For pendulums at each oscillation excite a motion in the fluid always contrary to the motion of the pendulum in its return; and the resistance arising from this motion, as also the resistance of the thread by which the pendulum is suspended, makes the whole resistance of a pendulum greater than the resistance deduced from the experiments of falling bodies. For by the experiments of pendulums described in that Scholium, a globe of the same density as water in describing the length of its semidiameter in air would lose the $\frac{1}{3342}$ part of its motion. But by the theory delivered in this seventh Section, and confirmed by experiments of falling bodies, the same globe in describing the same length would lose only a part of its motion equal to $\frac{1}{4586}$, supposing the density of water to be to the density of air as 860 to 1. Therefore the resistances were found greater by the experiments of pendulums (for the reasons just mentioned) than by the experiments of falling globes; and that in the ratio of about 4 to 3. But yet since the resistances of pendulums oscillating in air, water, and quicksilver, are alike increased by like causes, the proportion of the resistances in these mediums will be rightly enough exhibited by the experiments of pendulums, as well as by the experiments of falling bodies. And from all this it may be concluded, that the resistances of bodies, moving in any fluids whatsoever, though of the most extreme fluidity, are, other things being equal, as the densities of the fluids.

These things being thus established, we may now determine what part of its motion any globe projected in any fluid whatsoever would nearly lose in a given time. Let D be the diameter of the globe, and V its velocity at the beginning of its motion, and T the time in which a globe with the velocity V can describe in a vacuum a space that is to the space $\frac{8}{3}$D as the density of the globe to the density of the fluid; and the globe projected in that fluid will, in any other time t lose the part $\frac{tV}{T+t}$, the part $\frac{TV}{T+t}$ remaining; and will describe a space, which will be to that described in the same time in a vacuum with the uniform velocity V, as the logarithm of the number $\frac{T+t}{T}$ multiplied by the number 2.302585093 is to the number $\frac{t}{T}$ by Cor. VII,

Prop. xxxv. In slow motions the resistance may be a little less, because the figure of a globe is more adapted to motion than the figure of a cylinder described with the same diameter. In swift motions the resistance may be a little greater, because the elasticity and compression of the fluid do not increase as the square of the velocity. But these little niceties I take no notice of.

And though air, water, quicksilver, and the like fluids, by the division of their parts *in infinitum,* should be subtilized, and become mediums infinitely fluid, nevertheless, the resistance they would make to projected globes would be the same. For the resistance considered in the preceding Propositions arises from the inactivity of the matter; and the inactivity of matter is essential to bodies, and always proportional to the quantity of matter. By the division of the parts of the fluid the resistance arising from the tenacity and friction of the parts may be indeed diminished; but the quantity of matter will not be at all diminished by this division; and if the quantity of matter be the same, its force of inactivity will be the same; and therefore the resistance here spoken of will be the same, as being always proportional to that force. To diminish this resistance, the quantity of matter in the spaces through which the bodies move must be diminished; and therefore the celestial spaces, through which the globes of the planets and comets are continually passing towards all parts, with the utmost freedom, and without the least sensible diminution of their motion, must be utterly void of any corporeal fluid, excepting, perhaps, some extremely rare vapors and the rays of light.

Projectiles excite a motion in fluids as they pass through them, and this motion arises from the excess of the pressure of the fluid at the fore parts of the projectile above the pressure of the same at the hinder parts; and cannot be less in mediums infinitely fluid than it is in air, water, and quicksilver, in proportion to the density of matter in each. Now this excess of pressure does, in proportion to its quantity, not only excite a motion in the fluid, but also acts upon the projectile so as to retard its motion; and therefore the resistance in every fluid is as the motion excited by the projectile in the fluid; and cannot be less in the most subtile ether in proportion to the density of that ether, than it is in air, water, and quicksilver, in proportion to the densities of those fluids.

SECTION VIII

The motion propagated through fluids.

PROPOSITION XLI. THEOREM XXXII

A pressure is not propagated through a fluid in rectilinear directions except where the particles of the fluid lie in a right line.

If the particles *a, b, c, d, e* lie in a right line, the pressure may be indeed directly propagated from *a* to *e*; but then the particle *e* will urge the obliquely posited particles *f* and *g* obliquely, and those particles *f* and *g* will not sustain this pressure, unless they be supported by the particles *h* and *k* lying beyond them; but the particles that support them are also pressed by them; and those particles cannot sustain that pressure, without being supported by, and pressing upon, those particles that lie still farther, as *l* and *m,* and so on *in infinitum.* Therefore the pressure, as soon as it is propagated to particles that lie out of right lines,

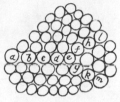

begins to deflect towards one hand and the other, and will be propagated obliquely *in infinitum;* and after it has begun to be propagated obliquely, if it reaches more distant particles lying out of the right line, it will deflect again on each hand; and this it will do as often as it lights on particles that do not lie exactly in a right line. Q.E.D.

Cor. If any part of a pressure, propagated through a fluid from a given point, be intercepted by any obstacle, the remaining part, which is not intercepted, will deflect into the spaces behind the obstacle. This may be demonstrated also after the following manner. Let a pressure be propagated from the point A towards any part, and, if it be possible, in rectilinear directions; and the obstacle NBCK being perforated in BC, let all the pressure be intercepted but the coniform part APQ passing through the circular hole BC. Let the cone APQ be divided into frustums by the transverse planes, *de, fg, hi.* Then while the cone ABC, propagating the pressure, urges the conic frustum *degf* beyond it on the surface *de,* and this frustum urges the next frustum *fgih* on the surface *fg,* and that frustum urges a third frustum, and so *in infinitum;* it is manifest (by the third Law) that the first frustum *defg* is, by the reaction of the second frustum *fghi,* as much urged and pressed on the surface *fg,* as it urges and presses that second frustum. Therefore the frustum *degf* is compressed on both sides, that is, between the cone A*de* and the frustum *fhig;* and therefore (by Case 6, Prop. xix) cannot preserve its figure, unless it be compressed with the same force on all sides. Therefore with the same force with which it is pressed on the surfaces *de,*

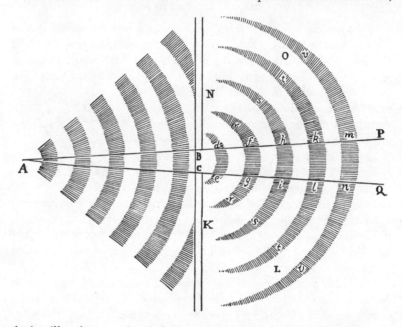

fg, it will endeavor to break forth at the sides *df, eg;* and there (being not in the least tenacious or hard, but perfectly fluid) it will run out, expanding itself, unless there be an ambient fluid opposing that endeavor. Therefore, by the effort it makes to run out, it will press the ambient fluid, at its sides *df, eg,* with the same force that it does the frustum *fghi;* and therefore, the pressure will be propagated as much from the sides *df, eg,* into the spaces NO, KL this way and that way, as it is propagated from the surface *fg* towards PQ. Q.E.D.

PROPOSITION XLII. THEOREM XXXIII

All motion propagated through a fluid diverges from a rectilinear progress into the unmoved spaces.

CASE I. Let a motion be propagated from the point A through the hole BC, and, if it be possible, let it proceed in the conic space BCQP according to right lines diverging from the point A. And let us first suppose this motion to be that of waves in the surface of standing water; and let *de, fg, hi, kl,* &c., be the tops of the several waves, divided from each other by as

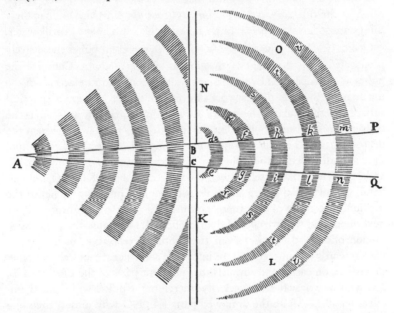

many intermediate valleys or hollows. Then, because the water in the ridges of the waves is higher than in the unmoved parts of the fluid KL, NO, it will run down from off the tops of those ridges, *e, g, i, l,* &c., *d, f, h, k,* &c., this way and that way towards KL and NO; and because the water is more depressed in the hollows of the waves than in the unmoved parts of the fluid KL, NO, it will run down into those hollows out of those unmoved parts. By the first deflux the ridges of the waves will dilate themselves this way and that way, and be propagated towards KL and NO. And because the motion of the waves from A towards PQ is carried on by a continual deflux from the ridges of the waves into the hollows next to them, and therefore cannot be swifter than in proportion to the celerity of the descent; and the descent of the water on each side towards KL and NO must be performed with the same velocity: it follows that the dilatation of the waves on each side towards KL and NO will be propagated with the same velocity as the waves themselves go forwards with directly from A to PQ. And therefore the whole space this way and that way towards KL and NO will be filled by the dilated waves *rfgr, shis, tklt, vmnv,* &c. Q.E.D. That these things are so, anyone may find by making the experiment in still water.

Case 2. Let us suppose that *de, fg, hi, kl, mn* represent pulses successively propagated from the point A through an elastic medium. Conceive the pulses to be propagated by successive condensations and rarefactions of the medium, so that the densest part of every pulse may occupy a spherical surface described about the centre A, and that equal intervals intervene between the successive pulses. Let the lines *de, fg, hi, kl,* &c., represent the densest parts of the pulses, propagated through the hole BC; and because the medium is denser there than in the spaces on either side towards KL and NO, it will dilate itself as well towards those spaces KL, NO, on each hand, as towards the rare intervals between the pulses; and hence the medium, becoming always more rare next the intervals, and more dense next the pulses, will partake of their motion. And because the progressive motion of the pulses arises from the continual relaxation of the denser parts towards the antecedent rare intervals; and since the pulses will relax themselves on each hand towards the quiescent parts of the medium KL, NO with very near the same celerity; therefore the pulses will dilate themselves on all sides into the unmoved parts KL, NO with almost the same celerity with which they are propagated directly from the centre A; and therefore will fill up the whole space KLON. Q.E.D. And we find the same by experience also in sounds which are heard through a mountain interposed; and, if they come into a chamber through the window, dilate themselves into all the parts of the room, and are heard in every corner; and not as reflected from the opposite walls, but directly propagated from the window, as far as our sense can judge.

CASE 3. Let us suppose, lastly, that a motion of any kind is propagated from A through the hole BC. Then since the cause of this propagation is that the parts of the medium that are near the centre A disturb and agitate those which lie farther from it; and since the parts which are urged are fluid, and therefore recede every way towards those spaces where they are less pressed: they will by consequence recede towards all the parts of the quiescent medium, as well to the parts on each hand, as KL and NO, as to those right before, as PQ; and by this means all the motion, as soon as it has passed through the hole BC, will begin to dilate itself, and from thence, as from its principle and centre, will be propagated directly every way. Q.E.D.

PROPOSITION XLIII. THEOREM XXXIV

Every tremulous body in an elastic medium propagates the motion of the pulses on every side straight forwards; but in a nonelastic medium excites a circular motion.

CASE 1. The parts of the tremulous body, alternately going and returning, do in going urge and drive before them those parts of the medium that lie nearest, and by that impulse compress and condense them; and in returning suffer those compressed parts to recede again, and expand themselves. Therefore the parts of the medium that lie nearest to the tremulous body move to and fro by turns, in like manner as the parts of the tremulous body itself do; and for the same cause that the parts of this body agitate these parts of the medium, these parts, being agitated by like tremors, will in their turn agitate others next to themselves; and these others, agitated in like manner, will agitate those that lie beyond them, and so on *in infinitum*. And in the same manner as the first parts of the medium were condensed in going, and relaxed in returning, so will the other parts be condensed every time they go, and expand themselves every time they return. And therefore they will not be all going and all returning at the same instant (for in that case they would always maintain determined distances from each other, and there could be no alternate condensation and rarefaction); but since, in the places where they are condensed, they approach to, and, in the places where they are rarefied, recede from each other, therefore some of them will be going while others are returning; and so on *in infinitum*. The parts so going, and in their going condensed, are pulses, by reason of the progressive motion with which they strike obstacles in their way; and therefore the successive pulses produced by a tremulous body will be propagated in rectilinear directions; and that at nearly equal distances from each other, because of the equal intervals of time in which the body, by its several tremors, produces the several pulses. And though the parts of the tremulous body go and return in some certain and determinate direction, yet the pulses propagated from thence through the medium will dilate themselves towards the sides, by the foregoing Proposition; and will be propagated on all sides from that tremulous body, as from a common centre, in surfaces nearly spherical and concentric, as in waves excited by shaking a finger in water, which proceed not only forwards and backwards agreeably to the motion of the finger, but spread themselves in the manner of concentric circles all round the finger, and are propagated on every side. For the gravity of the water supplies the place of elastic force.

CASE 2. If the medium be not elastic, then, because its parts cannot be condensed by the pressure arising from the vibrating parts of the tremulous body, the motion will be propagated in an instant towards the parts where the medium yields most easily, that is, to the parts which the tremulous body would otherwise leave vacuous behind it. The case is the same with that of a body projected in any medium whatever. A medium yielding to projectiles does not recede *in infinitum,* but with a circular motion comes round to the spaces which the body leaves behind it. Therefore as often as a tremulous body tends to any part, the medium yielding to it comes round in a circle to the parts which the body leaves; and as often as the body returns to the first place, the medium will be driven from the place it came round to, and return to its original place. And though the tremulous body be not firm and hard, but every way flexible, yet if it continue of a given magnitude, since it cannot impel the medium by its tremors anywhere without yielding to it somewhere else, the medium receding from the parts of the body where it is pressed will always come round in a circle to the parts that yield to it. Q.E.D.

COR. Hence it is a mistake to think that the agitation of the parts of flame conduces to the propagation of a pressure in rectilinear directions through an ambient medium. Such a pressure must be derived not from the agitation only of the parts of flame, but from the dilatation of the whole.

PROPOSITION XLIV. THEOREM XXXV

If water ascend and descend alternately in the erected legs KL, MN of a canal or pipe; and a pendulum be constructed whose length between the point of suspension and the centre of oscillation is equal to half the length of the water in the canal: I say, that the water will ascend and descend in the same times in which the pendulum oscillates.

I measure the length of the water along the axes of the canal and its legs, and make it equal to the sum of those axes; and take no notice of the resistance of the water arising from its attrition by the sides of the canal. Let, therefore, AB, CD represent the mean height of the water in both legs; and when the water in the leg KL ascends to the height EF, the water will descend in the leg MN to the height GH. Let P be a pendulous body, VP

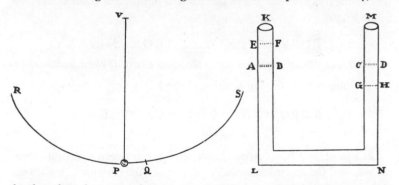

the thread, V the point of suspension, RPQS the cycloid which the pendulum describes, P its lowest point, PQ an arc equal to the height AE. The force with which the motion of the water is accelerated and retarded alternately is the excess of the weight of the water in one leg above the weight in the other; and, therefore, when the water in the leg KL ascends to EF, and in the other leg descends to GH, that force is double the weight of the water EABF, and therefore is to the weight of the whole water as AE or PQ to VP or PR. The force also with which the body P is accelerated or retarded in any place, as Q, of a cycloid, is (by Cor., Prop. LI, Book I) to its whole weight as its distance PQ from the lowest place P to the length PR of the cycloid. Therefore the motive forces of the water and pendulum, describing the equal spaces AE, PQ, are as the weights to be moved; and therefore if the water and pendulum are quiescent at first, those forces will move them in equal times, and will cause them to go and return together with a reciprocal motion. Q.E.D.

Cor. i. Therefore the reciprocations of the water in ascending and descending are all performed in equal times, whether the motion be more or less intense or remiss.

Cor. ii. If the length of the whole water in the canal be of 6⅑ feet of *French* measure, the water will descend in one second of time, and will ascend in another second, and so on by turns *in infinitum;* for a pendulum of 3¹/₁₈ such feet in length will oscillate in one second of time.

Cor. iii. But if the length of the water be increased or diminished, the time of the reciprocation will be increased or diminished as the square root of the length.

PROPOSITION XLV. THEOREM XXXVI

The velocity of waves varies as the square root of the breadths.

This follows from the construction of the following Proposition.

PROPOSITION XLVI. PROBLEM X

To find the velocity of waves.

Let a pendulum be constructed, whose length between the point of suspension and the centre of oscillation is equal to the breadth of the waves, and in the time that the pendulum will perform one single oscillation the waves will advance forwards nearly a space equal to their breadth.

That which I call the breadth of waves is the transverse measure lying between the deepest part of the hollows, or the tops of the ridges. Let ABCDEF represent the surface of stagnant water ascending and descend-

ing in successive waves; and let A, C, E, &c., be the tops of the waves; and let B, D, F, &c., be the intermediate hollows. Because the motion of the waves is carried on by the successive ascent and descent of the water, so that the parts thereof, as A, C, E, &c., which are highest at one time, become lowest immediately after; and because the motive force, by which the highest parts descend and the lowest ascend, is the weight of the elevated water, that alternate ascent and descent will be analogous to the reciprocal motion of the water in the canal, and will observe the same laws as to the times of ascent and descent; and therefore (by Prop. XLIV) if the distances between the highest places of the waves A, C, E and the lowest B, D, F be equal to twice the length of any pendulum, the highest parts A, C, E will become the lowest in the time of one oscillation, and in the time of another oscillation will ascend again. Therefore with the passage of each wave, the time of two oscillations will occur; that is, the wave will describe its breadth in the time that pendulum will oscillate twice; but a pendulum of four times that length, and therefore equal to the breadth of the waves, will just oscillate once in that time. Q.E.I.

COR. I. Therefore, waves, whose breadth is equal to $3\frac{1}{18}$ *French* feet, will advance through a space equal to their breadth in one second of time; and therefore in one minute will go over a space of $183\frac{1}{3}$ feet; and in an hour a space of 11000 feet, nearly.

COR. II. And the velocity of greater or less waves will be augmented or diminished as the square root of their breadth.

These things are true upon the supposition that the parts of water ascend or descend in a straight line; but, in truth, that ascent and descent is rather performed in a circle; and therefore I give the time defined by this Proposition as only approximate.

PROPOSITION XLVII. THEOREM XXXVII

If pulses are propagated through a fluid, the several particles of the fluid, going and returning with the shortest reciprocal motion, are always accelerated or retarded according to the law of the oscillating pendulum.

Let AB, BC, CD, &c., represent equal distances of successive pulses; ABC the line of direction of the motion of the successive pulses propagated from A to B; E, F, G three physical points of the quiescent medium situate in the right line AC at equal distances from each other; E*e*, F*f*, G*g* equal spaces of extreme shortness, through which those points go and return with a reciprocal motion in each vibration; ε, φ, γ any intermediate places of the same points; EF, FG physical short lines, or linear parts of the medium lying between those points, and successively transferred into the places εφ, φγ, and *ef, fg.* Let there be drawn the right line PS equal to the right line E*e.* Bisect the same in O, and from the centre O, with the radius OP, describe the circle SIP*i.* Let the whole time of one vibration, with its proportional parts, be represented by the whole circumference of this circle and its parts, in such sort, that, when any time PH or PHS*h* is completed, if there be let fall to PS the perpendicular HL or *hl,* and there be taken Eε equal to PL or P*l,* the physical point E may be found in ε. A point, as E, moving according to this law with a reciprocal motion, in its going from E through ε to *e,* and returning again through ε to E, will perform its several vibrations with the same degrees of acceleration and retardation with those of an oscillating pendulum. We are now to prove that the several physical points of the medium will be agitated with such a kind of motion. Let us suppose, then, that a medium hath such a motion excited in it from any cause whatsoever, and consider what will follow from thence.

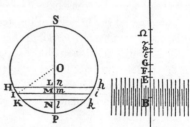

In the circumference PHS*h* let there be taken the equal arcs, HI, IK, or *hi, ik,* having the same ratio to the whole circumference as the equal right lines EF, FG have to BC, the whole interval of the pulses. Let fall the perpendiculars IM, KN, or *im, kn;* then because the points E, F, G are successively agitated with like motions, and perform their entire vibrations composed of their going and return, while the pulse is transferred from B to C; if PH or PHS*h* be the time elapsed since the beginning of the motion of the point E, then will PI or PHSi be the time elapsed since the beginning of the motion of the point F, and PK or PHSk the time elapsed since the beginning of the motion of the point G; and therefore Eε, Fϕ, Gγ will be respectively equal to PL, PM, PN, while the points are going, and to Pl, Pm, Pn, when the points are returning. Therefore $\varepsilon\gamma$ or EG + Gγ − Eε will, when the points are going, be equal to EG − LN, and in their return equal to EG + ln. But $\varepsilon\gamma$ is the breadth or expansion of the part EG of the medium in the place $\varepsilon\gamma$; and therefore the expansion of that part in its going is to its mean expansion as EG − LN to EG; and in its return, as EG + ln or EG + LN is to EG. Therefore since LN is to KH as IM to the radius OP, and KH to EG as the circumference PHShP to BC; that is, if we put V for the radius of a circle whose circumference is equal to BC the interval of the pulses, as OP is to V; and, multiplying together corresponding terms of the proportions, we obtain LN to EG as IM to V; the expansion of the part EG, or of the physical point F in the place $\varepsilon\gamma$, is to the mean expansion of the same part in its first place EG, as V − IM is to V in going, and as V + im is to V in its return. Hence the elastic force of the point F in the place $\varepsilon\gamma$ is to its mean elastic force in the place EG as $\dfrac{1}{V - IM}$ is to $\dfrac{1}{V}$ in its going, and as $\dfrac{1}{V + im}$ is to $\dfrac{1}{V}$ in its return. And by the same reasoning the elastic forces of the physical points E and G in going are as $\dfrac{1}{V - HL}$ and $\dfrac{1}{V - KN}$ is to $\dfrac{1}{V}$; and the difference of the forces is to the mean elastic force of the medium as $\dfrac{HL - KN}{VV - V \cdot HL - V \cdot KN + HL \cdot KN}$ is to $\dfrac{1}{V}$; that is, as $\dfrac{HL - KN}{VV}$ is to $\dfrac{1}{V}$, or as HL − KN is to V; if we suppose (by reason of the very short extent of the vibrations) HL and KN to be indefinitely less than the quantity V. Therefore since the quantity V is given, the difference of the forces is as HL − KN; that is (because HL − KN is proportional to HK, and OM to OI or OP; and because HK and OP are given), as OM; that is, if Ff be bisected in Ω, as $\Omega\phi$. And for the same reason the difference of the elastic forces of the physical points ε and γ, in the return of the physical short line $\varepsilon\gamma$, is as $\Omega\phi$. But that difference (that is, the excess of the elastic force of the point ε above the elastic force of the point γ) is the very force by which the intervening physical short line εγ of the medium is accelerated in going, and retarded in returning; and therefore the accelerative force of the physical short line εγ is as its distance from ☊, the middle place of the vibration. Therefore (by Prop. xxxviii, Book 1) the time is rightly represented by the arc PI; and the linear part of the medium εγ is moved according to the law above mentioned, that is, according to the law of a pendulum oscillating; and the case is the same of all the linear parts of which the whole medium is compounded. Q.E.D.

Cor. Hence it appears that the number of the pulses propagated is the same with the number of the vibrations of the tremulous body, and is not multiplied in their progress. For the physical short line εγ as soon as it returns to its first place is at rest; neither will it move again, unless it receives a new motion either from the impulse of the tremulous body, or of the pulses propagated from that body. As soon, therefore, as the pulses cease to be propagated from the tremulous body, it will return to a state of rest, and move no more.

PROPOSITION XLVIII.[1] THEOREM XXXVIII

The velocities of pulses propagated in an elastic fluid are in a ratio compounded of the square root of the ratio of the elastic force directly, and the square root of the ratio of the density inversely; supposing the elastic force of the fluid to be proportional to its condensation.

Case 1. If the mediums be homogeneous, and the distances of the pulses in those mediums be equal amongst themselves, but the motion in one medium is more intense than in the other, the contractions and dilatations of the corresponding parts will be as those motions; not that this proportion is perfectly accurate. However, if the contractions and dilatations are not exceedingly intense, the error will not be sensible; and therefore this proportion may be considered as physically exact. Now the motive elastic forces are as the contractions and dilatations; and the velocities generated in the same time in equal parts are as the forces. Therefore equal and corresponding parts of corresponding pulses will go and return together, through spaces proportional to their contractions and dilatations, with velocities that are as those spaces; and therefore the pulses, which in the time of one going and returning advance forwards a space equal to their breadth, and

[1 Appendix, Note 38.]

are always succeeding into the places of the pulses that immediately go before them, will, by reason of the equality of the distances, go forwards in both mediums with equal velocity.

CASE 2. If the distances of the pulses or their lengths are greater in one medium than in another, let us suppose that the correspondent parts describe spaces, in going and returning, each time proportional to the breadths of the pulses; then will their contractions and dilatations be equal; and therefore if the mediums are homogeneous, the motive elastic forces, which agitate them with a reciprocal motion, will be equal also. Now the matter to be moved by these forces is as the breadth of the pulses; and the space through which they move every time they go and return is in the same ratio. And, moreover, the time of one going and returning is in a ratio compounded of the square root of the matter and the square root of the space; and therefore is as the space. But the pulses advance a space equal to their breadths in the times of going once and returning once; that is, they go over spaces proportional to the times, and therefore are equally swift.

CASE 3. And therefore in mediums of equal density and elastic force, all the pulses are equally swift. Now if the density or the elastic force of the medium were augmented, then, because the motive force is increased in the ratio of the elastic force, and the matter to be moved is increased in the ratio of the density, the time which is necessary for producing the same motion as before will be increased as the square root of the ratio of the density, and will be diminished as the square root of the ratio of the elastic force. And therefore the velocity of the pulses will be in a ratio compounded of the square root of the inverse ratio of the density of the medium, and the square root of the direct ratio of the elastic force. Q.E.D.

This Proposition will be made clearer from the construction of the following Problem.

PROPOSITION XLIX. PROBLEM XI

The density and elastic force of a medium being given, to find the velocity of the pulses.

Suppose the medium to be pressed by an incumbent weight after the manner of our air; and let A be the height of an homogeneous medium, whose weight is equal to the incumbent weight, and whose density is the same with the density of the compressed medium in which the pulses are propagated. Suppose a pendulum to be constructed whose length between the point of suspension and the centre of oscillation is A: and in the time in which that pendulum will perform one entire oscillation composed of its going and returning, the pulse will be propagated right onwards through a space equal to the circumference of a circle described with the radius A.

For, letting those things stand which were constructed in Prop. XLVII, if any physical line, as EF, describing the space PS in each vibration, be acted on in the extremities P and S of every going and return that it makes by an elastic force that is equal to its weight, it will perform its several vibrations in the time in which the same might oscillate in a cycloid whose whole perimeter is equal to the length PS; and that because equal forces will impel equal corpuscles through equal spaces in the same or equal times. Therefore since the times of the oscillations are as the square root of the lengths of the pendulums, and the length of the pendulum is equal to half the arc of the whole cycloid, the time of one vibration would be to the time of the oscilla-

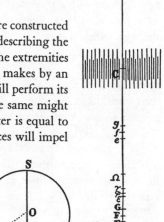

tion of a pendulum whose length is A as the square root of the length ½PS or PO to the length A. But the elastic force with which the physical short line EG is urged, when it is found in its extreme places P, S, was (in the demonstration of Prop. XLVII) to its whole elastic force as HL−KN is to V, that is (since the point K now falls upon P), as HK to V; and all that force, or, which is the same thing, the incumbent weight by which the short line EG is compressed, is to the weight of the short line as the altitude of the incumbent weight is to EG the length of the short line; and therefore, taking the product of corresponding terms, the force with which the short line EG is urged in the places P and S is to the weight of that short line as HK · A is to V · EG; or as PO · A is to VV; because HK was to EG as PO to V. Therefore, since the times in which equal bodies are impelled through equal spaces are inversely as the square root of the forces, the time of one vibration, produced by the action of that elastic force, will be to the time of a vibration, produced by the impulse of the weight, as the square root of the ratio of VV to PO · A, and therefore to the time of the oscillation of a pendulum whose length is A as the square root of the ratio of VV to PO · A, and as the square root of the ratio of PO to A conjointly; that is, in the entire ratio of V to A. But in the time of one vibration composed of the going and returning of the pendulum, the pulse will be propagated right onwards through a space equal to its breadth BC. Therefore the time in which a pulse runs over the space BC is to the time of one oscillation composed of the going and returning of the pendulum as V is to A, that is, as BC is to the circumference of a circle whose radius is A. But the time in which the pulse will run over the space BC is to the time in which it will run over a length equal to that circumference in the same ratio; and therefore in the time of such an oscillation the pulse will run over a length equal to that circumference. Q.E.D.

Cor. i. The velocity of the pulses is equal to that which heavy bodies acquire by falling with an equally accelerated motion, and in their fall describing half the altitude A. For the pulse will, in the time of this fall, supposing it to move with the velocity acquired by that fall, run over a space that will be equal to the whole altitude A; and therefore in the time of one oscillation composed of one going and return, will go over a space equal to the circumference of a circle described with the radius A; for the time of the fall is to the time of oscillation as the radius of a circle to its circumference.

Cor. ii. Therefore since that altitude A is directly as the elastic force of the fluid, and inversely as the density of the same, the velocity of the pulses will be in a ratio compounded of the square root of the ratio of the density inversely, and the square root of the ratio of the elastic force directly.

PROPOSITION L.   PROBLEM XII

To find the distances of the pulses.

Let the number of the vibrations of the body, by whose tremor the pulses are produced, be found to any given time. By that number divide the space which a pulse can go over in the same time, and the part found will be the breadth of one pulse. Q.E.I.

SCHOLIUM

The last Propositions respect the motions of light and sounds; for since light is propagated in right lines, it is certain that it cannot consist in action alone (by Prop. XLI and XLII). As to sounds, since they arise from tremulous bodies, they can be nothing else but pulses of the air propagated through it (by Prop. XLIII); and this is confirmed by the tremors which sounds, if they be loud and deep, excite in the bodies near them, as we experience in the sound of drums; for quick and short tremors are less easily excited. But it is well known that any sounds, falling upon strings in unison with the sonorous bodies, excite tremors in those strings. This is also confirmed from the velocity of sounds; for since the specific gravities of rain water and quicksilver are to one another as about 1 to 13⅔, and when the mercury in the barometer is at the height of 30 inches of our measure, the specific gravities of the air and of rain water are to one another as about 1 to 870, therefore the specific gravities of air and quicksilver are to each other as 1 to 11890. Therefore when the height of the quicksilver is at 30 inches, a height of uniform air, whose weight would be sufficient to compress our air to the density we find it to be of, must be equal to 356700 inches, or 29725 feet of our measure; and this is that very height of the medium, which I have called A in the construction of the foregoing Proposition. A circle whose radius is 29725 feet is 186768 feet in circumference. And since a pendulum 39⅕ inches in length completes one oscillation, composed of its going and return, in two seconds of time, as is commonly known, it follows that a pendulum 29725 feet, or 356700 inches in length will perform a like oscillation in 190¾ seconds. Therefore in that time a sound will go right onwards 186768 feet, and therefore in one second 979 feet.

But in this computation we have made no allowance for the crassitude of the solid particles of the air, by which the sound is propagated instantaneously. Because the weight of air is to the weight of water as 1 to 870, and because salts are almost twice as dense as water; if the particles of air are supposed to be of about the same density as those of water or salt, and the rarity of the air arises from the intervals of the particles; the diameter of one particle of air will be to the interval between the centres of the particles as 1 to about 9 or 10, and to the interval between the particles themselves as 1 to 8 or 9. Therefore to 979 feet, which, according to the above calculation, a sound will advance forwards in one second of time, we may add $\frac{979}{9}$, or about 109 feet, to compensate for the crassitude of the particles of the air: and then a sound will go forwards about 1088 feet in one second of time.

Moreover, the vapors floating in the air being of another spring, and a different tone, will hardly, if at all, partake of the motion of the true air in which the sounds are propagated. Now if these vapors remain unmoved, that motion will be propagated the swifter through the true air alone, and that as the square root of the defect of the matter. So if the atmosphere consist of ten parts of true air and one part of vapors, the motion of sounds will be swifter as the square root of the ratio of 11 to 10, or very nearly in the entire ratio of 21 to 20, than if it were propagated through eleven parts of true air: and therefore the motion of sounds above discovered must be increased in that ratio. By this means the sound will pass through 1142 feet in one second of time.

These things will be found true in spring and autumn, when the air is rarefied by the gentle warmth of those seasons, and by that means its elastic force becomes somewhat more intense. But in winter, when the air is condensed by the cold, and its elastic force is somewhat remitted, the motion of sounds will be slower as the square root of the density; and, on the other hand, swifter in the summer.

Now by experiments it actually appears that sounds do really advance in one second of time about 1142 feet of *English* measure, or 1070 feet of *French* measure.

The velocity of sounds being known, the intervals of the pulses are known also. For M. *Sauveur,* by some experiments that he made, found that an open pipe about five *Paris* feet in length gives a sound of the same tone with a viol string that vibrates a hundred times in one second. Therefore there are near 100 pulses in a space of 1070 *Paris* feet, which a sound runs over in a second of time; and therefore one pulse fills up a space of about 10⁷⁄₁₀ *Paris* feet, that is, about twice the length of the pipe. From this it is probable that the breadths of the pulses, in all sounds made in open pipes, are equal to twice the length of the pipes.

Moreover, from the Corollary of Prop. XLVII appears the reason why the sounds immediately cease with the motion of the sonorous body, and why they are heard no longer when we are at a great distance from the sonorous bodies than when we are very near them. And besides, from the foregoing principles, it plainly appears how it comes to pass that sounds are so mightily increased in speaking-trumpets; for all reciprocal motion tends to be increased by the generating cause at each return. And in tubes hindering the dilatation of the sounds, the motion decays more slowly, and recurs more forcibly; and therefore is the more increased by the new motion impressed at each return. And these are the principal phenomena of sounds.

SECTION IX
The circular motion of fluids.

HYPOTHESIS
The resistance arising from the want of lubricity in the parts of a fluid, is, other things being equal, proportional to the velocity with which the parts of the fluid are separated from one another.

PROPOSITION LI. THEOREM XXXIX
If a solid cylinder infinitely long, in an uniform and infinite fluid, revolves with an uniform motion about an axis given in position, and the fluid be forced round by only this impulse of the cylinder, and every part of the fluid continues uniformly in its motion: I say, that the periodic times of the parts of the fluid are as their distances from the axis of the cylinder.

Let AFL be a cylinder turning uniformly about the axis S, and let the concentric circles BGM, CHN, DIO, EKP, &c., divide the fluid into innumerable concentric cylindric solid orbs of the same thickness. Then, because the fluid is homogeneous, the impressions which the contiguous orbs make upon each other will be (by the Hypothesis) as their translations from each other, and as the contiguous surfaces upon which the impressions are made. If the impression made upon any orb be greater or less on its concave than on its convex side, the stronger impression will prevail, and will either accelerate or retard the motion of the orb, according as it agrees with, or is contrary to, the motion of the same. Therefore, that every orb may continue uniformly in its motion, the impressions made on both sides must be equal and their directions contrary. Therefore since the impressions are as the contiguous surfaces, and as their translations from one another, the translations will be inversely as the surfaces, that is, inversely as

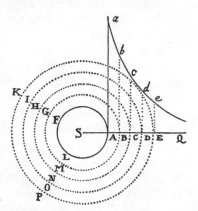

the distances of the surfaces from the axis. But the differences of the angular motions about the axis are as those translations applied to the distances, or directly as the translations and inversely as the distances; that is, joining these ratios together, inversely as the squares of the distances. Therefore if there be erected the lines A*a*, B*b*, C*c*, D*d*, E*e*, &c., perpendicular to the several parts of the infinite right line SABCDEQ, and inversely proportional to the squares of SA, SB, SC, SD, SE, &c., and through the extremities of those perpendiculars there be supposed to pass an hyperbolic curve, the sums of the differences, that is, the whole angular motions, will be as the correspondent sums of the lines A*a*, B*b*, C*c*, D*d*, E*e*, that is (if to constitute a medium uniformly fluid the number of the orbs be increased and their breadth diminished *in infinitum*), as the hyperbolic areas A*a*Q, B*b*Q, C*c*Q, D*d*Q, E*e*Q, &c., analogous to the sums; and the times, inversely proportional to the angular motions, will be also inversely proportional to those areas. Therefore the periodic time of any particle, as D, is inversely as the area D*d*Q, that is (as appears from the known methods of quadratures of curves), directly as the distance SD. Q.E.D.

Cor. i. Hence the angular motions of the particles of the fluid are inversely as their distances from the axis of the cylinder, and the absolute velocities are equal.

Cor. ii. If a fluid be contained in a cylindric vessel of an infinite length, and contain another cylinder within, and both the cylinders revolve about one common axis, and the times of their revolutions be as their semidiameters, and every part of the fluid continues in its motion, the periodic times of the several parts will be as the distances from the axis of the cylinders.

Cor. iii. If there be added or taken away any common quantity of angular motion from the cylinder and fluid moving in this manner, yet because this new motion will not alter the mutual attrition of the parts of the fluid, the motion of the parts among themselves will not be changed; for the translations of the parts from one another depend upon the attrition. Any part will continue in that motion, which, by the attrition made on both sides with contrary directions, is no more accelerated than it is retarded.

Cor. iv. Therefore if there be taken away from this whole system of the cylinders and the fluid all the angular motion of the outward cylinder, we shall have the motion of the fluid in a quiescent cylinder.

COR. V. Therefore if the fluid and outward cylinder are at rest, and the inward cylinder revolve uniformly, there will be communicated a circular motion to the fluid, which will be propagated by degrees through the whole fluid; and will go on continually increasing, till such time as the several parts of the fluid acquire the motion determined in Cor. IV.

COR. VI. And because the fluid endeavors to propagate its motion still farther, its impulse will carry the outmost cylinder also about with it, unless the cylinder be forcibly held back; and accelerate its motion till the periodic times of both cylinders become equal with each other. But if the outward cylinder be forcibly held fast, it will make an effort to retard the motion of the fluid; and unless the inward cylinder preserve that motion by means of some external force impressed theron, it will make it cease by degrees.

All these things will be found true by making the experiment in deep standing water.

PROPOSITION LII. THEOREM XL

If a solid sphere, in an uniform and infinite fluid, revolves about an axis given in position with an uniform motion, and the fluid be forced round by only this impulse of the sphere; and every part of the fluid continues uniformly in its motion: I say, that the periodic times of the parts of the fluid are as the squares of their distances from the centre of the sphere.

CASE I. Let AFL be a sphere turning uniformly about the axis S, and let the concentric circles BGM, CHN, DIO, EKP, &c., divide the fluid into innumerable concentric orbs of the same thickness. Suppose those orbs to be solid; and, because the fluid is homogeneous, the impressions which the contiguous orbs make one upon another will be (by the supposition) as their translations from one another, and the contiguous surfaces upon which the impressions are made. If

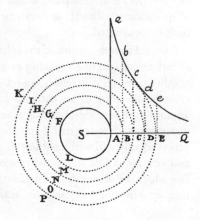

the impression upon any orb be greater or less upon its concave than upon its convex side, the more forcible impression will prevail, and will either accelerate or retard the velocity of the orb, according as it is directed with a conspiring or contrary motion to that of the orb. Therefore that every orb may continue uniformly in its motion, it is necessary that the impressions made upon both sides of the orb should be equal, and have contrary directions. Therefore since the impressions are as the contiguous surfaces, and as their translations from one another, the translations will be inversely as the surfaces, that is, inversely as the squares of the distances of the surfaces from the centre. But the differences of the angular motions about the axis are as those translations applied to the distances, or directly as the translations and inversely as the distances; that is, by compounding those ratios, inversely as the cubes of the distances. Therefore, if upon the several parts of the infinite right line SABCDEQ there be erected the perpendiculars A*a*, B*b*, C*c*, D*d*, E*e*, &c., inversely proportional to the cubes of SA, SB, SC, SD, SE, &c., the sums of the differences, that is, the whole angular motions, will be as the corresponding sums of the lines A*a*, B*b*, C*c*, D*d*, E*e*, &c., that is (if to constitute an uniformly fluid medium the number of the orbs be increased and their thickness diminished *in infinitum*), as the hyperbolic areas A*a*Q, B*b*Q, C*c*Q, D*d*Q, E*e*Q, &c., analogous to the sums; and the periodic times being inversely proportional to the angular motions, will be also inversely proportional to those areas. Therefore the periodic time of any orb DIO is inversely as the area D*d*Q, that is (by the known methods of quadratures), directly as the square of the distance SD. Which was first to be demonstrated.

CASE 2. From the centre of the sphere let there be drawn a great number of indefinite right lines, making given angles with the axis, exceeding one another by equal differences; and, by these lines revolving about the axis, conceive the orbs to be cut into innumerable annuli; then will every annulus have four annuli contiguous to it, that is, one on its inside, one on its outside, and two on each hand. Now each of these annuli cannot be impelled equally and with contrary directions by the attrition of the interior and exterior annuli, unless the motion be communicated according to the law which we demonstrated in Case 1. This appears from that demonstration. And therefore any series of annuli, taken in any right line extending

itself *in infinitum* from the globe, will move according to the law of Case
1, except we should imagine it hindered by the attrition of the annuli on
each side of it. But now in a motion, according to this law, no such is, and
therefore cannot be, any obstacle to the motions continuing according to
that law. If annuli at equal distances from the centre revolve either more
swiftly or more slowly near the poles than near the ecliptic, they will be
accelerated if slow, and retarded if swift, by their mutual attrition; and so
the periodic times will continually approach to equality, according to the
law of Case 1. Therefore this attrition will not at all hinder the motion
from going on according to the law of Case 1, and therefore that law will
take place; that is, the periodic times of the several annuli will be as the
squares of their distances from the centre of the globe. Which was to be
demonstrated in the second place.

CASE 3. Let now every annulus be divided by transverse sections into
innumerable particles constituting a substance absolutely and uniformly
fluid; and because these sections do not at all respect the law of circular
motion, but only serve to produce a fluid substance, the law of circular motion will continue the same as before. All the very small annuli will either
not at all change their asperity and force of mutual attrition upon account
of these sections, or else they will change the same equally. Therefore the
proportion of the causes remaining the same, the proportion of the effects
will remain the same also; that is, the proportion of the motions and the
periodic times. Q.E.D. But now as the circular motion, and the centrifugal
force thence arising, is greater at the ecliptic than at the poles, there must be
some cause operating to retain the several particles in their circles; otherwise the matter that is at the ecliptic will always recede from the centre, and
come round about to the poles by the outside of the vortex, and from thence
return by the axis to the ecliptic with a continual circulation.

COR. 1. Hence the angular motions of the parts of the fluid about the axis
of the globe are inversely as the squares of the distances from the centre
of the globe, and the absolute velocities are inversely as the same squares
applied to the distances from the axis.

COR. 11. If a globe revolve with an uniform motion about an axis of a given
position in a similar and infinite quiescent fluid with an uniform motion,
it will communicate a whirling motion to the fluid like that of a vortex, and

that motion will by degrees be propagated onwards *in infinitum;* and this motion will be increased continually in every part of the fluid, till the periodical times of the several parts become as the squares of the distances from the centre of the globe.

COR. III. Because the inward parts of the vortex are by reason of their greater velocity continually pressing upon and driving forwards the external parts, and by that action are continually communicating motion to them, and at the same time those exterior parts communicate the same quantity of motion to those that lie still beyond them, and by this action preserve the quantity of their motion continually unchanged, it is plain that the motion is continually transferred from the centre to the circumference of the vortex, till it is quite swallowed up and lost in the boundless extent of that circumference. The matter between any two spherical surfaces concentric to the vortex will never be accelerated; because that matter will be always transferring the motion it receives from the matter nearer the centre to that matter which lies nearer the circumference.

COR. IV. Therefore, in order to continue a vortex in the same state of motion, some active principle is required from which the globe may receive continually the same quantity of motion which it is always communicating to the matter of the vortex. Without such a principle it will undoubtedly come to pass that the globe and the inward parts of the vortex, being always propagating their motion to the outward parts, and not receiving any new motion, will gradually move slower and slower, and at last be carried round no longer.

COR. V. If another globe should be swimming in the same vortex at a certain distance from its centre, and in the meantime by some force revolve constantly about an axis of a given inclination, the motion of this globe will drive the fluid round after the manner of a vortex; and at first this new and small vortex will revolve with its globe about the centre of the other; and in the meantime its motion will creep on farther and farther, and by degrees be propagated *in infinitum,* after the manner of the first vortex. And for the same reason that the globe of the new vortex was carried about before by the motion of the other vortex, the globe of this other will be carried about by the motion of this new vortex, so that the two globes will revolve about some intermediate point, and by reason of that circular motion mutually fly from each other, unless some force restrains them. Afterwards, if the constantly impressed forces, by which the globes continue in their motions, should cease, and everything be left to act according to the laws of mechanics, the motion of the globes will languish by degrees (for the reason assigned in Cor. III and IV), and the vortices at last will quite stand still.

COR. VI. If several globes in given places should constantly revolve with determined velocities about axes given in position, there would arise from them as many vortices going on *in infinitum*. For upon the same account that any one globe propagates its motion *in infinitum,* each globe apart will propagate its motion *in infinitum* also; so that every part of the infinite fluid will be agitated with a motion resulting from the actions of all the globes. Therefore the vortices will not be confined by any certain limits, but by degrees run into each other; and by the actions of the vortices on each other, the globes will be continually moved from their places, as was shown in the last Corollary; neither can they possibly keep any certain position among themselves, unless some force restrains them. But if those forces, which are constantly impressed upon the globes to continue these motions, should cease, the matter (for the reason assigned in Cor. III and IV) will gradually stop, and cease to move in vortices.

COR. VII. If a similar fluid be inclosed in a spherical vessel, and, by the uniform rotation of a globe in its centre, be driven round in a vortex; and the globe and vessel revolve the same way about the same axis, and their periodic times be as the squares of the semidiameters: the parts of the fluid will not go on in their motions without acceleration or retardation, till their periodical times are as the squares of their distances from the centre of the vortex. No constitution of a vortex can be permanent but this.

COR. VIII. If the vessel, the inclosed fluid, and the globe, retain this motion, and revolve besides with a common angular motion about any given axis, because the mutual attrition of the parts of the fluid is not changed by this motion, the motions of the parts among themselves will not be changed; for the translations of the parts among themselves depend upon this attrition. Any part will continue in that motion in which its attrition on one side retards it just as much as its attrition on the other side accelerates it.

Cor. ix. Therefore if the vessel be quiescent, and the motion of the globe be given, the motion of the fluid will be given. For conceive a plane to pass through the axis of the globe, and to revolve with a contrary motion; and suppose the sum of the time of this revolution and of the revolution of the globe to be to the time of the revolution of the globe as the square of the semidiameter of the vessel to the square of the semidiameter of the globe; and the periodic times of the parts of the fluid in respect of this plane will be as the squares of their distances from the centre of the globe.

Cor. x. Therefore if the vessel move about the same axis with the globe, or with a given velocity about a different one, the motion of the fluid will be given. For if from the whole system we take away the angular motion of the vessel, all the motions will remain the same among themselves as before, by Cor. viii, and those motions will be given by Cor. ix.

Cor. xi. If the vessel and the fluid are quiescent, and the globe revolves with an uniform motion, that motion will be propagated by degrees through the whole fluid to the vessel, and the vessel will be carried round by it, unless forcibly held back; and the fluid and the vessel will be continually accelerated till their periodic times become equal to the periodic times of the globe. If the vessel be either restrained by some force, or revolve with any constant and uniform motion, the medium will come little by little to the state of motion defined in Cor. viii, ix, x, nor will it ever continue in any other state. But if then the forces, by which the globe and vessel revolve with certain motions, should cease, and the whole system be left to act according to the mechanical laws, the vessel and globe, by means of the intervening fluid, will act upon each other, and will continue to propagate their motions through the fluid to each other, till their periodic times become equal among themselves, and the whole system revolves together like one solid body.

SCHOLIUM

In all these reasonings I suppose the fluid to consist of matter of uniform density and fluidity; I mean, that the fluid is such, that a globe placed anywhere therein may propagate with the same motion of its own, at distances from itself continually equal, similar and equal motions in the fluid in the same interval of time. The matter by its circular motion endeavors to recede from the axis of the vortex, and therefore presses all the matter that lies beyond. This pressure makes the attrition greater, and the separation of the parts more difficult; and by consequence diminishes the fluidity of the matter. Again; if the parts of the fluid are in any one place denser or larger than in the others, the fluidity will be less in that place, because there are fewer surfaces where the parts can be separated from each other. In these cases I suppose the defect of the fluidity to be supplied by the smoothness or softness of the parts, or some other condition; otherwise the matter where it is less fluid will cohere more, and be more sluggish, and therefore will receive the motion more slowly, and propagate it farther than agrees with the ratio above assigned. If the vessel be not spherical, the particles will move in lines not circular, but answering to the figure of the vessel; and the periodic times will be nearly as the squares of the mean distances from the centre. In the parts between the centre and the circumference the motions will be slower where the spaces are wide, and swifter where narrow; nevertheless, the particles will not tend to the circumference at all the more because of their greater swiftness; for they then describe arcs of less curvity, and the tendency to recede from the centre is as much diminished by the lessening of this curvature as it is augmented by the increase of the velocity. As they go out of narrow into wide spaces, they recede a little farther from the centre, but in doing so are retarded; and when they come out of wide into narrow spaces, they are again accelerated; and so each particle is retarded and accelerated by turns forever. These things will come to pass in a rigid vessel; for the state of vortices in an infinite fluid is known by Cor. VI of this Proposition.

I have endeavored in this Proposition to investigate the properties of vortices, that I might find whether the celestial phenomena can be explained by them; for the phenomenon is this, that the periodic times of the planets revolving about Jupiter are as the 3/2th power of their distances from Jupiter's centre; and the same rule obtains also among the planets that revolve about the sun. And these rules obtain also with the greatest accuracy, as far as has been yet discovered by astronomical observation. Therefore if those planets are carried round in vortices revolving about Jupiter and the sun, the vortices must revolve according to that law. But here we found the periodic times of the parts of the vortex to be as the square of the distances from the centre of motion; and this ratio cannot be diminished and reduced to the $\frac{3}{2}$th power, unless either the matter of the vortex be more fluid the farther it is from the centre, or the resistance arising from the want of lubricity in the parts of the fluid should, as the velocity with which the parts of the fluid are separated goes on increasing, be augmented with it in a greater ratio than that in which the velocity increases. But neither of these suppositions seems reasonable. The more gross and less fluid parts will tend to the circumference, unless they are heavy towards the centre. And though, for the sake of demonstration, I proposed, at the beginning of this Section, an Hypothesis that the resistance is proportional to the velocity, nevertheless, it is in truth probable that the resistance is in a less ratio than that of the velocity; which granted, the periodic times of the parts of the vortex will be in a greater ratio than the square of the distances from its centre. If, as some think, the vortices move more swiftly near the centre, then slower to a certain limit, then again swifter near the circumference, certainly neither the $\frac{3}{2}$th power, nor any other certain and determinate power, can obtain in them. Let philosophers then see how that phenomenon of the $\frac{3}{2}$th power can be accounted for by vortices.

PROPOSITION LIII. THEOREM XLI

Bodies carried about in a vortex, and returning in the same orbit, are of the same density with the vortex, and are moved according to the same law with the parts of the vortex, as to velocity and direction of motion.

For if any small part of the vortex, whose particles or physical points continue a given situation among themselves, be supposed to be congealed, this particle will move according to the same law as before, since no change is made either in its density, inertia, or figure. And again; if a congealed or solid part of the vortex be of the same density with the rest of the vortex, and be resolved into a fluid, this will move according to the same law as before, except so far as its particles, now become fluid, may be moved among themselves. Neglect, therefore, the motion of the particles among themselves as not at all concerning the progressive motion of the whole, and the motion of the whole will be the same as before. But this motion will be the same with the motion of other parts of the vortex at equal distances from the centre; because the solid, now resolved into a fluid, is become exactly like the other parts of the vortex. Therefore a solid, if it be of the same density with the matter of the vortex, will move with the same motion as the parts thereof, being relatively at rest in the matter that surrounds it. If it be more dense, it will endeavor more than before to recede from the centre; and therefore overcoming that force of the vortex, by which, being, as it were, kept in equilibrium, it was retained in its orbit, it will recede from the centre, and in its revolution describe a spiral, returning no longer into the same orbit. And, by the same argument, if it be more rare, it will approach to the centre. Therefore it can never continually go round in the same orbit, unless it be of the same density with the fluid. But we have shown in that case that it would revolve according to the same law with those parts of the fluid that are at the same or equal distances from the centre of the vortex.

Cor. i. Therefore a solid revolving in a vortex, and continually going round in the same orbit, is relatively quiescent in the fluid that carries it.

Cor. ii. And if the vortex be of an uniform density, the same body may revolve at any distance from the centre of the vortex.

SCHOLIUM

Hence it is manifest that the planets are not carried round in corporeal vortices; for, according to the *Copernican* hypothesis, the planets going round the sun revolve in ellipses, having the sun in their common focus; and by radii drawn to the sun describe areas proportional to the times. But the parts of a vortex can never revolve with such a motion. For, let AD, BE, CF represent three orbits described about the sun S, of which let the outmost circle CF be concentric to the sun; let the aphelions of the two innermost be A, B; and their perihelions D, E. Hence a body revolving in the orb CF, describing, by a radius drawn to the sun, areas proportional to the times, will move with an uniform motion. And, according to the laws

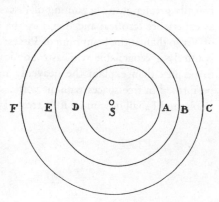

of astronomy, the body revolving in the orbit BE will move slower in its aphelian B, and swifter in its perihelion E; whereas, according to the laws of mechanics, the matter of the vortex ought to move more swiftly in the narrow space between A and C than in the wide space between D and F; that is, more swiftly in the aphelion than in the perihelion. Now these two conclusions contradict each other. So at the beginning of the sign of Virgo, where the aphelion of Mars is at present, the distance between the orbits of Mars and Venus is to the distance between the same orbits, at the beginning of the sign of Pisces, as about 3 to 2; and therefore the matter of the vortex between those orbits ought to be swifter at the beginning of Pisces than at the beginning of Virgo in the ratio of 3 to 2; for the narrower the space is through which the same quantity of matter passes in the same time of one revolution, the greater will be the velocity with which it passes through it. Therefore if the earth being relatively at rest in this celestial matter should be carried round by it, and revolve together with it about the sun, the velocity of the earth at the beginning of Pisces would be to its velocity at the beginning of Virgo in the ratio of 3 to 2. Therefore the sun's apparent diurnal motion at the beginning of Virgo ought to be above 70 minutes, and at the beginning of Pisces less than 48 minutes; whereas, on the contrary, that apparent motion of the sun is really greater at the beginning of Pisces than at the beginning of Virgo, as experience testifies; and therefore the earth is swifter at the beginning of Virgo than at the beginning of Pisces; so that the hypothesis of vortices is utterly irreconcilable with astronomical phenomena, and rather serves to perplex than explain the heavenly motions. How these motions are performed in free spaces without vortices, may be understood by the first Book; and I shall now more fully treat of it in the following Book.